IS EARTH REALLY SOLID?

The Evidence Re-Examined

Oliver Milatovic

Adventures Unlimited Press

Other Books of Interest:

VIMANA
ARK OF GOD
THE LOST WORLD OF CHAM
OBELISKS: TOWERS OF POWER
ANCIENT TECHNOLOGY IN PERU & BOLIVIA
THE MYSTERY OF THE OLMECS
PIRATES AND THE LOST TEMPLAR FLEET
TECHNOLOGY OF THE GODS
A HITCHHIKER'S GUIDE TO ARMAGEDDON
LOST CONTINENTS & THE HOLLOW EARTH
ATLANTIS & THE POWER SYSTEM OF THE GODS
THE FANTASTIC INVENTIONS OF NIKOLA TESLA
LOST CITIES OF NORTH & CENTRAL AMERICA
LOST CITIES OF CHINA, CENTRAL ASIA & INDIA
LOST CITIES & ANCIENT MYSTERIES OF AFRICA & ARABIA
LOST CITIES & ANCIENT MYSTERIES OF SOUTH AMERICA
LOST CITIES OF ANCIENT LEMURIA & THE PACIFIC
LOST CITIES OF ATLANTIS, ANCIENT EUROPE & THE MEDITERRANEAN
LOST CITIES & ANCIENT MYSTERIES OF THE SOUTHWEST
YETIS, SASQUATCH AND HAIRY GIANTS
THE ENIGMA OF CRANIAL DEFORMATION
THE CRYSTAL SKULLS
ALICE'S WONDERLAND
PYTHAGORAS OF SAMOS
THE GODS IN THE FIELDS
JACK THE RIPPER'S NEW TESTAMENT
THE LANDING LIGHTS OF MAGONIA
U-33: HITLER'S SECRET ENVOY
OTTO RAHN AND THE QUEST FOR THE HOLY GRAIL
CRYSTALS, MOTHER EARTH & THE FORCES OF LIVING NATURE

IS EARTH REALLY SOLID?

The Evidence Re-Examined

Adventures Unlimited Press

Is Earth Really Solid?

ISBN 978-1-948803-77-9

Published by:
Adventures Unlimited Press
One Adventure Place
Kempton, Illinois 60946 USA
auphq@frontiernet.net

AdventuresUnlimitedPress.com

Cover by Terry Lamb

For my mother and Geneviéve

About the author

Oliver Milatovic earned his BSc in Engineering Geology from the University of the Andes, Venezuela. He worked in the oil industry, analysing geophysical data to assess the petrophysical conditions of the subsurface and point out the places that are likely to contain oil; using techniques ranging from Gamma Ray (GR) Logs that measure radioactivity of the rocks to determine the amount of shale in a formation through a borehole or drill hole, to Sonic (or Borehole Compensated) Logs, that measure porosity by measuring how fast sound waves travel through rocks. Also, he worked in the laboratory, performing testing of candidate drilling, completion, remedial, treatment or workover fluids; under reservoir conditions. He is a former consultant of environmental issues, ensuring hazards such as subsidence; seismic risks and unstable slopes are identified. Former professor of Geology showing students how to apply academic knowledge and skills to a range of Earth science problems, mainly those related to Civil Engineering. Technical advisor in design and/or improvement of laboratory test procedures for the oil and construction industry. Amateur astronomer. His main interest is the practical application of geological/engineering/astronomical knowledge either conventional or cutting-edge, to a wide range of engineering and scientific problems. He lives and works in the city of Aberdeen, Scotland.

For an insightful woman

Come with me, I invite you forth
On a voyage to the centre of the Earth
Unravelling what's inside Earth is
like trying to find something pretty much
invisible
Trust me, not one but several
truthful ways and solutions
are possible
Unbelievable, inconceivable
fantastic they might seem
Nature decided that this must have been
But you will be the first
The very first
To quench your thirst
For this new knowledge, that you always dreamed about
The current answers are based upon
traditional and obsolete researchers'
pretensions of Nature's knowledge
Errors, uncertainties and arrogance are always present
but they don't want this to acknowledge
You may wonder: what is the
reliability/uncertainty of the current
Earth's inner model?
Now you will see what is wrong and who is right
In the end, you will see the light
On our voyage to the centre of the Earth.

Contents

Introduction

Truth, bitter truth.
Danton

One of popular science book's primary functions is to make what would otherwise be inaccessible, specialist knowledge accessible to the lay reader. A basic level of scientific literacy strikes the right balance between scientific accuracy and universal appeal. Generally, anything that inspires people to be curious about the world and to enjoy the pursuit of knowledge is a pretty great thing, any popular scientific text must accomplish. This is the aim of this book.

However, this is not an ordinary popular science book which talks about science. It is off the beaten path. This book talks science. It is aimed for the layman (and laywoman) either an accountant or a blacksmith. The book was written to be easily understood and technical terms are kept at minimum, though maintaining the scientific rigor of the book.

These terms are explained whenever it is necessary. A basic knowledge of high school mathematics by the reader would suffice, especially in chapter five. Some paragraphs in this book were written as introductory lectures so the untrained reader can grasp the basic scientific knowledge; which is necessary when discussing a specific topic.

Despite twenty-first century technology that boggles the mind, the *terra firma* we take for granted walking around on what we call Earth is largely an *unknown* entity. We've been taught to believe it's a solid spherical mass with semisolid mantle and a molten iron core. But where's the actual proof? Without empirical evidence, what conventional science merely postulates as the truth really is nothing more than a model or hypothesis.

And what if Earth is not a solid as commonly believed? What if instead it is a spherical thick *shell*, defying the current academic establishment. Because realistically, based on the previous historic track record, initially any *new* proposed theories and conclusions about truth and reality were always vehemently resisted and rejected by the established conventions of the day.

But gradually as stronger evidence emerges, the truth is supposed to win out over the established false consensus. Indeed, as it will be discussed trough this book, part of current science had become a big magic trick, you can't believe anything you see.

Counterfeit science is like that. To the untrained eye, it looks real, but when you examine the evidence, it's simply f*abricated*, taken out of context, or extrapolated beyond what the evidence indicates. It is worthless information. The details do not support the conclusions being drawn from the evidence. It's understandable how counterfeit science and money prevail so often in our culture. The general population simply are not trained to identify the details of either, so they can be easily taken in. Popular science exists merely to entertain the masses. Those who have the capacity of a *deeper understanding* know it.

Usually popular science books present the so-called bright side of the scientific endeavour and some of its advocates acquire a *cult status*. Seldom these popular science books mention the fact that for instance, current physics is burdened by unnecessary pitfalls and owes many of its troubles to unclear or false definitions, inconsistent modelling, untenable assumptions, neglected conditions, carelessly applied mathematics, careless simplifications and misunderstood or even *fake* experimental results.

This embodies the *crisis* modern science is currently facing and this book, emphasizes this trend in regards to the current accepted model of Earth's interior. The ideals of the crowd have *usurped* research and experience, and we are each left clinging to the beliefs that best confirm our biases. This decadent Western culture elevates subjectivity over factuality, *scientific methodology* and common sense. Therefore, *real* scientific truth becomes an *endangered species*.

Given the above, the problem is the inevitable spreading of *misinformation*. Most of the public is certainly getting a cartoon version of almost everything that science says. Some with sound scientific background might be immune to it. In general, the public tend to give scientists the power by trusting them more than their common sense. It is rightfully to ask: does popular science books generate an enlightened population or a bunch of misinformed 'know-it-all's'?

Science frequently makes choices between alternatives. Once the choice is made, however, scientists tend to unify (groupthink) behind the *accepted* alternative to the extent of denying and eventually forgetting that there was any "real" choice made. Subsequent textbooks gloss over any possible *alternatives*, depicting science as a straightforward march up the one correct path toward truth.

Since it is forgotten and denied that such choices existed, the results of these choices are rarely reviewed. Not only is there no provision, or incentive, for such a review. There is positive, and powerful, peer pressure against any such *questioning* of basic premises. This book dares to question these basic premises.

Average popular science books should be *truthful* and enlightening, so people can read a given science topic and be confident in what they are reading. This book was written with this purpose, aiming to generate debate, influence culture and inspire future researchers.

This book is divided in five chapters. A brief description of each one is given below. The reader, especially the one with no scientific background is encouraged to thoroughly, digest the first chapter as it explains what science is and how *real* scientific knowledge is acquired. This enables the reader to keep the scientific cheaters off and comprehend what this book is all about. Students will find how often they are being told *lies*. A complete list of sources for further reading, is added at the end.

Chapter One: Is focused in explaining exactly what is known as science and the only valid way to acquire scientific knowledge: The *scientific method*, stressing the importance of observables and experiments. Its five parts are briefly discussed with examples. Key concepts such as law, hypothesis and theory, among others are rigorously defined. *Mistakes* in applying the scientific method as well logical reasoning flaws, that distort the scientific endeavour are highlighted. The importance of a *concise language* in science and the use of clear and precise concepts to avoid ambiguities and misunderstandings, that are misleading the scientific research are stressed out. Scientific *fraud* as common practice is described. A brief discussion of the importance in understanding the concept of *scientific paradigm* and how it hinders the advance of science, from past to present, is included. This chapter explains the old and outdated paradigms still in use and the importance of the emerging new ones, that are used to support the alternative view discussed in this book.

Finally, there is a description of the current division in the scientific endeavour: one going *underground* and almost kept off the public (and is related to some points of this book) the other known as *mainstream* or conventional science, running in universities and colleges; where the public is aware of its development through popular science media outlets such as TV, books and magazines.

Chapter Two: Is dedicated to examine how the current model or paradigm of the Earth's interior was developed and why it is *questionable*, applying several points discussed in chapter one and the latest scientific research. This chapter addresses the methodology, limitations and issues faced by Seismology when interpreting raw data to figure out a model of the Earth's interior. Basic concepts of Geology and Seismology are briefly explained for a better understanding of the book's main topic.

Prevailing ideas of pressure and temperature present in the centre of Earth along its alleged inner iron core are proven to be *untenable*. The accepted origin of the Earth's magnetic field is scientifically *impossible*. Finally, during the drilling of the 12-km depth *Kola Deep Borehole* in Russia, analysis of samples showed that several geological and geophysical ideas held by mainstream science proved to be *false*, casting serious *doubts* of the current model of the Earth's interior.

Chapter Three: Deals with Astronomy because Earth being a planet, must follow the same principles that gave the *raison d'être* to other planets in the Solar System and the Universe in general. It addresses the difficulties, flaws and *inconsistencies* of the conventional science to explain the origin of the Solar System and its planets, especially Earth.

Contrary to mainstream science, which hinges gravitation as the driving force in the formation and organization of galaxies, stars and planets; the importance of *three* variables: gravitation, electro-magnetism and vortexes/rotational physics, that indeed shape the Universe, are highlighted.

This unconventional knowledge, helps to dispel misconceptions such as dark matter among others. Based in recent scientific research and the application of the *new emergent paradigms*, detailed in chapter one; this chapter explains the reasons of the *unfeasibility* of both: the nebula collapse model for the origin of the Solar System and the accepted chaotic-accretion model of planetary formation that originates a solid spherical planet with a hot inner core.

Finally, here is presented a more realistic star and planetary origin, based on *observables* and the *latest* scientific research, that lead to a *spherical shell* planetary formation. With this novel approach, the origin of Earth and its spherical shell distribution of mass is explained. Other aspects of this alternative Earth formation are detailed in chapter five.

Chapter Four: Discusses the observed *anomalies* (mostly in the North Pole) that are inconsistent with the definition of polar regions, pointing out a different conception of the Earth's interior, other than the current paradigm.

Conventional science dismisses some of those and gives unsatisfactory explanations for the rest. Similar anomalies are observed in other planets in the same locations. Antarctica is a special and restricted area. Ignored scientific research is brought into light to explain the *true* cause of Earth's magnetic field along notable anomalies within the Solar System.

Chapter Five: Some *inconsistencies* in the current gravity models that affect Earth's mass, are explained. The concept of Moment of Inertia (MoI) which defines the internal mass distribution of a rotating body is applied in the case of Earth. Conventional science, with a *flawed* procedure, misapplied the concept of *moment of inertia* concluding wrongly, a solid Earth. A more accurate and less artificial MoI technique shows otherwise, the distribution of Earth's mass is indeed a *spherical shell*, mostly void which is in phase with *recent* studies by a space probe sent to Jupiter, that shows this same feature.

Earth has an inner spinning body which radiates varied types of energy and can explain the observed anomalies in the polar regions. This radiated energy comes out through apertures or *polar openings*. Information of polar regions is mostly filtered and edited before becoming public. A brief exposition of the expanding Earth along the *unfeasibility* of convection currents and inconsistencies of Plate Tectonics, is included.

Finally, the aim of this book is not to persuade the reader but to encourage a critical thinking. Quoting German scientist and philosopher Immanuel Kant (1724-1804): "Dare to know! Have the courage to use your own intelligence! It is never too late to become reasonable and wise".

CHAPTER I

What's *real* science and scientific approach.

Today's scientists have substituted mathematics for experiments, and they wander off through equation after equation, and eventually build a structure which has no relation to reality.
Nikola Tesla (1856-1943)

Nowadays, people have an unshakable faith in science. Our lives are permeated by science and technology as never before. The reader should be familiar with newspaper articles often start with: "Scientists say...," without ever actually naming the "scientists" as if the word is coming down from on high, through the scientific priesthood. When involved in a debate, people often resort to: "The science says…" or "What does the science say?" to support their argument.

Science it seems, is the Ultimate Authority. People believe that science can give us truth. Not so. Part of it has little or no scientific substance. We are told that we cannot trust ourselves, we must bow to the higher wisdom of science: "We need to know that instinct is no substitute for the neutral evaluation of a hypothesis".

However, we have come to a place where scientific theories and data are completely *ignored* unless they have been published in a peer-reviewed, scientific journal. As if getting a handful of other scientific priests to agree somehow, grants holiness to the idea.

The current system of peer review publication has been in place for several decades. It makes it impossible for anyone who is *not* a member of the scientific priesthood to be heard. It also makes it nearly impossible for even a member of the priesthood, to be heard if his or her ideas are considered scientific heresy. Meanwhile the people, not being conversant in scientific language and largely unable to decode the scientific literature, are dependent on a go-between, a translator: meaning the popular science books and mainstream media broadcasts.

This book might be considered a heresy to the guild mentioned before. But its content is indeed a rightful criticism and an *alternative* view, which the people should know and judge by themselves. For the purposes of this book and the fact it covers different branches of science and reaches an unorthodox conclusion; it is important to clarify to the layman, before going to the matter, the true meaning and scope of what's commonly known as *real* science. It will make possible to the reader, the dispelling of some misconceptions about what really science and acquiring scientific knowledge is all about and how it must be carried out.

This would prove to be key when discussing several issues along this book. The aim is to get a yardstick that enables us to compare real science with pseudoscience. Several "scientific facts" that are sold to the public are indeed, either speculations or *unproven* hypothesis. Modern physics continues to be more *fantasy* than reality. It supports highly speculative theories about the nature of space, time, and reality itself; theories which to date have NO experimental, real world *evidence*.

What is science *really* and how it is done? Science is a special body of knowledge which is elaborated by certain persons called scientists, either professionals or amateurs. In fact, some comets were discovered by amateur astronomers. An amateur fossil hunter and bus driver in Scotland, has found a unique fossil which is considered, the remains of the oldest creature known to have lived on land.

Science in itself, is a human endeavour and we can define it, operationally:

Science is the pursuit and application of knowledge and understanding of the natural and social world following a systematic methodology based on <u>evidence</u>.

Science is a social activity and has an impact upon society. People are curious and many want to know real facts. Usually people are objective. Specifically, we base our judgements on observations and verified facts.

However, often we realize to what extent one and everyone else can be *biased* by individual's perspective. Science as a social activity, should be driven by the following five trends:

Communalism (not communism): knowledge should be accessible for all people.

Universalism: everyone should have the right to contribute.

Disinterestedness: science should be objective and not ruled by special interests.

Originality: the results should be new.

Scepticism: scientists should be open to criticism.

The key in any branch of science is to understand the term *scientific methodology*, because often, scientists seem to forget following procedure. Scientific methodology basically means that the foundations of any science lays in *evidence* or empirical facts not *opinion,* preferences or dogmas, let alone pure mathematics. Science starts and ends in facts given by *experience*. However, several individuals wrongly labelled as scientists, the so-called "theoretical physicists" or theorists for short; often disregard experience and conclude that whatever they think or calculate, automatically is the "truth". Therefore, in order to any knowledge can be considered "scientific" it must be the fruit of a *scientific methodology*. In other words, it can't just pop out from the thin air or pure reasoning independent of experience, so to speak.

Finally, the aim of science is to understand Nature in all its physical manifestations and how it works. Until the beginning of the 19th century, individuals engaged in any scientific research were known as *natural philosophers* and science was still known as *Natural Philosophy*. The term *scientist* wasn't used yet, as such.

The word "science" is derived from the Latin words *Scientia* and *Scire,* which means "to know" and is knowledge based on *demonstrable* and *reproducible* empirical data, nothing else. True to this definition, science aims for measurable results through *testing* and analysis. We must emphasize true science is *empirical*, essentially meaning it deals with what can be observed with the accepted five senses and is enhanced with instruments such as telescopes, microscopes, radiation detectors, etc. Also, it uses common sense and logical reasoning, glued to the universal law of *cause* and *effect.*

In contrast, some modern "scientific" theories aren't built upon empirical data but mathematical assumptions of what reality ought to be, have several untestable arguments, disregard common sense and not necessarily; these theories follow logic reasoning which gives them a lack of consistency and coherence. That's why many people consider these theories weird and confusing, giving the false impression that only a few can truly "understand" the Universe.

Due to the vast and broad range of aspects we identify in Nature, science has many disciplines and subdivisions. The natural manifestations of Nature are known as *phenomena* which is the plural of phenomenon. The word phenomenon comes from the Greek *φαινόμενον* (phainómenon) which means "that which appears or is seen,"

A natural phenomenon is not an engineered event manufactured by humans, although it may affect them. In general, all phenomena do occur in our three-dimensional (3-D) space during a specific interval of time. Phenomena involves objects. For the purposes of science, an object is that which has shape in a background or region.

Science is ultimately concerned only with objects that do *exist*. If a concept or statement regarding any object, doesn't have its counterpart in the physical reality "out there" then, it doesn't exist and thus does not concern science. Plain and simple.

Due to the diverse phenomena studied, science developed a wide range of basic concepts and units to measure them such as time (second), length (metre), mass (kilogram), electrical charge (coulomb), kinetic energy (joule), etc. As a rule, the more measurements and quantities are involved in a scientific study, more reliable scientific knowledge is attained.

Common examples of natural phenomena include sunrise, the natural weather patterns, decomposition, free fall and erosion. Most natural phenomena, such as fog, are relatively harmless so far as humans are concerned. Bear in mind scientific knowledge is produced dealing *only* with phenomena observable in the *present*, in studying physical processes that actually do occur, in real time.

In fact, the huge success of science as body of knowledge and its technical applications; was cumulatively achieved, by dealing phenomena that were and still are observable in the present such as eclipses, bodies expansion due to heat or radioactive decay.

While a good scientific theory can predict a phenomenon not yet known, it can be verified in the future. However, there is no way to return to the past or bring it back and verify any new phenomenon "predicted" by theory. On the other hand, the attempt to explain a present observable such as the Solar System, being the result of alleged past processes or phenomena, that are *not* observable in the present or inexistent, is a nonsense. This issue will be discussed in chapter three.

Even though the knowledge of the natural world was aimed since Antiquity, by several cultures; modern science as we know it today, started in the Seventeenth Century's Europe through the Scientific Revolution pioneered mainly by Nicolas Copernicus, Galileo Galilei, Johannes Kepler and Isaac Newton among others.

They contributed to lay down the methodological procedure to attain *true* scientific knowledge and demolished ancient ideas of how nature works such as the body's free fall and the geocentric concept of the Solar System. This methodological procedure is known as the *scientific method.* Any true scientific knowledge *must* be produced through this method to be deemed as such. An honest scientist always will follow the scientific method to produce reliable scientific knowledge.

Due to the many facets of the natural word and the scope of its activity, science is divided in branches such as Physics, Biology, Chemistry, Geology, Anthropology, etc. Each branch also is subdivided in Biochemistry, Geophysics, Ecology, Seismology and so on. Thus, scientists are often, restricted to their reduced areas of specialization which makes them to think deeply but not clearly. Scientists, working in a specific field are often not familiar enough with theories of other disciplines, involved in their studies/research.

Science comprises individual disciplines that reflect historical developments and the organization of natural and social phenomena for study. Social scientists may have methods for recording research data that differ from the methods of biologists or physicists, and scientists who depend on complex instrumentation may have authorship practices different from those of scientists who work in small groups or carry out field studies. Even within a discipline, experimentalists engage in research practices that differ from the procedures followed by theorists, who are often divorced from reality.

The scientific method

Before discussing the scientific method, we must introduce some basic concepts. Usually, when scientists are faced with an inquiry or issue of a specific sector of *physical reality* within Nature, they must isolate this sector from its surroundings to conduct a research. This isolated sector is known as *system*. A system is a portion of the Universe that has been chosen for studying the changes, that take place within it in response to varying conditions.

A system may be complex, such as a planet, or relatively simple, as the liquid within a glass. Also, some assumptions must be done, especially when some variables are unknown (which is questionable), in order to carry out the study satisfactorily. However, this approach is workable to some extent and with reliable outcomes. But in fact, this isolation for a specific study, is artificial.

Nothing in Nature exists independently of its surroundings. Everything in the Universe can be considered an *interconnecting* network of things; a complex whole.

A common example of a system is the Solar System or a living cell. Both can be isolated from its surroundings to allow a better study but this doesn't mean these are separated from its environment. In most real situations, many different effects are operating. Let's consider the elementary physics problem of a falling object. One normally calculates how long it takes to hit the ground considering only gravity, height and maybe air resistance.

In reality, however, its course will be influenced by radiation pressure, the Earth's magnetic field, the gravitational pull of Jupiter, Coriolis force, cosmic ray bombardment and the perfume being worn by the person who dropped it, to name but a few. Most of these effects are tiny: far too small to have a measurable effect.

There are two types of systems:

An *open system* is a system that freely exchanges energy and matter with its surroundings. For instance, when you are boiling soup in an open saucepan on a stove, energy and matter are being transferred to the surroundings through steam. Putting a lid on the saucepan makes the saucepan a closed system.

A *closed system* is a system that exchanges only energy with its surroundings, not matter. Due to its methodology and high specialization, science has put more emphasis in closed systems and tend to ignore its surroundings. This approach is misleading as we will discuss it later.

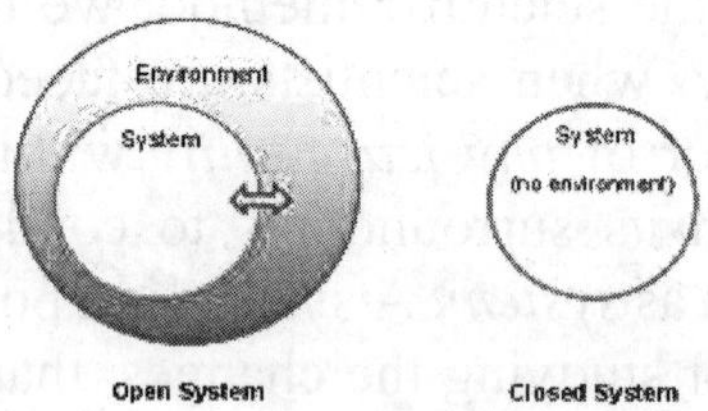

Fig. 1- An essential procedure in scientific research is to distinguish between an open sysatem and a closed system. An open system interacts with its environment through giving and receiving information/energy/matter. Many natural phenomena are open systems. A closed system only exchanges energy and it is typical in laboratories.

When conducting research of a determinate system, scientists use the scientific method to collect measurable, empirical evidence or *variables* in a system under *controllable conditions*, usually known as experiments, either in the laboratory or field. A variable is a factor or element that can be changed and manipulated in ways that are observable and measurable.

An essential feature in any science is the concept of *interaction*. It is commonly defined as reciprocal action, effect, or influence between two objects or things. Technically it is defined as force, is a push or pull upon an object resulting from the object's interaction with another object.

Whenever there is an interaction between two objects, there is a force upon each of the objects. When the interaction ceases, the two objects no longer experience the force. Forces only exist because of an interaction.

Conventional or mainstream science acknowledges only four basic forces: gravitational, electromagnetic, strong, and weak; that govern how objects or particles interact and how certain particles decay. These forces are the base of the known Standard Model of Physics. According to this theoretical framework, all the known forces of Nature can be traced to these fundamental interactions.

However, since a few decades ago, *new phenomena* and anomalies were discovered in several branches of science, that *cannot* fit into the Standard Model. This led some scientists to either re-consider it properly or introduce a *new force* or interaction (fifth force), as it will be discussed in chapter four. This does show that mainstream science and its advocates still don't fully understand these fundamental forces of Nature.

Cause and *effect* are often interchangeable with interaction. Cause and effect is a type of relationship between events whereby a cause creates an effect. It is the *foundation* of any science. Is important to mention that in many cases, an effect can result from many causes and the exact nature of these relationships can be difficult to determine. In some cases, cause and effect are straightforward. Scientists attempt to uncover and understand cause and effect relationships through well-designed experiments and investigations.

Nonetheless, scientists depending of their specialization (or indoctrination), often focus in just one or perhaps two causes and dismiss the other equally important ones. This necessarily would lead to incomplete and flawed scientific theories, as it will be discussed through this book.

Measurement is an integral part of modern science as well as of engineering, commerce, and daily life. It is often considered a hallmark of the scientific enterprise and a privileged source of knowledge relative to qualitative modes of inquiry. Measurement is an activity that involves interaction with a concrete system with the aim of representing aspects of that system in abstract terms (e.g., in terms of classes, numbers, vectors etc.) determining a target's size, length, weight, capacity, or other aspect.

Due the vast number of scientific disciplines and subtypes, it's not easy to whittle down a simple definition of the scientific method. Each discipline has its own scope and issues. However, there are a few fundamentals that are common to all.

Scientific method procedure

Usually it starts when a specific phenomenon poses a problem of what is happening or how to explain it.

Step 1: Make observations

Observations can be qualitative or quantitative. Qualitative observations describe properties or occurrences in ways that do not rely on numbers. Examples of qualitative observations include the following: the outside air temperature is cooler during the winter season, table salt is a crystalline solid, sulphur crystals are yellow, and dissolving a penny in dilute nitric acid forms a blue solution and a brown gas.

Quantitative observations are measurements, which by definition, consist of both a number and a unit. Examples of quantitative observations include the following: the melting point of crystalline sulphur is 115.21° Celsius, and 35.9 grams of table salt whose chemical name is sodium chloride dissolve in 100 grams of water at 20° Celsius.

Often in this step, key factors involve measurement and data (possibly although not necessarily using mathematics as a tool).

Step 2: Formulate a hypothesis

A hypothesis is a *tentative statement* about the relationship between two or more *variables*. It is a specific, testable prediction about what you expect to happen in a study. There are two types of variables involved in any hypothesis. The *dependen*t variable is what is being studied and measured in the experiment. It's what *changes* because of the changes to the *independent* variable.

After deciding to learn more about an observation or a set of observations, scientists generally begin an investigation by forming a hypothesis, a tentative explanation for the observation(s) or in other words, an educated guess about what Nature is going to do, or about why Nature does what it does.

Most of the time, the hypothesis begins with a question which is then explored through background research. It is only at this point that researchers begin to develop a testable hypothesis. The hypothesis may not be correct, but it puts the scientist's understanding of the system being studied into a form that can be tested. A hypothesis often follows a basic format of "If (this happens) then (this will happen)."

For example: assuming constant temperature, if *moisture* levels in the air changes, bacterial *growth* will be affected. In this case the independent variable is moisture and the dependant variable is growth. The *experiment* will have the last word in deciding the *validity* of this hypothesis. This is a crucial step for scientists branded as experimentalists. However, theorists simply just don't care.

Step 3: Design and perform experiments

Without exceptions, any scientific hypothesis *must* be "testable". However, this is often overlooked in some branches of science, regardless of the sophistication of equipment used and/or scientist's "expertise". Science proceeds by making careful observations of Nature, often with experiments. If a hypothesis does not generate any *observational tests*, there is nothing that a scientist can do with it.

Arguing back-and-forth about what should happen, or what ought to happen, is *not* the way science makes progress. Unfortunately, many scientists, often in the branch of "theorists" make this mistake, over and over.

Consider the following hypothesis:

Hypothesis X:

"*Dark matter is matter (or mass) in the Universe that we cannot see directly using any of our telescopes. Our telescopes see not only visible radiation (constituting the spectrum of colours that our own eyes can detect), but other types of radiation as well. Dark matter does not reveal its presence by emitting any type of electromagnetic radiation.*

It emits no infrared radiation, nor does it give off radio waves, ultraviolet radiation, X-rays or gamma rays. It is truly dark. Cosmologists believe we can only see about 5 percent of the matter in the Universe".

This statement may or may not be true, but it is NOT a scientific hypothesis. By its very nature, it is *not testable* and bluntly, it *breaches* the scientific method. There are no observations that a scientist could make to tell whether or not, the hypothesis is correct. Notice the word *believe* in it.

In real science, there is no room for *beliefs,* only and only facts. Building a skyscraper, sending artificial satellites into space or performing a successful open-heart surgery are not based on beliefs whatsoever, only in well proven scientific and technical *facts.*

Ideas such as Hypothesis X are interesting to think about, but real science has nothing to say about them. Hypothesis X is a *speculation,* not a scientific hypothesis whatsoever. Sadly, several "scientific" facts sold to the public are indeed, nothing more and nothing less than speculations.

Technically speaking, it is *pseudoscience*. In fact, pseudoscience is theory or speculation which has the trappings and rhetoric of science, and is presented as science, but does NOT follow the scientific method.

One of the hallmarks of a pseudoscience is that it makes claims that cannot be verified or proven false.

After a hypothesis, has been formed, scientists conduct experiments to test its validity. Experiments are systematic *observations* or measurements, preferably made under *controlled conditions* or variables that is, under conditions in which a single variable change.

In the case of bacterial growth mentioned before, the experiment to test the hypothesis, would be made as follow: two samples of the *same* known bacterial culture and moisture content, will be taken. One sample is kept under the same moisture and temperature as the original culture.

Technically this is known as *control sample.* The other sample will be put under the same temperature but with higher or lower moistures.

Then its growth is checked out. If there is any considerable *variation* in the growth, compared with the control sample, the hypothesis is proven to be correct. If not, the hypothesis is proven to be false.

Scientists like detectives, must have the *talent* for creating or choosing hypotheses that are correct. They must have a sound *intuition.*

One word of warning about the so called "thought experiments". These are NOT part of the scientific method because the "scientist" engages in armchair or wheelchair speculation, as far removed from reality as possible. Thought experiments, which true scientists sneer at precisely, because it doesn't require one to get one's hands dirty. Also, thought experiments usually lead to *misleading* conclusions.

Step 4: Accept or modify the hypothesis

Even the best and most elegant scientific hypothesis *must* be subject to being either proven correct or wrong by experimental or observational data. If the experiment or observation agrees with the hypothesis's prediction, then the hypothesis passed the test and has basis for a theory. Scientists will tentatively accept the hypothesis as being correct but continue to carry further tests. The most important identifying feature of *real science* however is what happens if the hypothesis *fails* the test. For instance:

At the end of the 18th century, many scientific thinkers accepted the caloric hypothesis as an explanation for the phenomena associated with heat. *Caloric* was considered to be a "massless" fluid that could flow between bodies, a fluid not unlike the electric fluid proposed by US physicist Benjamin Franklin or a comparable magnetic fluid, also in vogue.

In some versions of the theory, the fluid was composed of discrete particles. The caloric theory was in accord with many observations, including the existence of heat capacities that governed the exchange of heat between bodies, the latent heat absorbed by substances in phase changes, and the release of heat in chemical reactions. Hot bodies had a specific "excess" of caloric. When it cooled down, is because the same quantity of caloric was "transferred" to the surroundings. Although some phenomena, such as the production of heat by friction, *lacked* a satisfactory explanation within the hypothesis.

Then in 1798, US physicist and inventor Benjamin Thompson (1753-1814) Count Rumford, while drilling out cannons in the Munich munitions works, he noticed that the canon became hot as long as the friction of boring continued. Furthermore, Rumford observed, the amount of heat released would be sufficient to completely melt the canon if it could be returned to the metal. Since more heat was being released than could have been originally contained in the metal, these *observations* were an outright *contradiction* to the caloric hypothesis.

Rumford thought there was no separate caloric fluid and that the heat content of an object was associated with motion or internal vibrations, motion which in the case of the cannon was bolstered by the friction of the tools.

He had recognized the relationship between heat energy and the physicists' concept of 'work' the transfer of energy from a system into the surroundings, caused by the work done, results in a difference in temperature. Thus, the hypothesis of caloric was discarded and replaced for a new one that could *account* for the *contradictions* observed in the old one. In other words, it is a superseded scientific theory or better said hypothesis.

Fig. 2- During the manufacturing process of canons, teams of horses harnessed to a large-toothed wheel supplied the power needed to drill a hole several inches in diameter straight down the center of a solid brass or bronze cylinder, which was cooled by water. Based on his observations, Rumford found that heat and work are equivalent ways of transferring energy. This fact proved the caloric theory (hypothesis) to be false. Rumford concluded that heat is the result of the interaction between the motions of the particles of ordinary matter and the vibrations of an ambient aether, acting as a mediator.

This example shows that if the data disagree with whatever the hypothesis predicted, then the data proved the hypothesis wrong. Data falsified the hypothesis. Scientists must then *change* the hypothesis. Changes to the hypothesis might range from minor tweaks to a completely *new* hypothesis that incorporates and agrees with the new data. A properly designed and executed experiment enables a scientist to determine whether the original hypothesis is valid. In which case, he can proceed to step 5. In other cases, experiments often demonstrate that the hypothesis is incorrect or that it must be modified thus requiring further experimentation.

However, this procedure is *not* always followed in the scientific endeavour, because a hypothesis might become entrenched in the scientists' minds. Despite overwhelming evidence disagreeing with a specific hypothesis, scientists are *reluctant* to admit that it is wrong and nevertheless, they cling to it as we will explore it later. This entrenched hypothesis usually becomes what is known as a paradigm or scientific dogma.

Step 5: Development into a conclusion, law and/or theory

The final step of the scientific method is developing a *conclusion*. When more experimental data is collected, and analysed making sure there are *no* discrepancies between observations and hypothesis, a scientist may begin to think that the results are sufficiently *reproducible* to merit being summarized in a law, a verbal or mathematical description of a phenomenon that allows for general predictions. A law simply says what happens; it does not address the question of why. The same applies to scientific discoveries. It is the starting point which enable scientists to develop broad general explanations, or *scientific theories*.

Bear in mind that the *reproducibility* of published experiments is the foundation of science. No reproducibility no science. When German chemist Otto Hahn discovered and published the phenomenon of nuclear fission in 1938, independent scientists around the world could *confirm* his discovery, because they could *repeat* or replicate consistently the experiment and *report* nuclear fission, which later lead scientists to take advantage of the atomic energy.

Another example is the discovery of X rays and its applications. An isolated finding, without corroboration or replication from other sources, is a dubious science.

This is the issue of how much *consistency* should be expected in replicated experiments. Scientists often make a mistake when one or maybe two "successful" studies or experiments can convince them that a claimed hypotheical fact is real. For example, in particle physics, some "fundamental" particles or fuzzy and microscopic vibrational clouds, lasting only nanoseconds are to be considered "found" based on only one or two events out of a total of more than 100.000 experimental trials carried out by gigantic and costly particle accelerators.

In other words, an event with an extremely *poor* replication rate, observed once in a hundred thousand times, was considered enough to convince most physicists that a "fundamental" particle was real. This can be considered cherry-picking. Later, this is found not to be true at all, like the case of the Higgs Boson or "God's Particle" now considered: "the particle that wasn't". It melted away. Even an entire book has been written by a particle physicist (Unzicker, 2013), exposing this fakery.

However, some top universities like to show off in their brochures, that one of its Physics Department professors, was involved in the research that led to the "discovery" of this notorious particle. One must ask: are we supposed to take these physicists seriously and whatever they might say through the media?

An appendix to the scientific method, includes the device known as *Occam's Razor*. It states that:

Essentially, when faced with competing explanations for the same phenomenon, the simplest is likely the correct one.

The scientific method strives in letting 'reality' show itself, in the *simplest* possible way. In fact, so far, *successful* and proven scientific theories are characterized for being coherent and *simple.* This of course, doesn't mean and shouldn't never be confused with *oversimplification*, at all. However, some scientists don't follow this procedure and tend to elaborate complicated and often unverifiable claims of how Nature should work, which also offer more questions than answers.

For example: the many-worlds interpretation (MWI) is *criticized* for this reason. It suggests that we live in a near-infinity of universes, all superimposed in the same physical space but mutually isolated and evolving independently. In many of these universes there exist replicas of you and me, all but indistinguishable yet leading other lives. If this claim is to be taken seriously, then it is unthinkable and by the scientific method *procedure*, it doesn't fulfil the function of science. It falls in the realm of science-fiction.

The *scientific method* is critical to the development of *any* scientific theory. Only through this method, *true* scientific knowledge is produced and indeed, true scientists stick to it. Because scientists can enter the cycle shown in the figure at any point, the actual application of the scientific method to different topics can take many different forms.

For example, a scientist may start with a hypothesis formed by reading about work done by others in the field, rather than by making direct observations. Any finding reached through the scientific method is (in theory) faced with critical exposure to scrutiny, peer review and assessment, to ensure reliability.

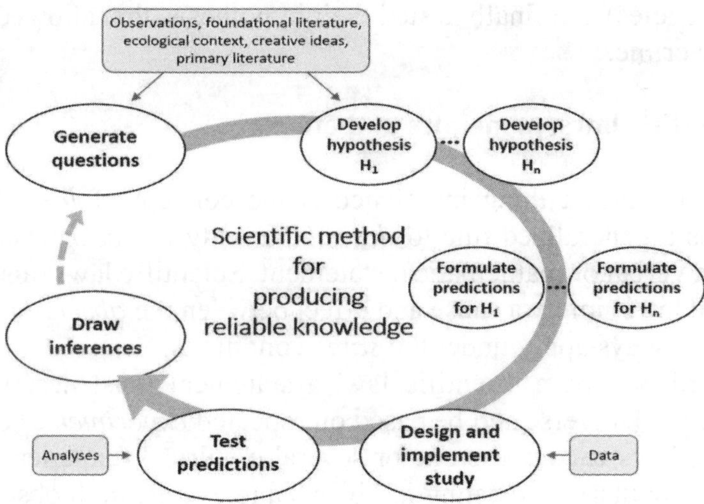

Fig. 3- Flowchart of the scientific method. True scientific knowledge is produced when <u>all</u> steps have been carried out, thoroughly. The scientific method can only answer questions that can be proven or disproven through testing.

Recent well-publicized cases of scientific fraud prove that scientists can be as susceptible to the allures of wealth, power and fame as politicians, the group that enjoys the lowest public trust. Glaring recent cases have included falsified results in the development of an HIV vaccine and new techniques for producing stem cells. Such breaches prove that scientists do not always base their work strictly on rigorous experimentation, data collection and analysis, and hypothesis testing.

On the other hand, scientists frequently *disagree* with one another, both as individuals and as representatives of competing schools of thought. Some of these debates rage on for years. Superstring Theory, sometimes called apparently the "theory of everything," has been a topic of vigorous *controversy* for over three decades. It has become a *degenerative* research project, getting increasingly complicated and, at the same time, *removed* from empirical reality. A total waste of time.

The work of a scientist can be compared to that of the detectives portrayed in literature and movies. They come upon the scene of a crime after all the interesting events have happened. Then looked carefully at all data and collected subtle and seemingly unrelated clues, making sure to avoid to jumping to conclusion. Later, these detectives attempted to build up a coherent story of data (being sensitive to slightest inconsistencies) and finally tested their hypothesis, that allowed them to solve the crime.

Scientific laws, principles and effects

One important thing in science is the concept of *law.* A law in science is a generalized rule to *describe* a body of *observations* in the form of a verbal or mathematical statement. Scientific laws (also known as natural laws) imply a cause and effect between the *observed* elements and must always apply under the *same* conditions.

In order to be a scientific law, a statement must describe some aspect of the Universe and be based on repeated *experimental* evidence. Scientific laws encompass one or several empirical (experiential) laws regarding regularities existing in objects and events, both observed and posited. Scientific laws may be stated in words, but many are expressed as mathematical equations.

A law is simply an equation describing a relationship between physical quantities or variables. Laws are widely accepted as true, but new data can lead to changes in a law or to exceptions to the rule. Sometimes laws are found to be true under certain conditions, but not others. For example, Newton's Law of Gravity holds true for most situations, but it breaks down at the sub-atomic level.

Irish chemist Robert Boyle studied the compressibility of gases in 1660. In his *experiments*, he observed: "At a fixed temperature, the volume of a gas is inversely proportional to the pressure exerted by the gas."

Mathematically:

$$P \propto 1/V$$

$$\text{or } PV = \text{constant}$$

For an initial and final conditions of the gas sample, the Boyle's Law formula is:

$$P_1V_1 = P_2V_2$$

This law is applicable to some extent until pressure is too big and the change in volume deviates from the Boyle's Law. Often in science, laws are called by the scientist who first discovered or elaborated them: Newton's laws, Faraday's laws, Ohm's law, etc.

Pressure p(kPa)	Volume (cm^3)
200	31
180	34
140	44
100	62
85	73
70	88
60	103

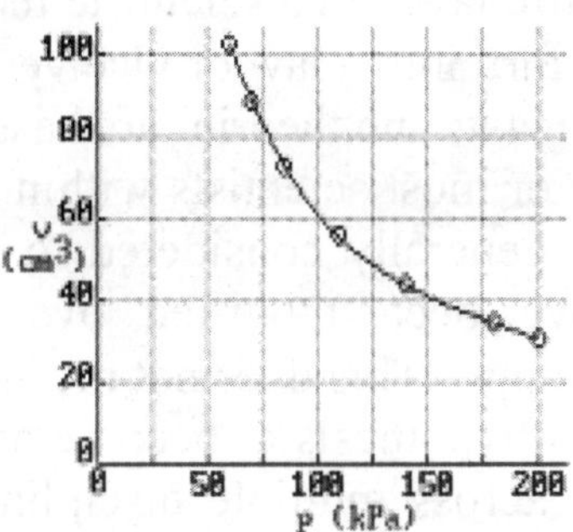

Fig. 4- Scientific laws come from either controlled experiments or careful field observations and data often is expressed in graphs or equations, describing the relationship between the variables involved. This graph shows the inverse proportion between pressure and gas volume. (Boyle's law).

Another concept often used in science is *principle.* It is a basic statement regarding a natural fact, based on an empirical controlled observation, which cannot be reducible or based on more simple facts.

For instance:

Principle of Archimedes: "A body immersed in a fluid is buoyed up by a force equal to the weight of the displaced fluid"

Other well know principles in science are: the principle of least action and the conservation of energy, among others. Scientific laws and principles are the building blocks of any science. Often, we can come across with the term *effect.* When using the term effect in science, it is the description of a specific natural phenomenon of interest and often bears the name of its discoverer.

For instance:

Photoelectric effect: "Phenomenon defined as the ejection of electrons from a metal plate, when light falls on it"

Other notable effects are known as: Döppler Effect, Coriolis Effect, Butterfly Effect, etc.

Scientific Theory and Scientific Model

Scientific laws do not try to explain 'why' an observed event happens, but only that the event actually occurs the same way over and over. The *explanation* of how a phenomenon works is a *scientific theory*. A scientific law and a scientific theory are not the same thing, a theory does not turn into a law or vice versa.

Both laws and theories are based on *empirical* data and are accepted by many or most scientists within the appropriate discipline. Scientific laws are generally considered to be without exception, though some laws have been modified over time after further testing found discrepancies. This does not mean theories are not meaningful.

For a hypothesis to become a theory, *rigorous testing* must occur, typically across multiple disciplines by separate groups of scientists. Saying something is "just a theory" is a layperson's term that has no relationship to science. One must be *cautious* with contemporary "scientists" who claim they are happy to say, that a theory never becomes a fact. This is very misleading. It is pseudoscience.

To most people a theory is a hunch. In science, a theory is the theoretical framework to *explain* the observed facts and *predict* new ones. Laws are a starting place, from there, scientists can then ask the questions, why and how?

For example, Newton's Law of Gravity is a mathematical relation that describes how two bodies interact with each other in space. This law does not explain *why* gravity works or even what gravity *is*. Scientific laws can be used to make new discoveries and perform calculations.

In the case of the AC (alternating current) generator, it was invented utilizing a spinning coil at a constant rate, in a magnetic field to induce an oscillating electromotive force. This was possible applying the *Faraday's law of induction.*

Technically in science, a true *theory* is a systematic ideational structure of broad scope, conceived by the human imagination, that encompasses a family of *empirical* (experiential) laws regarding regularities existing in objects and events, regarded as phenomena.

A scientific theory is a structure suggested by these laws and is devised to explain them in a scientifically *rational* (logical) manner. Thus, when empirical laws can satisfy curiosity by uncovering an orderliness in the behaviour of objects or events, the scientist may advance a systematic scheme, or scientific theory, to provide an accepted explanation of *why* these laws do occur.

An example of scientific theory is the Atomic Theory in chemistry, which states that all matter is made up of tiny indivisible particles (atoms). According to modern interpretations of the theory, the atoms of each element are effectively identical, but differ from those of other elements, and unite to form compounds in fixed proportions. Atoms of the same kind (element) are uniform in size, weight, and other properties.

Atomic theory *explains* chemical laws such as the law of definite proportion, (or constant composition) sometimes called *Proust's law*. It states that a given chemical compound always contains its component elements in fixed ratio (by mass) and does not depend on its source and method of preparation. For instance, any sample of pure water contains 11.19% hydrogen and 88.81% oxygen by mass. It does not matter where the sample of water came from or how it was prepared. Despite being demonstrated atoms are divisible and made of subatomic particles, this doesn't affect the Atomic Theory as such and its use in chemistry. Scientific theories have much greater scope, explaining a variety of such laws and predicting new phenomena as yet undiscovered.

Let's face it, the *validity* of any science lies in its power to *predict*. A prediction is a statement suggesting what will happen in the future, based on *observation,* experience or a hypothesis. Scientists abstract hypotheses from observations of the world, which they then deploy to test their reliability. The best way to test reliability is to predict an effect before it occurs. If we can manipulate the independent variables (the efficient causes) that make it occur, then ability to predict makes it possible to control.

For example: According to Newtonian mechanics, the *spinning* of a planet on its axis should cause it to bulge at the equator. This is the content of Proposition 18 of Book III of the *Principia*, and in Proposition 19 Newton estimates that the equatorial diameter of the Earth differs from the polar diameter by about 1 part in 230 (modern estimates are about 1 part in 297.4). With this statement, Newton predicted the Earth could not be spherical, but ellipsoid instead slightly flattened at the poles. It was matter of *verifying* whether that prediction was true or not.

If Newton was correct then to account for 90° between the equator and the poles, the *distance* corresponding to one degree of latitude should be *shorter* at the equator than at the poles, as illustrated in the following figure:

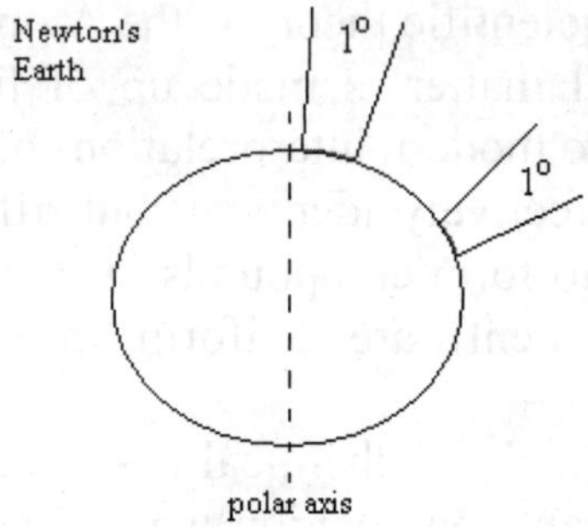

Fig. 5- Newton's prediction of an Earth slightly flattened at the poles, meaning the distance corresponding to one degree of latitude should be shorter at the equator than at the poles, was confirmed by two French expeditions in 1735-1736.

In 1735 and 1736, two French-led expeditions set out with the goal of the verification of Newton's hypothesis on the flattening of the terrestrial globe in the polar regions. One expedition which included the notable French mathematician and astronomer Alexis Clairaut, headed poleward to Lapland to measure several degrees of meridian at the arctic circle.

The other went equatorward, to the Viceroyalty of Peru, by then a territory of the Spanish crown. They established their headquarters in Quito, present-day Ecuador, to make the same measurement in the vicinity of the equator. The Quito team included Pierre Bouger, a French mathematician and surveyor alongside two Spanish scientists, Antonio de Ulloa and Jorge Juan.

After each expedition measured the length of a degree of latitude at their respective locations, the scientific result was clear: The Earth is indeed a spheroid flattened at the poles, as Newton had *predicted* and made his theory a success. However, current alleged scientific theories such as String Theory or the Big Bang, have *failed* for decades, to predict new phenomena in their respective areas. In other cases, an observed phenomenon "predicted" by theory was in fact, a *fake* one such as the notorious Higgs boson, mentioned before or the gravitational waves, which will be discussed later.

Scientific model is the generation of a physical, conceptual, or mathematical *representation* of a real phenomenon or physical object that is impossible/difficult to observe *directly*. This is key to understand as it tends to misleading. Scientific models are used to explain and predict the behaviour of real objects or systems and are used in a variety of scientific disciplines, ranging from physics and chemistry to ecology and Earth sciences. Although modelling is a central component of modern science, scientific models at best are *approximations* of the objects and systems that they represent; they are not exact replicas.

As a rule of thumb, all models are wrong but some are useful. Thus, scientists constantly are working to improve and refine models. However, people often *mistake* scientific models with *reality*. For instance, what's inside Earth cannot be observed "directly" thus, with the study of how seismic waves propagate trough Earth, scientists can devise a model of the Earth's interior, which might *approximate* to what really is inside it but often this model is mistaken as a *fact* and thus is widely thought in schools and universities. But this model is *questionable* as we will see through this book.

Following scientific laws and models, a *scientific explanation* is an *account* of a natural or social phenomenon, that is couched in terms of an established set of scientific principles, laws and facts, that have been repeatedly confirmed through *observation* and *experimentation*. A scientific explanation must be a valid *deductive argument* whose conclusion is the event to be explained. It is not an "educated guess." Statements such as "because X person says so, that's why;" is not a scientific explanation. It is a *dogma*.

One last thing is the mention of *computer simulations*. Computer models and simulations are used for state-of-the art research, testing and planning in areas ranging from the design of commercial jets to the simulation of crowd movements in new sports complexes. A computer simulation is a *computer program* that attempts to simulate an abstract model of a specific real, physical system. Computer models have been most of the time, very successful at assisting scientists and engineers with solving complex problems. While these models represent crucial research tools and instruments for analysis, models have also been known to *misrepresent* data and present *false* information as well.

A computer simulation is related to experiment, but all the action is running inside microprocessors, not in the *physical* reality of the laboratory nor the field. Furthermore, many 'experiments' alluded to a specific research, were computer simulations in which the necessary values were assumed, in order to ensure the prevailing ideas. Too much approximation and simplification coupled with wrong assumptions of the conditions (variables) assigned to the physical system, lead to *false* conclusions. An example of this will be shown in chapter five.

However, scientists tend to assume that whatever a computer simulation says, it is ultimately, the truth. Finally, there is a common misconception that scientists "prove" a hypothesis/theory is true. In reality, no number of experiments can ever prove a hypothesis is true beyond *all* doubt; which is the point. They can only show or validate that it's *consistent* with the *evidence,* this is what really matters in science. As evidence accumulates and competing explanations are disproven, of course, it becomes more and more reasonable to believe the hypothesis is the best explanation.

At this point scientists, will refer to it as a theory (for example, the Quantum Theory). Nonetheless, if a theory and its predictions have been repeatedly verified by *experiment*, it will be generally accepted, unless there is sufficient evidence to show it should be discarded in favour of a new theory. *Untestable* predictions and hypotheses lie *outside* the realm of science. These belong to either philosophy or pseudoscience.

Observations versus interpretations in science

An observation is any report from your five senses, that's it. It does not involve a statement of what had happened. An observation can also involves measurements. Such an observation is a quantitative one, as opposed to a qualitative one (no measurements). On the other hand, an *interpretation* is an *attempt* to figure out what has been observed (raw data) and often it is not *directly* verifiable. The model of what's inside Earth trough seismological interpretation is an example.

When doing laboratory/field research it is important not to *confuse* observations and raw data with interpretations. For instance, if you are asked to observe, you should not identify gases. "Hydrogen gas was produced when zinc was added to acid" is not an observation. What you see is the liquid getting cloudy with bubbles rising from the area surrounding the metal. The latter is what you should be reporting as an observation. Later if you collect the gas and it has hydrogen's characteristic *property*, then it will be time to *conclude* that the bubbles contained hydrogen gas.

Whereas everybody agrees that heat dilates the bodies or water is made of two parts of hydrogen and one of oxygen, as any chemical analysis can prove; with *scientific interpretations*, things are not so easy. Scientific explanations and scientific interpretations are NOT the same but many do think so. In scientific research either in laboratory or field, lab reports *observations* belong in the data section, whereas *interpretations* are part of the analysis.

This gives someone looking at your data the opportunity to interpret the *same* results *differently*, and it is a reminder to the writer of the report that the observing and interpreting are two different processes. But too often than not, data can be *misconstrued* and *misunderstood,* leading to false "scientific" conclusions.

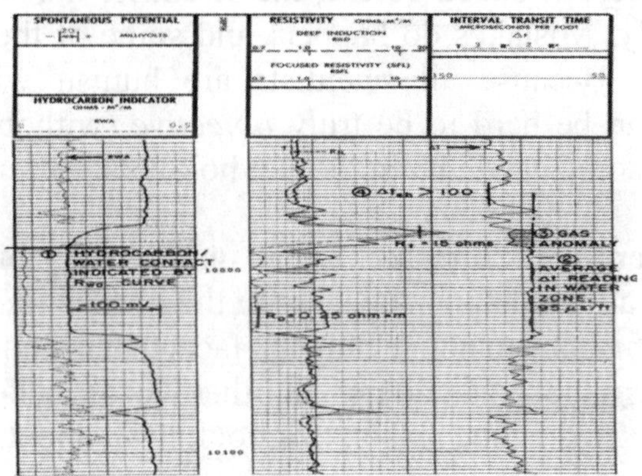

Fig. 6- Well log record interpretation. Detailed lithological information and precise determination of reservoir properties (lithology, porosity, permeability and hydrocarbon saturation) of the drilled section, are obtained from cuttings and side-wall cores. However, the number and length of cored intervals are generally restricted and reservoir properties cannot be obtained from cuttings. Thus, getting key information in the vast majority of wells rely on interpreting the data stored in a wide range of petrophysical well logs. A correct interpretation depends of the expertise and knowledge of the geologist in charge and it is not always 100% correct.

As within any human endeavour, scientists can make mistakes or even intentionally *deceive* their peers. In fact, some scientists misrepresent the way the Universe works. Scientific *interpretations* (not facts) are neither absolute truth nor personal opinion:

These are inferences, suggestions, or hypotheses about what the experimental data mean, based on a foundation of scientific knowledge and individual expertise. That's why different scientists can interpret the *same* data in *different* ways.

Also, *misinterpretation* of scientific data often happens. Therefore, scientific interpretations are NOT facts nor explanations whatsoever, as many believe. Given all this, one should start to show a little more mistrust towards some of the pronouncements of science.

Interpretation is a *tricky* business and this is crucial when discussing Earth's structure trough seismological interpretation, which ignores the context and values the quantities more than the qualities. This will be dealt thoroughly in the next chapter.

Common mistakes in applying the scientific method

Unfortunately, as stated before, the scientific method isn't always applied correctly. Mistakes do happen, and some of them are actually fairly common. Because all scientists are human with *biases* and *prejudices*, it can be hard to be truly *objective* (without opinions and preferences) in some cases. Even those who oversee peer reviews, can't escape.

It's important that all results are as untainted by bias as possible, but that doesn't always happen because of the *confirmation bias:* People have a natural *tendency* to notice only the facts that support their position while *discounting* those that do not. In other words, believing what you *want* to believe. This is one reason this book is all about.

Misunderstanding the difference between scientific hypothesis and theory can lead to *flawed* research. For example, some researchers might treat an untested hypothesis as if it were a well-established theory, leading to *overconfidence* in their conclusions. Similarly, they might fail to update or revise their hypotheses when new data *contradicts* their initial assumptions.

A common mistake is taking something as *obvious* or deciding that something is so *logical* that it doesn't need to be tested. Also, mistaking the hypothesis for an explanation of a phenomenon, *without* performing experimental tests or backed by field data. Scientists branded as "theorists" such as cosmologists, are caught in this trap and their "discoveries" sold to the gullible public as scientific "truth". For instance, in the following statement:

"If the galaxies are flying apart, the Universe is expanding then clearly; at some earlier time, the Universe was smaller than at present.

Following back logically, like a movie played in reverse it must ultimately, have had some beginning when it was very tiny indeed"

Scientists tend to forget that everything must be *tested* before it can be considered a solid hypothesis, let alone a scientific fact. Only careful *experimentation* can advance the understanding of the structure of the Universe. This is the reason why remarkable experimentalists such as Michael Faraday, Louis Pasteur or Nikola Tesla among others; were able to make wonderful discoveries of how Nature works, that no theorist sitting in an armchair or wheelchair, could have ever dreamed of. Scientists also must be willing to look at *every* piece of data, even those which *invalidate* the hypothesis.

Often this doesn't happen. Some scientists so strongly *believe* their hypothesis that they try to *explain away* data that disproves it. They want to find some reason as to why that data or experiment must be wrong instead of looking at their hypothesis again. All data must be considered in the *same* way, even if it goes *against* the hypothesis.

However, instead of reconsider their faulty hypothesis, scientists come up with weird, outlandish concepts or ideas to bridge the increasing *gap* between the observed data and the hypothesis. One example of such concept is the one of "dark matter" mentioned before, which is a flagrant *breach* of the scientific method. This sort of misconduct is more prevalent than anyone would like to think.

Another common mistake arises from the failure to estimate quantitatively systematic errors (and all errors). There are many examples of discoveries which were missed by experimenters whose data contained a new phenomenon, but who explained it away as a systematic background (noise).

Conversely, there are many examples of alleged "new discoveries" which later proved to be due to systematic errors not accounted for by the "discoverers." The alleged *red shift* of galaxies, which is the foundation of an expanding Universe, falls in the later.

Also, a common mistake when applying the scientific method is that scientists, generally disregard inconsistent processes as processes that are simply not yet *understood*. This entirely misses and dismisses as fluke effects that may be caused by systems or processes that are much larger than the hypothesis may cover.

Essentially, the limited scope and rigidity of such over-specialized and controlled environments precludes a *deeper*, more varied understanding of phenomena.

For instance, in the nineteenth century, several researchers based on experiments, had to conclude the existence of a sort of *subtle plenum*, endowed with special characteristics, that filled ordinary space, and could explain certain physical phenomena.

This cosmic plenum received the name of *aether*. However, in the 1880's, several experiments designed to detect the notable "aether wind," conducted by the Michelson-Morley team; in fact, *showed* values of aether wind tough *below* the expectations. Trouble is that, the scientific establishment *downgraded* those values and *wrongly* concluded that such aether didn't exist; despite several scientists warned *not* to do so.

This mistake was officially accepted in mainstream science as "truth", this apparently non-detection of aether wind for many years. Nowadays, cutting-edge scientific research, vindicates this cosmic plenum or aether; though with a different name: *quantum vacuum* or the *Zero Point Field* (ZPF) which pervades, shapes and sustains our physical reality; including ourselves.

Finally, the controlled environments and instruments that are used in scientific inquiry only really produce answers that are *relevant* to those same controlled environments and instruments. Science, as any human endeavour is fundamentally subjective, there is no way for it be truly 'objective'.

All of its verifications are performed from within itself, and are being generally *biased* towards the types of perspectives that are drawn to the method. To put it plainly, a scientist is focused on the consistently repeatable and reliable, and has a blind spot for what appears to be *inconsistent* on the limited and controlled scale that it operates. This blind spot applies to the accepted "knowledge" of the interior of the Earth, as we will see through this book.

Science as intellectual or logical endeavour

Since antiquity, *mathematics* had been an exemplar of certain knowledge. Reasoning like a mathematician meant reasoning reliably from assumptions to conclusions. Geometry, for example, was taught for centuries as a model for reasoning in general.

However, since seventeenth century, the pioneers of the Scientific Revolution, showed that geometry's role in the search for truth has been replaced by *data science*. In fact, it was Italian pioneering physicist, astronomer and inventor of the experimental concept, Galileo Galilei (1564-1642) who introduced scientific data and *experiments* to prove *false* ancient ideas made only by logic, mathematical reasoning and superficial observations. Reality *cannot* be conveyed by mathematics; they are distinct from each other and they just won't integrate properly.

Mathematics and logic are considered *formal sciences*. In natural sciences, these are important *tools* to organize, describe and derive conclusions based on empirical scientific data. Because of their *non-empirical* nature, formal sciences are construed by outlining a set of axioms and definitions from which other statements (theorems) are constructed based upon them.

Whereas the natural sciences and social sciences seek to characterize *physical* systems and *social* systems, respectively, using empirical methods, the formal sciences are language tools concerned with characterizing *abstract* structures described by *sign* systems, which are totally *independent* of physical reality. Therefore, Mathematics is *not* the language of Physics as it is wrongly believed. In fact, it has absolutely nothing to do with Physics. Math is Math and Physics is Physics, both terms cannot be confounded. Science deals with *explanations* not descriptions.

A description is not science, though is helpful. Mathematical descriptions are *only tools*. What really matters in science is the qualitative, physical interpretation which is rooted in experience. Only then do we have true understanding of Nature. The formal sciences aid the natural and social sciences by providing information about the structures the latter use to describe the world, and what inferences may be made about them.

Bear in mind that formal sciences are *not* concerned with the validity of theories based on observations in the *real world*, instead follows the established rules of the formal (abstract) systems. Since seventeenth century, the ideal of science was to express with mathematics, all scientific knowledge as physics does.

However, the world is too complex and hard to be captured by any single tool, so we need multiple tools. A formal model, such mathematics, might be used as a map of the world, *not* the world itself. It never was and never will. Nevertheless, often scientists, do confound the map with the world as one and the same thing, which is misleading and unrealistic. In fact, many mathematical "physical" theories do lack *physicality.* There is a lot of talk, a lot of smoke but no fire, anywhere.

It is often neglected that during the Scientific Revolution, mathematics began to be used as *tool* to describe dynamic situations or *changes* pertinent to physical systems. English physicist Isaac Newton and German mathematician Gottfried Wilhelm Leibnitz, invented *Calculus* for that purpose. Therefore, in science, mathematics is a language of symbols used exclusively to *describe* dynamic situations.

The scientific purpose of mathematics is to synthesize in an equation, *how* something moved or changed whether it is an orbiting planet, a charged particle traveling within an electric/magnetic field or the growth of a bacteria culture. A mathematical expression *cannot* tell us what an invisible or visible entity IS or looks like. An equation or a function or a bunch of numbers will not tell us what a car is or what it looks like. It can at best tell us how a car moved. This is *crucial* to understand when studying the current alleged Earth's inner's structure, which will be discussed in chapter two.

In several scientific disciplines, *empirical data* is feed or modelled into equations to explain current phenomena and successfully predict new ones. However, with their characteristic contempt for reality and experimentation, theorists do assume or ideate data to be feed or modelled into their equations; which apparently explain certain phenomena but *cannot* predict new ones, as happens with String Theory or the Big Bang Theory.

Mathematics does indeed become a *problem* when scientists attempt to provide a physical interpretation to their equations, *independently* of experience. Even worse, theoretical physicists for instance, swear that they can "visualize" their alleged 4-D Universe when the truth is that *no one* can visualize the 4-D objects of their mathematical theories, because they are not objects.

All mathematical objects invented since the Babylonians are *abstract* concepts. There is absolutely *nothing* to visualize in mathematics. To visualize something, one must to look at it, in the real 3-D world. Because often, these mathematical theories cannot point to a physical entity or *observable* as required by the scientific method, these theories hardly qualify as scientific. On the other hand, science is made by *statements* derived from the scientific method, which in itself, involves *scientific arguments*.

These arguments are made through *logic reasoning*. Whether we are consciously aware of it or not, our arguments all follow a certain basic structure. They begin with one or more *premises*, which are *facts* that the argument *takes for granted* as the starting point. Then a principle of logic is applied in order to come to a conclusion. This structure is often illustrated symbolically with the following example:

Premise 1: If A = B,
Premise 2: and B = C
Logical connection: Then (apply principle of equivalence)
Conclusion: A = C

In order for an argument to be considered valid, the *logical* form of the argument must work, it must be *valid.* A valid argument is one in which, if the premises are true, then the conclusion must be true also. However, if one or more premise is *false* then a valid logical argument may still lead to a false conclusion. A sound argument is one in which the logic is valid and the premises are true, in which case the conclusion must be true. \During the scientific process, *deductive* reasoning is used to reach a logical true conclusion. Another type of reasoning, *inductive*, is also used.

Where deductive reasoning proceeds from general premises to a specific conclusion, inductive reasoning proceeds from specific premises to a general conclusion. Scientists use inductive reasoning to formulate hypothesis and theories, and deductive reasoning when applying them to specific situations. Deductive reasoning is a *logical* process in which a conclusion is based on the *concordance* of several premises that are generally *assumed* to be true.

First premise: Noble gases are stable.
Second premise: Neon is a noble gas.
Inference: Therefore, neon is stable.

Inductive reasoning, also called induction or bottom-up logic, constructs or evaluates general propositions that are derived from specific examples. It progresses from observations of individual cases to the development of a *generality*.

In science, this generality is known as *universality*. For instance, based upon the observations and measurements of planets by Kepler and his own work on Earth, Newton could elaborate his universal law of gravitation which states:

"Every particle in the Universe attracts every other particle with a force along a line joining them. The force is directly proportional to the product of their masses and inversely proportional to the square of the distance between them"

There is simply no way to verify if each particle in the Universe, behaves in a such manner. A metal bar will be lengthened proportionally to the force applied, but not indefinitely. At a certain point, it will break.

A general conclusion made from many specific observations may not always be true. Exceptions are indeed found. Inductive reasoning which uses facts to extrapolate conclusions, always involves uncertainty. By often ignoring this fact, scientists are very skilled throwing extrapolations everywhere, making their "truths" shaky or unreliable. Though scientists do their best to shirk off the flawed modes of thinking that are associated with the common people, they are not perfect. They make errors just like the rest of us.

A fallacy is a kind of error in reasoning. Such logical fallacies result in sloppy, substandard research, work that's stamped with the seal of science, but ultimately does very little to elucidate the true nature of things. This is often ignored by the public.

Consider this argument:

If it rains, then the pavement will be wet
The pavement is wet
Therefore, it must have rained

This is of the very same form as several scientific arguments that have been made. But the conclusion does not logically follow.The pavement could be wet for some other reason than rain. Someone might have turned on the sprinklers. It might be wet because some children were playing in water.

There are many other reasons the pavement could be wet, which is why this form of reasoning is usually considered suspect. Numerous scientific theories have been victimized by this fallacy: what the A that the scientists thought was being caused by the B was really being caused by X or Y or Z.

Consider the following statement:

"Scientists do think that the only way of what can produce such large quantities of energy, might be the gravitational collapse of the whole central region of a galaxy along stars.

This gives the idea that a black hole is the most likely explanation of the observations."

As we will see later in chapter three, this statement is false. Fortunately, some scientists are able to think outside the box and point out, to *other* causes that can explain the same phenomenon, in a more efficient way.

On the other hand, assumptions which are often wrong, might be considered a special case of fallacy. An *assumption* is the act of taking something for granted, accepting that it is true or will happen, *without* any real proof. We all want the arguments and evidence to be based on *facts*, not assumptions. Nevertheless, when it comes to science and many aspects of our daily lives (such as politics), people are often more than happy to accept assumptions, and people frequently state them as if they are facts. Therefore, let's have a closer look on how to deal with assumptions.

In science, assumptions are commonly made for two reasons: one, to *simplify* the study of a specific phenomenon. In the Kinetic Theory of Gases for instance, one assumption is that gases are made of particles in constant, random motion which can be *considered* as point masses with no volume. In this case, *experience* shows this assumption, among others, are a valid *approximation* of a real gas.

Another reason is when there are variables in a study, that are *unknown* or *unattainable*. In this case, scientists are tempted to make assumptions on how those variables, might behave. For instance, the alleged temperatures of the Earth's core. However, due to the unattainability of those variables, in the first place; it makes impossible to test those assumptions. In other words, the alleged temperature of Earth's core is a guess disguised as a fact. This breaches the scientific method and thus it's not a scientific knowledge, as it will be discussed in chapter two. A variant of an assumption is a *conjecture*. In science, the word conjecture is defined as an opinion based on *incomplete* information. The word can be taken to be slightly pejorative, but given that conjecture also involves imagination and creative effort, it can be reasonable to say that in scientific research there is a natural progression from conjecture to hypothesis and after its experimental verification, to a theory. A very close relative of assumption is the concept of *postulate.* It comes from Geometry.

A postulate is a proposition or statement, that is assumed to be true *without* any proof. For example: "a line is infinitely long", "right angles are all equal". It is an *agreement* (consensus) by everyone to be correct. In Geometry and Mathematics, postulates are the fundamental propositions used to prove other statements known as theorems like the Pythagorean Theorem. Postulate is synonymous with *axiom*. An axiom is something we accept to be "obviously true." It doesn't have to be true in the real world, but Mathematics isn't necessarily about the real world.

If it is used in science as a "fundamental principle," a postulate *must* be based on *reliable* and repeatable *experimental evidence,* not guesswork or assumptions as it is often the case. Before 1905 no respectable, competent physicist would dare resort to postulates, in order to explain a natural phenomenon. Only to *empirical* laws and principles. To resort to postulates is being mediocre. An example in physics, are the two well-known postulates of Special Theory of Relativity:

First postulate: *All laws of physics should be the same for all observers in a referential frame with uniform motion.*

Second postulate: *As measured in any inertial frame of reference, light is always propagated in empty space with a constant velocity c that is independent of the state of motion of the emitting body*

The first postulate is nothing new. It is essentially Galileo's notion of relativity. It wasn't about space or time when it was considered by Galileo, but instead about a much simpler concept: the detectability of *motion.* Galilean relativity states that the laws of motion (not all physics) are the same in all inertial frames. Galileo described this *principle* in 1632 using the example of a ship traveling at *constant* velocity, without rocking, on a smooth sea; any observer doing experiments below the deck would not be able to tell whether the ship was moving or stationary. There is no *mechanical experiment* by which one can distinguish whether a system is at rest or is moving with a constant speed in a straight line. The first postulate is also known as principle of relativity. It claims motion is relative, there is no absolute rest or absolute motion.

The term "All laws of physics" in the first postulate was concocted to eliminate an *apparent* problem. Galilean relativity exists only for mechanics laws, but not for *electromagnetic* laws embodied in Maxwell's equations (there is an asymmetry). Electromagnetism was not "invariant" under Galilean transformation: meaning that Maxwell's equations don't maintain the same forms (symmetry) for different inertial frames. Maxwell's equations are valid *only* in a well-defined reference frame, identified as the all-pervading continuum of *luminiferous aether,* a space-filling imponderable medium supporting the wave propagation itself, traveling at the speed of light *c*.

There is an *absolute* inertial frame (the aether) as demonstrated by the work of Faraday, Maxwell, Tesla and other scientists. Later on, Maxwell's equations were *forced* to become artificially "invariant" in the mathematical form under certain assumptions, to be symmetric (Lorentz transformations, 1904). Tough these still were based on the aether. Finally, Special Relativity came and being annoyed because it was the aether that gave rise to this asymmetry between Galilean relativity and Electromagnetism, just for convenience, the first postulate *wrongly* considers aether as a "superfluous" agent in physics, rejecting it. One wonders why this fuss about "invariance" wasn't applied to the laws of Thermodynamics. This indeed, would have included "All laws of physics."

The second postulate is also nothing new. This statement of the velocity of light from a moving source has never, as a matter of fact, been taken seriously. We know from the phenomenon of interference that light is a wave motion. The *velocity* of waves is a matter of the physical properties of the *region* they are traversing; once away from the source they look after themselves. If the properties of this region are *not* constant, the speed of the wave *changes*.

The second postulate *doesn't* hold with the several *experimental* researches done all that long. In fact, the 1961 interplanetary radar contact with Venus presented the first opportunity to overcome technological limitations and perform direct experiments of Special Relativity's second postulate of a constant light speed of *c* in space.

When the radar calculations were based on the postulate, the observed-computed residuals ranged to over 3 milliseconds of the expected error of 10 microseconds from the best fit, the Massachusetts Institute of Technology Lincoln Laboratory could generate, a variation range of over 30,000%. An analysis of the data showed a component that was relativistic in a c+v Galilean (classical) sense and matched Newton's emission theory of light (the speed of light relative to the observer varies with the speed of the emitter, c'=c+v). The Lincoln Laboratory made a complete c analysis of all the radar data up to 1966. The velocity of the object relative to the observer including c is the *sum* of these velocity vectors, as any high school leaver should recall.

This anomaly led to the publication of *Radar Testing of the Relative Velocity of Light in Space*, Bryan G. Wallace, *Spectroscopy Letters*, 1969. In December 1974, during the American Astronomical Society, Dynamical Astronomy Meeting, it was reported that significant unexplained systematic variations existed in all the interplanetary data, and that they are forced to use *empirical* correction factors that have no theoretical foundation. Thus, NASA Jet Propulsion Laboratory was basing their analysis of signal transit time in the Solar System on Newtonian Galilean c+v, and not c as predicted by Special Relativity Theory.

On the other hand, French physicist Alain Aspects' experiment in 1982, proved a Quantum Theory prediction that it's possible for two particles to become "entangled," meaning they will retain a sort of causal relationship with each other, no matter their distance in time and space. If you measure one particle and it spins clockwise, for example, then its entangled companion would *instantly* collapse into a counter clockwise spin, even if it's on the other side of the Universe. This collapse occurs at a speed far *faster* than speed of light c meaning: instantaneously. This is technically known as *non-locality* and it shows that instantaneous transfer of information is possible. Quantum entanglement has been experimentally confirmed time and again and it is a fundamental feature of Nature.

Bear in mind that the postulates of Special Relativity are postulates (assumptions) *not* laws. We must give *priority* to (scientific) laws that are proven, rather that unproven assumptions. This shows that ultimately, real science is based on facts, *experimental evidence*. Assumptions can be *misleading*. Much delusion can be done by confusing assumptions with the truth, so scientists should always proceed with caution when making assumptions. Unfortunately, the scientific establishment routinely do *confound* assumptions with facts. There are other logical fallacies that are often made in science:

False analogy: This is a comparison between two things that are not really similar. "Swiss watches are prized for their precision; therefore, if you want to buy a precision automobile, purchase a Swiss car." The problem with this statement is that watches and cars are *different*. It is false because the two items do not have strong enough similarities to predict that what happens in one will happen in the other. The same applies to light and sound. Both behave as waves, but this doesn't mean light and sound have the same qualities. Light do possess *quantum properties* whereas sound doesn't. An analogy must do with conduct and not with structure. However, in science this is often ignored.

Circular reasoning: "The fossils and rocks are interpreted by the Theory of Evolution". The scientist might also have argued: "The Theory of Evolution is proven by the interpretation given to the fossils and rocks." However, neither statement actually *explains* why or how. String Theory holds that every particle, photons, electrons, quarks, protons, etc., is really a vibrating string. However, String Theory also holds that every string is in turn made of points (particles). This is a circular argument. Circular reasoning neglects to explain itself and is very common in academia. The circle that has been constructed may be large and confusing, and thus the logical mistake goes unseen. Therefore, many people, including scientists are fooled or deceived, without knowing it.

Close to the above, an elementary logical *mistake* in scientific reasoning is to conclude that if A implies B and B is observed to be true, then A is true. Not necessarily. B might have *other* causes than A. Let's illustrate:

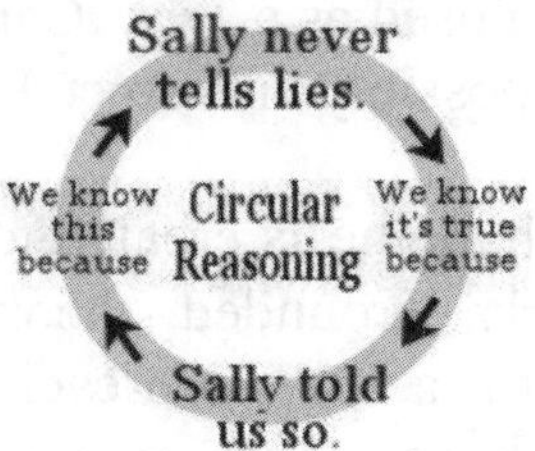

Fig. 7- Circular reasoning, or "begging the question," occurs when an argument doesn't go anywhere because the reasoner begins with what he or she is trying to end up with and in circular reasoning, there is no progress. A very common fallacy found in "scientific research". Instead of offering evidence, it simply repeats the conclusion, rendering the argument logically incoherent.

A = You bang your head into a wall
B = You have a headache

One might say that A implies B. Head bang leads to a sore headache, at least theoretically. Now, suppose that B is true, you've got a headache. Can we conclude that it is due to A? Not necessarily, you might have a headache from *other* causes, like drinking too much alcohol.

Also, it might be that the implication is even *incorrect*. So, there is no connection at all. You might have an unusually strong skull. Yet, this type of logic is *trademark* in modern physics and science in general:

If we assume (without proof) that a proton consists in three quarks, then we can theoretically derive a formula for the observed mass of a proton. Hence a proton consists in three quarks.

If we assume (without proof) a solid Earth, which is the result of small particles, sticking together after tens of millions of years of condensation and accretion; then we can theoretically explain the inner Earth's structure. Hence Earth is solid.

If we assume (without proof) there is a black hole in the centre of a galaxy, then we can theoretically explain the observed shape of a galaxy. Hence there is a black hole in the centre of a galaxy.

The fact that a certain phenomenon is observed, which can be theoretically explained from a certain *assumption,* is used in science not only as an assumption but posed as a *fact*. Can the reader see now, the potential danger of such possibly incorrect logic? Does it represent science or pseudoscience?

Ad hoc hypothesis: a hypothesis created to explain away facts that *refute* a shaky or poorly grounded theory, often it cannot be independently tested and it is a flagrant breach of the *scientific method.* As a trick amongst pseudo-scientists, an ad hoc hypothesis is often appended, in an attempt to justify why the expected results were *not* obtained.

The scientific method dictates that, if a hypothesis/theory is flawed, then that is final. The research needs to be redesigned or refined before the hypothesis can be tested again.

Advancing ad hoc hypothesis is a *dishonest* conduct to justify *failure* and deflect criticisms. The example of "dark matter" mentioned before, falls into this category and it will be discussed in chapter three.

Oversimplification: When a contributing factor is assumed to be the cause, or when a complex array of causal factors is *reduced* to a single cause. This is to simplify to the point of error, distortion, or misrepresentation.

Oversimplification is saying that "A" caused "B" while *ignoring* the other factors that lead to "B." It is a form of simplistic thinking that implies something is either a cause, or it is not.

It overlooks the important fact that, especially when referring to complex systems, causes are very *complex* and multidimensional. In the *real* world, events typically have *multiple* intersecting causes which together; produce the events or phenomena we see.

Sometimes that isn't so bad, but sometimes it can be disastrous an in science it often leads to dead ends. "People end up in jail because they are lazy and have no morals." While being, lazy and having, no morals may increase someone's chances of ending up in jail, there are many *other* factors involved in one's life that can lead to an arrest. Through this book, we will come across with some examples of oversimplification in science.

Weak analogy: The fallacy of weak analogy occurs when a presenter compares two or more things that really are *not* compatible in relevant respects. “Lettuce is leafy and green, and tastes great in a salad. Poison ivy is leafy and green; therefore, it should taste great in a salad”. This sort of analogy is frequently used by theoretical physicists to “explain” how their abstract mathematical concepts apparently, do affect reality.

An example is the rubber sheet analogy in teaching gravitational fields. It is a popular way to visualize the effect mass has on the curvature of space-time (a mathematical abstraction). Indeed, it’s straightforward to set up a demonstration in which a marble rolls around a central mass deforming a rubber trampoline. The deformed rubber trampoline represents the curvature of space-time, where the marble rolls or is “orbiting” the central mass.

But here’s the problem. Physicists have long suspected that the orbit of such a marble is *not* at all like a planet orbiting a star. Indeed, several years ago, US researchers Chad A. Middleton and Michael Langston in a paper published in the *American Journal of Physics* (2013) showed that there a rubber sheet deformed by a central mass simply *cannot* adopt a shape that reproduces the *real* orbits of objects observed in space within a span of time.

The motion of a marble on a deformed rubber sheet is entirely unlike the motion of a plant around star (or any small object moving around a large one under the force of gravity).

Indeed, physicists have known for nearly a decade that a rubber sheet deformed by a central mass can *never* take on a shape that reproduces the gravitational effects of space-time. So, the bottom line is that the rubber sheet weak analogy with space-time is fundamentally flawed. It simply *cannot* reproduce the motion that General Relativity predicts.

Science as a language

When people are engaged in *science*, the language of communication they use tries to be more *precise* and *consistent*. Methodologically, science *must* use clear and distinct concepts coupled with its correspondent clear and defined counterparts in the *real word*, in other words the 3-D objects we see involved within phenomena. This allow us through the scientific method, to elaborate specific arguments we understand with our minds. This is key in order to work effectively and elaborate *sound* scientific definitions and statements, to communicate ideas.

In this regard, a scientific definition is one that can be used *consistently* along an argument. Science often introduces *technical* words with *specific* meanings and also gives scientific meaning to words which may have a different usage in everyday language.

To a physicist, *energy* has a different meaning than to an athlete. In science, it is of utmost importance to avoid *vague* statements or *misleading* concepts and fully understand what a scientific argument entails. In other words, it is a matter of *semantics*.

Semantics is the study of the meaning of linguistic expressions. To assign meanings to the sentences of a language, you need to know what they are. It is the job of another area of linguistics, called syntax, to answer this question, by providing rules that show how sentences and other expressions are built up out of smaller parts, and eventually out of words.

The meaning of a sentence depends not only on the words it contains, but on its syntactic makeup. The sentence:

That can hurt you

For instance, is *ambiguous* it has two distinct meanings. These correspond to two distinct syntactic structures. In one structure 'That' is the subject and 'can' is an auxiliary verb (meaning "able"), and in the other 'That can' is the subject and 'can' is a noun (indicating a sort of container).

Real science has no room for *flawed* wording and imprecise or *vague* statements/arguments. Many of the *problems* of contemporary science, are of a *semantic* nature. The theorists along other scientists, relying heavily upon mathematics, have *confused* distance with displacement, real physical dimensions with abstract mathematical dimensions, and solids with volumes, to name but a few.

Bear in mind that mathematics only studies *adverbs* (relations, how something moves, etc.) and science deals with the remaining grammatical categories such as nouns, adjectives and verbs. Also, scientifically, adjectives can only be used to qualify *objects*. We can perhaps say that a table, chair or a rock (noun) is finite. Only an ignorant would say that running or swimming (verb) is finite.

Running is not a noun, but a verb. Running can perhaps be constant, perpetual, incessant, or tiring, but hardly infinite, straight, or massive. In science, the former is exclusively used as adverbs and the latter exclusively as adjectives. This is not arbitrary.

It is because this is the *only* way we can use these categories consistently (i.e., scientifically). Whenever a mediocre scientist uses the adjectives finite or infinite to qualify speed, one instantly notices that he/she is quite wrong. This scientist is invoking ordinary speech rather than scientific terminology.

Science requires precise definitions of the *key words* that make or break the theory. A theory is *not* mathematical description. A theory is a physical interpretation. The *inconsistent* language often used by the theorists only shows that they have no idea what they're talking about. They are confusing nouns with verbs in the same arguments. How can they draw sweeping conclusions about the speed of light or gravity if they have never in defined the word motion?

Then, to make matters worse, they use adjectives to qualify speed. Even in science, the word "energy" lacks clarity and distinction. A standard or generic definition is:

"From the perspective of physics, every physical system contains (alternatively, stores) a certain amount of a continuous, scalar quantity

called energy... There is no uniform way to visualize energy; it is best regarded as an abstract quantity useful in making predictions."

Most scientists would say that energy is: "the ability to do work". However, this is misleading because an *ability* is not what something *is*. Ability is what something indeed *does*. If energy is equal to 'ability', it clearly does not qualify as a physical object. Another meaning of energy in science is "capacity" It is used in names of quantities which express the relative amount of some quantity with respect to another quantity upon which it depends.

For example, heat capacity is: ΔU/ΔT, where U is the internal energy and T is the temperature. We are unable to neither *visualize* nor illustrate capacity. Capacity has no shape. Capacity is at best an *attribute* of an object rather than an *object* in its own right. The word capacity does not represent a structural *entity*, but functions rather as a qualifier.

For the purposes of physics, the word capacity is an adverb. It stands for 'latent or potential ability.' Therefore, it is absolutely incorrect for a scientist to use the phrase 'energy transfer', as if he were transferring a bottle of whisky from one shelf to another. We can transfer water. We cannot transfer the capacity of the bucket nor the ability to do something.

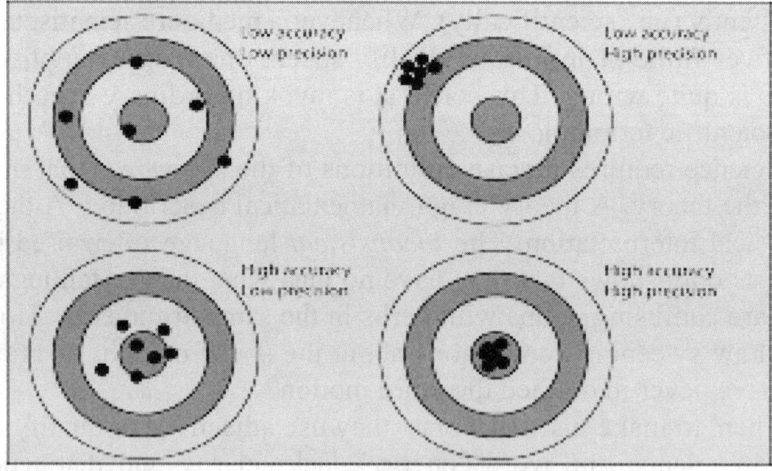

Fig. 8- Imprecise wording in research articles can mislead readers to believe that the findings are more powerful than they are. It common is for scholars to include statements in their articles that fail to accurately reflect their research design. As a result, the risk of misinterpreting and misquoting scientific findings is high.

This is an example of *misleading* language that the scientists have developed. It is wrong to say that a photon carries or transfers energy.

The scientist can at best argue that whatever light is, it induces an entity to vibrate (which is correct). Rather than saying that an atom vibrates, the scientists say that it has energy. These are two radically different explanations, and the scientists have yet to realize the distinction.

It is irrational to explain any phenomenon of Nature with this all-encompassing word "energy" until and unless science defines it *unambiguously*. Energy is considered a noun, verb, and adverb; all in one and this interferes with communication and understanding in science. In regards to Earth's inner structure, the phrase "convection currents" used by geoscientists has it issues as well, which will be discussed in chapter five.

There is more. It is usual to come across with scientific arguments that can be labelled as either misleading or half-truths. Consider the following statements:

It is widely accepted that the Earth's inner core formed about a billion years ago, when a solid, super-hot iron nugget spontaneously began to crystallize inside a 4,200-mile-wide ball of liquid metal at the planet's centre.

Elementary particles have intrinsic properties. Some of these properties are the same properties we associate with macroscopic objects, such as mass and charge. Some are purely quantum-mechanical and have no macroscopic analogue.

At this point, these clumped matters are called planetesimals, which just means a small, irregular-shaped body formed by colliding matter. Then eventually, the planetesimals grew larger by colliding and combining with other bodies of matter, growing larger to form planets.

When the word "spontaneous" is used in science, what this really means is that the underlying mechanism, is either *unknown* or poorly understood. There is always a cause-effect relationship in Nature. The same applies for "intrinsic" or that an object has an *inherent* characteristic, not caused externally. Precisely, the *ignorance* of that cause is what leads to the use of the word "intrinsic", to wash away that unknown cause. This issue will be discussed with more details in chapter four.

The word "eventually" is perhaps the worst. It is often used in physical processes that leads from the point or stage A to the point B without any real scientific knowledge of what's going on in *between.* It is the *belief* that what happens in A necessarily continues to B. In other words, it is *wishful thinking.* In science, this means intellectual mediocrity.

Finally, the reader should understand that in real science and engineering, we don't define words with mathematical symbols like theorists do. We define them with words and sentences rooted in *experience.* It is also necessary to define well, the crucial terms on which a theory holds. The *looser* we define a word, the *wider* the range of interpretations, and the more *dilute* the message that gets across will be.

Therefore, we need *rigorous* definitions to communicate a scientific theory, precisely

Science and Engineering

It is important to discuss, tough briefly, the *application* of scientific knowledge, which is known as *Engineering and Technology.* Engineering is the profession in which a knowledge of the mathematical and natural sciences gained by study, experience, and practice is applied with *judgment* to develop ways to utilize economically the materials and forces of Nature for the benefit of society.

Therefore, when at work an engineer just like a detective, is always concerned with *facts* and uses the best scientific theory that can handle these facts. If it can't, an engineer would do a research or might come up with an original theory that works.

And here's a key difference between Science and Engineering. Whereas for an engineer, scientific theory *must* always suit the facts, scientists wouldn't mind to *twist* facts to suit their theories; as they often do with their *ad hoc* hypothesis. Also, in contrast to many scientists, an engineer usually *never* commits the mistake of theorize in advance of the facts. Guessing is cheap, but guessing wrong can be expensive.

Whereas a scientist *believes* in a hypothetical particle that someday, will be found and fit in the so-called Supersymmetry Theory or when a massive star, more than a few solar masses, has exhausted its internal nuclear fuel and scientists *believe,* it would enter the stage of an endless gravitational collapse without having any final equilibrium state; the engineer moves from fact to fact, ensuring all the necessary technical and scientific knowledge is at hand and gets the job done. Moreover, the engineer must do it well. In engineering, knowledge and certainty are *paramount*. If a person doesn't like to think logically, basing conclusions on the *evidence* instead of professional biases towards that evidence as often scientists do, then one won't make a good engineer.

Because of being trained in *multidisciplinary* subjects, and engineer develops a *broad perspective* or vision in order to approach any given situation and tackle it from different angles. Scientists are taught to think that there is always *one* right way to approach an issue. Engineers are taught that there is *more* than one right way to tackle an issue. This includes available resources, physical, imaginative or technical aids, flexibility for future modifications and additions plus other factors. The approach works by optimizing the function, reduce the situation to *essentials* and strive to find the best solution.

By understanding the *constraints* of a scientific theory in regards to reality, engineers derive specifications for the *limits* within which a scientific theory is *applicable* and workable for any given *physical system*. In general, scientists however, don't care about this. For them, a scientific theory is far more important than the reality it represents or explains and thus, anything goes.

12. A horizontal circular jet of air strikes a stationary flat plate as shown. The jet velocity is 40 m/s and the jet diameter is 30 mm. If the air velocity magnitude remains constant as the air flows over the plate surface in the directions shown, determine: (a) the magnitude of F_A, the anchoring force required to hold the plate stationary, (b) the fraction of mass flow along the plate surface in each of the two directions shown, (c) the magnitude of F_A, the anchoring force required to allow the plate to move to the right at a constant speed of 10 m/s.

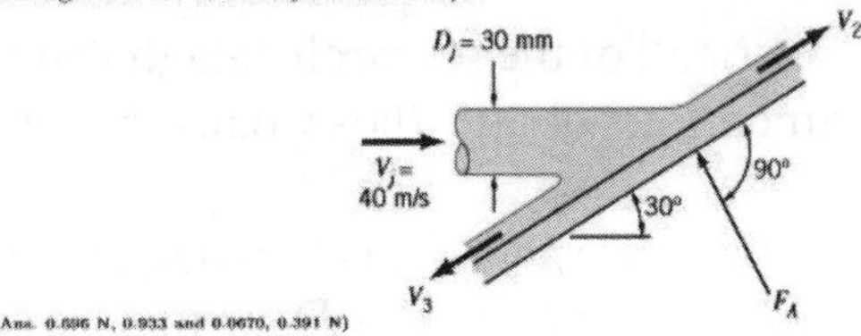

(Ans. 0.696 N, 0.933 and 0.0670, 0.391 N)

Fig. 9- Engineers must apply tested physical laws when dealing with real situations, that require valid and workable solutions. Scientists often indulge in theoretical scenarios that require hypothetical and untestable conclusions. Thus, their "scientific research" lacks physicality. Nowadays, scientists' success often isn't measured by the quality of their questions or the rigor of their methods. It's instead measured by how much grant money they win, the number of mediocre studies they publish, and how they spin their findings to appeal and deceive the public.

On the other hand, due to the *specialization* in a specific area, scientists often develop a *tunnel vision*, getting a selective perception which limits the ability to remain open and to see things clearly.

This leads to the narrowcasting of their professional reality, which can hinder an effective approach even in their specific fields, leading sometimes to a dead end. This situation will be closely examined in chapter three.

When the scientific endeavour goes wrong

Dishonest deceptions are not unusual in the history of science. A truth-seeker must always be concerned with how real science should be done. The number of scientific articles and journals being published around the world has grown so *large* that it is starting to confuse researchers, overwhelm the quality-control systems of science, encourage *fraud* and *distort* the dissemination of important findings.

There are many types of *science fraud*, from minor manipulation of results or incorrect causal connections to full-blown *fabrication* of results and plagiarism of the work of others.

The forms of scientific dishonesty that are practised and published, in *detriment* of society, are described in the following quartet:

Trimming: the smoothing of irregularities to make the data look extremely accurate and precise.

Cooking: retaining only those results that fit the theory and discarding others.

Stewing: hailing as confirming a scientific theory, an experiment that refutes it.

Forging: inventing some or all of the research data that are reported, and even reporting experiments to obtain those data that were never performed.

A notorious example of *scientific fraud* was the case of the Piltdown Man. You see, since the publication of Charles Darwin's *The Origin of Species* in 1859 and further books; Theory of Evolution "predicted" an intermediate specie o link between apes and humans.

For more than half a century, scientists were digging in different parts of the world without finding it yet. Then, between 1912 and 1914 among other remains, a fossilized skull apparently of the 'missing link' between apes and humans, was discovered in a quarry in Piltdown, Sussex, England. Two men astounded the scientific world by announcing they had found fossil evidence of the missing evolutionary link between apes and humans.

Palaeontologist Arthur Smith Woodward and amateur antiquarian Charles Dawson presented the skull of so-called Piltdown Man, dubbed *Eoanthropus Dawsoni*. Forty years later, it was discovered through fluorine testing, that the skull pieces had been *fraudulently* modified to appear ancient.

The Piltdown remains were only 50,000 years old at most not nearly old enough to be the missing link Dawson had claimed they were. It came to light that a jawbone of an orangutan with filed down teeth and at least two human specimens, possibly from the medieval period, were used to create the *fake* skull and planted in the quarry. More than a century later of this alleged discovery, chances are that the "missing link" is indeed, missing and never to be found.

The speed of sound ... 760 miles an hour at "standard temperature"—68 degrees above

(Continued On Page 18A)

Piltdown Man Named Hoax, Jolts Science

London, Nov. 21 (INS)—Three British scientists branded the famous "Piltdown man" a deliberate fake today and set off a controversy that may engage the scientific world for years to come.

Lengthy investigation ...

SECURITY ...ES CLAMPED ...N KIDNAP PAIR

Fig. 10- Piltdown Man was an audacious fake and a sophisticated scientific fraud. The frequency with which scientists fabricate and falsify data, or commit other forms of scientific misconduct is common. Journals are awash in a rising tide of scientific manuscripts from paper mills, secretive businesses that allow researchers to pad their publication records by paying for fake papers or undeserved authorship.

Recently, another *scientific fraud* which has been criticized by several scientists, has come to light: the so-called discovery of the allegedly "gravitational waves". Years after the US based Laser Interferometer Gravitational-Wave Observatory (LIGO) instrument discovered gravitational waves (GWs), the LIGO scientists still have not provided the answer for this question: What are the gravitational waves made of?

The explanation that three solar masses of black hole (BH) turned out to be the GWs didn't help, leading immediately to another question: So, what is the black hole made of?

And especially, how could a volume of BH material the size of three suns become waves that, after over one billion years of traveling in space, expanded itself big enough to fully cover every single point on the surface of a sphere with the radius 1.3 billion light-years long with enough *strength* to be detected?

It is interesting to observe that the LIGO teams themselves, report the presence of correlated noises in the two detectors and noise bursts (Blips) that appear continuously in the interval of a few minutes, in each day of operation, and that can be easily confused with gravitational waves, if by chance they happen at close intervals of time in the two LIGO detectors.

Actually, this widely and cheerfully accepted discovery creates many more mysteries and myths that need to be addressed. In fact, this alleged discovery has an *inconsistency*.

This GW, for example, is acting or moving, precisely *against* the laws of physics and logic. Throwing a big rock into water, we create waves that would certainly move toward the shore.

But if we dig a hole at the bottom of the lake that immediately sucks water in something similar to the black hole the newly created wave, if any, would move in the *opposite* direction, meaning toward the hole, not the shore.

But the LIGO's two black holes while merging and combining their sucking power that even light cannot escape suddenly generate waves that move *away* from the holes and toward the shore, or in this case, the LIGO instrument that is 1.3 billion light-years away from where they were born.

Another mystery is the unbelievable capacity of the LIGO instrument. The detected GWs were faint and lasted only *0.2 seconds*. But these weak waves, though they existed briefly, provided enough data for LIGO scientists to analyse and find out *everything* we need to know about them, including their origin, velocity, and the distance they have been traveling in space; in just one go. This looks more like a movie script than a genuine scientific report. As any honest experimentalist physicist, would know, just *one* faint observation isn't enough. But the LIGO scientists could confidently report in great detail that these GWs were born when two black holes merged, despite the fact that not a single black hole has ever been "observed". These have been for decades, *hypothetical* constructs.

Black Hole #1 has 29 solar masses and is 174 km wide; BH #2 has 36 solar masses and is 216 km wide. The new-born BH is supposed to be 29 + 36 = 65 solar masses, but is only 62 solar masses and 372 km wide because the third BH turned out to be the GWs that mysteriously worked against physical laws.

Instead of being sucked into the newly formed black hole, as everything else in the surrounding area, they ran away from it to reach the LIGO instrument located 1.3 billion light-years away. It is hard to believe that LIGO can detect a gravitational wave 1.3 billion light years away, but it has not, in four months of observational run, detected anything in our own galaxy.

It's like installing a microphone on the streets of a busy city and then detecting a horn beeping 1.3 billion miles away, but no sound detections from local traffic. (Our own milky-way galaxy is theorized to be full of gravitational wave sources). However, it has been claimed that a specific colliding neutron stars in our galaxy, have originated gravitational waves.

Trouble is that this collision produces colliding shockwaves not GWs. Like the case of the Piltdown's Man hoax, that was made up to "prove" the missing link required by Theory of Evolution, this discovery of LIGO was made up to prove the gravitational waves purportedly predicted by General Relativity. It is noteworthy to mention that there were no reliable *independent* sources confirming GWs alleged discovery.

In other words, this "finding" couldn't be verified. As per the scientific method rules: no replication no science. Very soon, a tsunami of LIGO's GWs will inundate our scientific textbooks, science classes, magazines, articles, papers, theses, etc., distributing a *fake* product into the treasure of common knowledge of mankind.

In this regard, we are failing to protect ourselves and future generations from consuming the scientifically *made-up* information. The same applies for the Higgs boson mentioned before and the recent alleged black hole "picture", captured for first time in space. Special mention deserves the fraud of the so-called *manmade climate change.*

Apparently, based on the comparison of atmospheric samples contained in ice cores and more recent direct measurements, some "scientists" lackeys of corporative and governmental interests, claim that this provides evidence that atmospheric CO_2 has increased since the Industrial Revolution to 400ppm.

Therefore, mankind is a threat to environment and we are forced into reducing our "carbon footprint". This enables governments a carbon dioxide (CO_2) tax on their citizens.

However other scientists have known for some time that a large amount of *volcanic activity* results in more CO_2 being released, but with previous analytical methods, it had been tricky to weigh up all the variables and to come up with an overall assessment of CO_2 concentrations.

In 2014, a new study made by University of Utrecht researchers and published in the *Proceedings of the National Academy of Sciences* suggest that for much of the Mesozoic, carbon dioxide levels in the Earth's atmosphere were much higher than they are today, even with global warming caused by greenhouse gases such as higher levels of CO_2.

Earlier studies had come to *similar conclusions* but this new research suggests that at the onset of the Triassic and into the Jurassic global CO_2 levels could have been as much as *five times* (2000ppm) their current levels, long *before* mankind ever walked on Earth.

The manmade climate change "crisis" is a political, not a scientific position. The ruling elites were using the U.N. as a platform to sell a global environment crisis and the global governance agenda.

This political agenda required "credibility" to accomplish the deception. It also required some *fake news* for momentum. Ideally, this would involve testimony from a "scientist" before a legislative committee.

Fraud scandals shake the *credibility* of science. The perpetrators are driven by personal motives such as the craving for recognition or greed for success. But the other reasons for fraud and manipulation are embedded in the system. Part of the responsibility for misconduct lies with the specialist publishing houses.

Researchers need to publish but only those with an impressive record are successful. Sounds logical but it is fatal, because it engenders misconduct. After all, the end justifies the means. So, science must break free from the tyranny of the luxury journals. The result will be better research that better serves science and society. The mechanisms that lead to fraud, deception or to looking the other way are universal.

The competition is huge. And, on top of this, today everyone works in teams. That makes it easier to cover up who is responsible for what, which favours deception. All disciplines are susceptible to scientific fraud.

However, in contrast to the scientific endeavour, there is little room for fraud in Engineering. This is because engineers when applying true scientific knowledge in areas of their competence, they must hold paramount the safety, health and welfare of the public.

Therefore, any engineering project is thoughtfully inspected to verify compliance with existing codes and regulations and if the work is deemed unsafe or it collapses, engineers by law will be sanctioned or prosecuted for misconduct.

Also, engineers are appraised for their practical professional performance on site, rather the corrupt peer-reviewed papers record. Engineers issue public statements only in an *objective* and *truthful* manner. From this perspective and in general, engineers do have far more *integrity* that a certain number of scientists. The author defies anyone to say otherwise.

The Scientific Revolution revisited

The Scientific Revolution in Europe of the seventeenth century, is something everybody should know from high school. It happened four and half centuries ago, starting with cosmology and astronomy with parallel developments in anatomy and physiology and then shifted to physics. The modern *scientific method* of observation, hypothesis, experimentation, analysis and conclusion was sculpted and refined in this era.

This historical event, started our current scientific and technological way of life. The Scientific Revolution is a thing from the past so, why bother to talk about it today?

This is because currently and to some extent, modern science has gone *backwards* to the ancient times, where scholars believed that abstract reasoning and sterile speculations lead to the true understanding of the Universe. The practical and *real world* was left to artisans, merchants and slaves, to deal with. The Scientific Revolution clearly ran contrary to tradition, to the authority of the Ancients and to established views in the universities and most church officials. Much of what was considered known about the natural world during the early middle ages in Europe dated back to the teachings of the Ancients. In fact, it was assumed that all significant knowledge, was already available there was no concept of progress; people looked for understanding to the past not the future.

There were a few exceptions in regards of scientific and mathematical findings made by medieval scholars. For centuries, after the downfall of the Roman empire, people still generally didn't question many of these long-held concepts or ideas, despite the many inherent *flaws*. Those concepts were based in *assumptions* and abstract (philosophical) reasoning with disregard to *experience*.

An example of a popular but *unproven* doctrine was the Aristotelian laws of physics. Aristotle taught that the rate at which an object falls, was determined by its weight since heavier objects fell faster than lighter ones.

However, Galileo trough *experimentation* proved him wrong. Another example was the flawed Ptolemaic or geocentric model, replaced by Copernicus and his heliocentric model, with further improvements, added later.

The Scientific Revolution, which emphasized careful *observations* and systematic *experimentation* as the most valid research method to acquire scientific knowledge, helped to throw into the trash bin the tradition of sterile speculations about Nature and how it works, that were in vogue in those days and defended by the Church authority.

It took Western civilization nearly 2000 years along with resistance and prosecutions, to recover from these *erroneous* physics and astronomic ideas among others, that came from Antiquity. Indeed, a sort of myths were replaced with *true* scientific knowledge, as stated by Kant:

"When Galileo let, his balls run down an inclined plane with a gravity which he had chosen himself... then a light dawned upon all natural philosophers".

However, this doesn't mean that today, in science, *myths* don't exist anymore. They do indeed. It's ironic that modern science helped to the resurrection of myths. They can be branded as *mathematical myths* and are furthered by a bunch of "scientists" that may be branded as *theorists,* who despise the scientific empiric approach, the only valid. Observation ability, tangibility and measurability are the basic *pillars* of any science.

These theorists instead of start working with observations, the first step of the *scientific method*; they begin with mathematical derivations from unquestionable assumptions which further are developed into theories that apparently, explain certain phenomena but cannot predict new ones. When further observations conflict the theory as they often do, new concepts are put forward to bridge the gap. For instance, the concept of "dark matter", discussed before.

A simple idea underpins science: "trust, but verify". Modern scientists are doing too much trusting and not enough *verifying* to the detriment of the whole of science, and of humanity.

Most theoretical physicists currently work (or better said wasting their time) on versions of superstring or M theory, which are complex mathematical structures based on the idea that there are 10 or 11 dimensions.

These theories have *inconsistencies* and are currently untestable and critics argue that they have therefore ceased to be scientific. In other words, it is *pseudoscience*.

Many other aspects of contemporary physics are also *untestable,* such as the multiverse theory and conventional cosmology depends on the postulation that 96% of the Universe is made up of dark matter and dark energy, whose nature is unknown and untestable.

Some physicists are indeed worried that physics has *lost* its way. Even one of them wrote a book (Hossenfelder, 2018). This is because modern physics, betrayed the spirit of the Scientific Revolution and its most precious gift to mankind: *the scientific method*.

In fact, particle physics is a decades long illogic-creep / rationale decay where poor assumptions are heaped on top of earlier assumptions until the whole endeavour, established as standard practice, ends up stuck in a bog, miles away from physics of *reality*, of actual phenomena in Nature. Because particle physics provides several "foundations" for Cosmology, this means that the usual explanations of cosmologists in regards of how the Universe works, are far from reality too.

All of the theoretical work that's been done since the 1970's has *not* produced a single successful prediction. That's a very shocking state of affairs for theoretical physics.

This doesn't mean physicists aren't busy; they write a lot of worthless papers, build a lot of (theoretical) models, hold a lot of conferences, cite each other and the journals are publishing more "research" than ever but all of this *lacks* scientific substance.

At least some of these theorists are honest enough to admit that research isn't doing much to advance our understanding of the Universe at least not the way physicists did in the last century.

Until the 1940-50's and with some exceptions, theoretical physicists used to explain what was *observed*, which by the way is their job and they should step on it.

Now theoretical physicists try to explain why they can't explain what was not observed. And they're not even good at that. This shortage in predicting success which is concerning modern physics, can lead to *fraudulent* discoveries, to justify their theories, as discussed before.

The credo of these theorists is that pure mathematical constructions enables them to "discover" the concept and the laws connecting them and thus the way to the understanding of Nature. Exactly as scholars from Antiquity thought and taught. In fact, they naively believe that a logical and mathematical thinking alone whilst sitting on an armchair or wheelchair, will lead them to the "truth".Not so.

This approach has *nothing* to do with physical reality. It never was and it never will. This platonic approach is absolutely unscientific and preposterous. The logical, however logical and mathematical that is, it has nothing to do with the real unless verified by *experience*. These theorists are not scientists but either philosophers or pseudoscientists.

Therefore, this sort of "scientists" what really are doing is hindering, instead of promoting the true advancement of science. They are fooling themselves and misleading the public. That's why the Scientific Revolution and its scientific method was aimed for. To emphasize careful *observations* and systematic *experimentation* as the valid research method to acquire scientific knowledge and dissipate *falsehood* based in philosophical speculations and untested theoretical "scientific knowledge".

Without *experimental verification*, one cannot be sure that the model and its mathematical analysis represents reality or *delusion*. In fact, many of the so-called "scientists" are actually never taught the scientific method. In an increasingly more radically *anti-science* and *pro-ideological* academic environment; as graduate students, they take oodles of courses in their chosen specialty; but their thesis advisors never sit them down and inculcate them on best practices. Consequently, these "scientists" will go off and make it worse. Instead of scientific progress, they indeed make a mess of it. They don't have the *insight* of the reality of the phenomena to be investigated and are unable to devise *valid* experimental arrangements, for the empirical formulation of their question to reality itself. These cranks end up parroting pure nonsense.

Fig. 11- Science is *corrupted* when it abandons the discipline of empirical validation or dis-confirmation. It is also weakened when it mistakes its mathematical assumptions for facts and its ready-made ideology for the way things are. This is what current theoretical physics stands for.

On the other hand, science is not a matter of opinion or consensus. There is only truth and reality. Also, one shouldn't be intimidated by any scientific theory. Bear in mind that *real* science is not difficult to understand, because is rooted in common sense and logic. Also, a consensus has absolutely nothing to do with truth or reality.

Organizing science per the dictates of consensus leads to corruption and bogus science. Most of this bogus science tends to be over complicated and confusing, mainly because its main objective is to confuse and complicate simple matters.

For example, the belief that the phenomena associated with the speed of light and gravitation cannot be perceived by ordinary people using their common sense, not because the mathematics or concepts are difficult, but because elementary logic must be abandoned. However, individuals with a strong sense of reality and the necessary talent needed to understand it, instinctively can recognize discrepancies and contradictions in these purported "scientific theories".

In science, authority and rhetoric must be kept aside. The motto of the Royal Society of London, when founded in 1660 was: *Nullius in Verba,* roughly, accept nothing based on words (or someone else's authority). Facts, proofs and experimental evidence are the *sole* sources of true scientific knowledge. Not empty words or theoretical but *untested* dissertations, often wrapped in mathematics and sold to the gullible public as "science".

The meaning of paradigm

Finally, this brief excursion in science would be incomplete without grasping the historical and conceptual meaning of *scientific paradigm*. A scientific paradigm is a framework containing all the commonly *accepted* views about a subject, *conventions* about what direction research should take and how it should be performed. It is crucial to understand that as history shows, scientific paradigms are often an *obstacle*, a ballast for the advancement of science.

This set of *assumptions* that supports a scientific domain and constitutes the whole theoretical framework within which they work is called a paradigm. The Oxford dictionary defines a paradigm as "a worldview underlying the theories and methodology of a particular scientific subject".

The paradigm or set of assumptions within which the enterprise of modern science operates was born approximately four hundred years ago, with the massive cultural transformation of the Renaissance and through the Scientific Revolution, that give us the cultural foundations of our modern world.

This new paradigm really came together and first found its most coherent full expression within the work of Sir Isaac Newton, whose work was extremely influential for centuries to come and laid the foundations for modern science, and of course, built into this foundation was a set of assumptions about how the world works. This whole set of assumptions is called the Newtonian paradigm or the *clockwork Universe*; in slightly more technical terms it can also be called *linear systems* theory.

Linear systems theory forms the backbone to virtually all of modern science. It is used in every domain from physics to biology to economics to psychology. Also, a specific branch of science can have its own paradigms. Catastrophism and uniformitarianism in geology, behaviourism and cognitivism in psychology, etc.

The modern usage of the term, however, began when Thomas Kuhn used it in his *Structure of Scientific Revolutions* (1962). Kuhn initially used the term "paradigm" in the contexts of history and philosophy of science. The term, however, was widely used in social sciences and human sciences and became a popular term in almost all disciplines. The body of pre-existing evidence in a field, conditions and shapes the collection and interpretation of all subsequent evidence.

The certainty that the current paradigm is reality itself is precisely what makes it so difficult to accept *alternatives*. In fact, Kuhn showed us that research in a deeply entrenched paradigm invariably ends up reinforcing that paradigm, since anything that contradicts it is *ignored* or else pressed through the present methods, until it conforms to already established dogma.

It is very common for scientists to discard certain models or ignore emerging theories. But every once in a while, enough *anomalies* accumulate within a field that the entire paradigm itself is required to *change* to accommodate them. Whereas scientists are well aware of this, not always is the case with the public.

A scientific revolution occurs, according to Kuhn, when scientists encounter anomalies which *cannot* be explained by the universally accepted paradigm within which scientific progress has thereto been made.

The paradigm, in Kuhn's view, is not simply the current theory, but the entire worldview in which it exists and all of the implications which come with it. There are anomalies for all paradigms, Kuhn maintained, that are brushed away as acceptable levels of error, or simply *ignored* and not dealt with.

Kuhn believed that science had periods of patiently gathering data within a paradigm, mixed in with the occasional revolution as the paradigm matured. A paradigm shift is not a threat to science, but rather the very manner in which it *progresses*.

Normal science is the step-by-step scientific process, which builds patiently upon previous research. Revolutionary science, often 'fringe science' *questions* the paradigm itself. Kuhn believed that a paradigm would make a sudden leap from one to the next, called a shift, where the new paradigm didn't build on the foundations of the old, but completely change the rules for that "building."

This book for instance, takes an alternative view of the current paradigm in regards to the Earth's interior structure. As it analyses, scientific data from *different* angles and proposes an *alternative* view, this book would fall in the category of fringe science. However, those scientists that cling to the current paradigm, would dismiss this book as pseudoscience. It's psychological.

An example of a paradigm shift

This example is a well-known story learned in school but it is worthy to refresh it because though contents and players are different, the basic script still applies *today*. In medieval Europe, it was generally accepted that the Earth lay at the centre of a finite Universe and that the Sun, planets and stars orbited around it in concentric spherical shells.

The framework in which this astronomy was set was established by Aristotle (384 - 322 BC) in the fourth century BC while in the second century AD Ptolemy (c. 100 - 170 AD) based on sloppy observations and too much imagination, he devised a detailed yet different geocentric astronomical system summarized in his work the *Almagest*.

Nonetheless, a heliocentric cosmos was proposed by Aristarchus of Samos (c. 310 - 230 BCE) but it wasn't widely accepted and fell into oblivion for nearly two millennia.

The Ptolemaic (Geocentric) model accounted for the apparent motions of the planets in a very direct way, by *assuming* that each planet moved on a small sphere or circle, called an epicycle, that moved on a larger sphere or circle, called a deferent.

The stars, it was assumed, moved on a celestial sphere around the outside of the planetary spheres. The Ptolemaic model was not seriously challenged for over 1,300 years. To make the planets appear to speed up and slow down, three tricks were used.

The epicycles we've just shown were the first trick. The second trick was to move the observer out of the centre of the circle, putting us into an "eccentric" position. The third trick was called the equant, placed on the same line but opposite to the centre and Earth at eccentric.

However, several problems were facing astronomers at the beginning of the sixteenth century. First, the tables (by means of which to predict astronomical events such as eclipses and conjunctions) were deemed not to be sufficiently accurate. Second, Portuguese and Spanish expeditions to the Far East and America sailed out of sight of land for weeks on end, and only astronomical methods could help them in finding their locations on the high seas but they observations not always matched theory. Third, the calendar, instituted by Julius Caesar in 44 BCE was no longer accurate.

The spring equinox, which at the time of the Council of Nicaea (325 CE) had fallen on the 21st, had now slipped to the 11th. Since the date of Easter (the celebration of the defining event in Christianity) was determined with reference to the equinox, and since most of the other religious holidays through the year were counted forward or backward from Easter, the slippage of the calendar with regard to celestial events was a very serious problem.

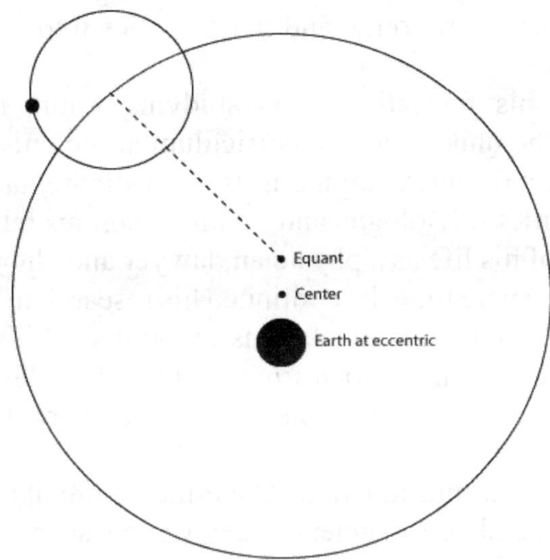

Fig. 12- Ptolemaic model to explain the apparent motion of planets, observed from Earth. A planet has two orbits: a large eccentric orbit around Earth called deferent and a small one called epicycle. The observed planet's speed variation was due to Earth moving from it eccentric to the equant position.

For the solution to all three problems, Europeans looked to the astronomers. Notice that the Geocentric model was not good enough to be *workable* and it started to crack.

This is a suggestion of either this model or paradigm needed a profound *revision* and update, or it had to be *discarded* for a new one which would fit smoothly all the observable data available by then.

Also, the many loops observed in the apparent motion of the planets implied many rearranges of the epicycles and deferents making this model, very *cumbersome* to work with.

In fact, a notable king of Spain, Alfonso X the Wise quoted: "had I been present at the Creation, I would have given some useful hints for the better ordering of the Universe". Nature operates with *simplicity*. The reader should recall the Occam's razor mentioned before.

Nicholas Copernicus or *Mikołaj Kopernik* (1473-1543) is often described as a lone Polish astronomer who defiantly argued that the Sun, not the Earth was at the centre of the Cosmos.

Copernicus' contributions to astronomy are so significant that they warrant their own term: *The Copernican Revolution*. Indeed, the Sun-centred model was a break with the past. Copernicus was the *first* to combine physics, astronomy, and mathematics into a fact-based model of the Universe.

He spent his formative years studying Latin, mathematics and astronomy in the undergraduate curriculum at the university of Cracow and then spent a decade studying in Italy medicine, law and astronomy at the Universities of Bologna and Padua. Upon his return to Poland, he spent the rest of his life as a physician, lawyer and church administrator.

During his spare time, he continued his research in astronomy. After years of observation and calculations he published his masterpiece *De Revolutionibus Orbium Coelestium* ("On the Revolutions of the Celestial Orbs"), which was published in Nuremberg, Germany, in 1543 the year of his death.

In the book's introduction Copernicus credits his heliocentric hypothesis to the ideas of ancient Greek writers such as Aristarchus and Philolaus. In the first section, Copernicus gave some basic mathematical rules, countered the old arguments about the fixity of the Earth, and discussed the order of the planets from the Sun.

He could no longer accept the old arrangement Earth, Moon, Mercury, Venus, Sun, Mars, Jupiter, and Saturn since this had been a consequence of a geocentric system.

He found it necessary to adapt it to his heliocentric system and adopted the following order from the stationary Sun: Mercury, Venus, Earth with the Moon orbiting around it, Mars, Jupiter, and Saturn. This was in direct *contradiction* of the teachings of the establishment of those days: The Catholic Church, which supported the Ptolemaic view of the Universe.

Scientifically, the Copernican theory or heliocentric model demanded two important *changes* in outlook. The first change had to do with the apparent *size* of the Universe. The stars always appeared in precisely the same fixed positions, but if the Earth were in orbit round the Sun, they should display a small periodic change.

Copernicus explained that the starry sphere was too far distant for the change to be *detected.* His theory thus led to the belief in a much larger Universe than previously conceived and in England, where the theory was openly accepted with enthusiasm, to the idea of an *infinite* Universe with the stars scattered throughout space.

The second change concerned the reason why bodies fall to the ground. Aristotle had taught that they fell to their "natural place," which was the centre of the Universe.

But because, according to the heliocentric theory, the Earth no longer coincided with the centre of the Universe, a *new* explanation was needed. This re-examination of the laws governing falling bodies led eventually to the Newtonian concept of universal gravitation.

In the Copernican system, the Sun is immobile at the centre of the Universe and the planets including the Earth, being the third planet rotate around it. The Earth is thus no longer a fixed hub but revolves around the Sun in one year. The daily motion of the Sun and stars is the result of the Earth turning on its axis in twenty-four hours.

Copernicus correctly accepted that the air (atmosphere) near the surface of the Earth, shares in its rotation thus preventing any "wind" from arising.

Copernicus reconciled placing the Sun at the centre of the Solar System with his religious beliefs by noting that the Sun is the source of light and life, a logical choice for the centre of the Universe.

For Copernicus, the motion of the heavenly bodies from east to west in the course of a day is only *apparent.* In fact, it is caused by the Earth's rotation in the opposite direction; also apparent is the Sun's annual travel along the ecliptic, actually due to terrestrial rotation; the retrograde movement of the planets is apparent as well.

For the ancients, who assumed the Earth to be immobile, each planet moved on a circle called the epicycle, whose centre, in turn, moved around the Earth on a larger circle known as the deferent.

In the Copernican system, the *cumbersome* and *unnecessary* epicycles where eliminated; instead, both motions are centred on the Sun, but only one is performed by the planet itself; the other is the revolution of the Earth.

The combination of the two motions generated the celestial body's apparent retrogression. A retrograde event then becomes the result of an inner planet overtaking and passing an outer planet. This is a *simpler* explanation of the retrogression.

Copernicus assumed that the shapes of the orbits are circles and that the planets move at constant speed (same as Ptolemy). This was later modified by Kepler. One more thing was the problem of the variation of the brightness of the planets.

This shouldn't happen if all planets are revolving, with the same distance, around Earth. However, the planets in the heliocentric system, naturally vary in brightness because they are not always the *same* distance from the Earth.

Nicholas Copernicus, in his book, quotes:

"For when a ship is floating calmly along, the sailors see its motion mirrored in everything outside, while on the other hand they suppose that they are stationary, together with everything on board. In the same way, the motion of the Earth can unquestionably produce the impression that the entire Universe is rotating."

In other words, Copernicus reasoning involves the later developed concept of *relative motion* as observed from or referred to some material system constituting a *frame of reference*. He was the first to highlight the need to define a frame of reference when studying *motion*. Reference point is too important in physics.

All-natural laws require a reference frame to maintain consistent relationships. 19th-century electrodynamics raised the question of a privileged frame of reference: the conception of light as an electromagnetic wave in the *aether* implied that the rest-frame of the aether itself plays a *distinguished* role in electrodynamical phenomena. Originally, (Maxwell, 1861) electromagnetic equations were based on the existence of an *absolute* reference system: the aether and successfully, *predicted* the existence of radio waves.

Physicists do all calculations according to the reference points. However, sometimes a reference frame is *wrongly* chosen which leads scientists to make mistakes, when interpreting certain celestial phenomena as we will see in chapter five.

Challenging a paradigm in current times

Whoever dares to *defy* the current paradigm faces dire consequences. Copernicus was well aware of that. He published his *De Revolutionibus Orbium Coelestium* in 1543 the year of his death. But he didn't publish it as early as 1529 for two likely reasons. One, he knew it would be very controversial and would question the Church's beliefs, and two, at that point, his model wasn't fully complete and couldn't accurately predict things like planetary positions.

Two Italians who lived decades after Copernicus, suffered for supporting his beliefs. Giordano Bruno not only agreed that the Earth revolved around the Sun, he even suggested space might be infinite, that our Solar System was but one of many, and that there were possibly other worlds inhabited by beings that might have intelligence equal to or even superior to man's.

In 1600 Bruno was condemned by the Papal Inquisition and burned at the stake for his views. Galileo Galilei, whose discovery of the moons of Jupiter in 1610 and claiming in 1632 that Earth orbited the Sun; was condemned by the Church in 1633; and forced to renounce all belief in the heliocentric system. He found himself under house arrest for committing *heresy* against the Catholic Church.

To show how the scientific endeavour is strongly affected by its *cultural* or social environment, one just must look at seventeenth century England where Newton was a hero, compared to Italy where Galileo was a criminal. More than three centuries later, another astronomer faced a similar fate of Galileo.

He was “condemned” not by the Church with its cardinals and popes but by the scientific establishment. The “heresy” was to prove that the grounds of a sacrosanct *dogma* in science: The Big Bang Theory, were flawed.

This was the case of a little known but remarkable US astronomer Halton “Chip” Arp (1927-2013) the Dean of US astronomy and former staff astronomer at the Hale Observatories in California, who faced *persecution* and was *blacklisted* in his own country, causing him to emigrate and continue his work at the Max-Planck-Institut für Astrophysik in by then, West Germany.

The reason: he found *experimental* evidence that made the Bing Bang Theory fundamentally wrong. Halton Arp did the right thing. He had to tell the truth as he saw it, especially about important things. No media coverage was made of this incident. That’s why Halton Arp is almost *unknown* by public.

As a rule, in science, new observations or discoveries the connections between objects and the new insights into the workings of the Universe are meant to be *communicated* and disseminated all; the primary obligations of academic science. However, it has generally tried to *suppress* or ignore such dissident information.

The Bing Bang Theory which indeed is a *hypothesis,* was grounded in what is known as the expanding Universe or expanding space. In astronomy, due to the vast and gigantic distances, the building blocks of the Universe are considered made by *galaxies*. Therefore, what the scientists mean is that galaxies are moving away from each other.

What fact if any, does support this expansion? Galaxies might be stationary or contracting. Even it could be a mixture of these three options. Let's examine now, the background of the events that led to the idea of the "expanding Universe" and why it is flawed.

In the 1920's very little was known *outside* our galaxy, The Milky Way. Extragalactic objects, which are those found outside our galaxy, haven't been fully observed yet. During those times astronomers were facing the problem of the so-called spiral nebulae.

The status of the spirals (as they were widely known) was then unclear. Were they distant star systems (galaxies in current terminology) comparable to the Milky Way Galaxy, or were they clouds of gas or sparse star clusters within, or close by, the Milky Way?

In 1923, US astronomer Edwin Hubble found Cepheid variable stars. An important fact is a characteristic variable of a Cepheid star's period or how often it *pulsates,* which is directly related to its luminosity or brightness in a predictable way. Therefore, these are useful in measuring interstellar and intergalactic distances in the Andromeda Nebula, a very well-known spiral. The fluctuations in light of these stars, enabled Hubble to determine the nebula's *distance* using the relationship between the period of the Cepheid fluctuations and its luminosity.

Although there was no clear consensus on the size of the Milky Way, Hubble's distance estimate placed the Andromeda Nebula approximately 900,000 light-years away. If Hubble was right, the Nebula clearly lay far *beyond* the borders of the Milky Way Galaxy (the largest estimates of its size put its diameter at around 300,000 light years). The Andromeda Nebula therefore had to be a galaxy and not a nebulous cloud or sparse star cluster within the Milky Way, as believed in those days.

Hubble's findings in the Andromeda Nebula and in other relatively nearby spiral nebulae swiftly convinced the great majority of astronomers that the Universe in fact contains a *myriad* of galaxies. Within a few years of this remarkable research, Hubble decided to tackle one of the outstanding puzzles about the external galaxies (or extragalactic nebulae, as Hubble always called them):

Why did the vast majority appear to be moving away from Earth? It is like the Universe *seems* to be expanding. Is noteworthy to mention that the word seems used in the previous sentence, is key for building the Big Bang Theory. Also, it contains the seed for its destruction.

Before going any further, is important to know the basis of an important tool used in astronomy to study distant objects through the *light* emitted by them: *Spectroscopy*.

If light from stars or galaxies is passed through a prism or grating, a spectrum is obtained, consisting of a series of lines and bands. These spectra can be used to identify the atomic elements present in the objects concerned, as each element has a distinct spectral 'signature' (spectral analysis).

Spectroscopy is this sort of study. This is how astronomers know the stuff, stars and other celestial objects, are made of. Also, depending of its energy, the light in the electromagnetic spectrum has two observable limits. High energy and shorter wavelengths are in the ultraviolet-blue limit. Low energy and larger wavelengths in the red-infrared limit.

However, the spectra studied in astronomy are not always static. There is an interesting phenomenon known as the *red shift*, which is often *believed* to be caused by the *Döppler effect* or the apparent difference between the frequency at which sound or light waves leave a source and that at which they reach an observer, caused by *relative motion* of the observer and the wave source.

This phenomenon is used in astronomical measurements. The following is an example of the Döppler effect: If you've ever heard a train whistle die away as it speeds past you, you've encountered the concept of recessional velocity. Similarly, the light from a star, observed from the Earth, shifts toward the red end of the spectrum (lower frequency or longer wavelength) if the Earth and star are receding or moving *away* from each other.

The galaxies observed by Hubble and other astronomers seemed to be moving away from Earth if the redshifts in their spectra are *interpreted* as the result of Döppler shifts. This is crucial to understand. In 1929, Hubble published his first paper on the relationship between redshift and *distance*. He tentatively concluded that there is a linear redshift-distance relationship; that is, if one galaxy is twice as far away as another, its redshift is twice as large. This is the basis of the *Hubble Law,* which by the way, has nothing to do with the so-called Hubble constant or the unit of measurement used to describe the "expansion" of the Universe. In astronomy redshift is symbolized by *z*, thus there are galaxies with low and high values of *z*.

By then, several scientists had applied General Relativity to the large-scale properties of the Universe. The redshift-distance relation established by Hubble was quickly meshed by various *theorists* with the General Relativity-based theory of an expanding Universe. In fact, one had absolutely nothing to do with the other. The result was that by the mid-1930's, the redshift-distance relationship was generally *interpreted* as a *velocity*-distance relationship such that the spectral shifts of the galaxies, were a consequence of their motions.

But Hubble throughout his career *resisted* the definite identification of the redshifts as *velocity* shifts. What Hubble really discovered was a redshift distance law *not* expansion of the Universe. He always stressed the *uncertainty* of the observational data but to a no avail. Astronomical data is the *most difficult* to acquire with precision.

Bear in mind that, due to the vast distances involved in astronomical observations, measurements are the least accurate and more prone to error that of any other science. Also, astronomers can *only* measure distances, they don't actually measure *time* to determine the age of the Universe.

Notice how some "scientists" *twisted* the scope from the redshift associated with *distance*, obtained from experience, to velocity so it can *fit* into General Relativity (GR). You see, around 1915-1916 it was proposed that GR's "field equations" which were NOT derived from careful *experimental observations* (Fig. 4) but from purely imaginary or abstract machinations; could be used to describe the diverse features of the Universe, originating the field study known as "modern cosmology".

However, it was found that these equations when solved, indicated two solutions: The Universe as a whole either must contract to, or expand from, a single point, a singularity. A *problem* with these "field equations" arose, showing the Universe as dynamic entity, which would *collapse* due to gravitational attraction among celestial bodies and thus, contradicting the (widely accepted) belief that the Universe is static.

Therefore, it was necessary to introduce a *fudge factor* a "cosmological constant" term Λ (lambda) which would hypothetically keep the Universe in equilibrium. It implied the existence of a repulsive force pervading space that counteracts the gravitational attraction holding matter together. This status quo of GR remained until 1929, when observations by Edwin Hubble were later *corrupted* to get rid of this cosmological constant and *falsely* claim that the Universe was not static but expanding. However, in real science, theory always must fit in with the *facts*. No fact shall be ever *distorted* in order to fit a theory, regardless of its grandeur, let alone who came up with that theory.

Due to this alleged velocity, a galaxy's redshift is proportional to its recession or escape velocity, which increases with its distance from Earth. Most astronomers and cosmologists *interpreted* the redshift to mean that all galaxies are flying apart at high speed and that the Universe is expanding. If the Universe is expanding it would mean the Universe had been in the past, somehow more "concentrated", less organized and more "primitive".

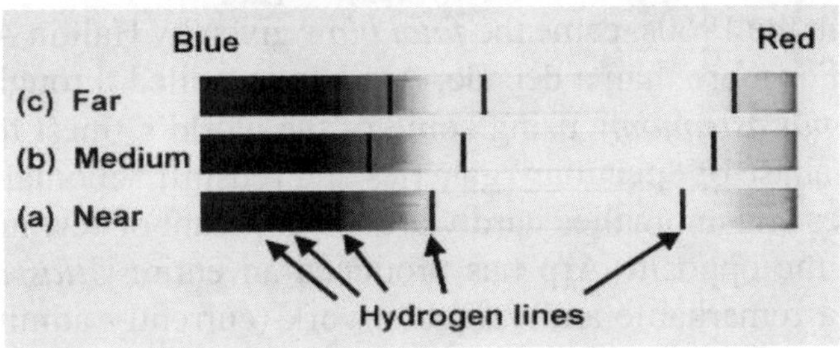

Fig. 13- Idealized galaxy spectra showing typical "absorption" lines (black against a rainbow-coloured background) produced by hydrogen atoms absorbing light. The more distant the galaxy, the more the lines are shifted to the red side (right) of the spectrum (log scale). The idea has now been extrapolated to <u>all</u> objects in the Universe. And because the class of objects called "quasars" have very large redshifts in general they are expected to be very distant.

These are the foundations upon the Big Bang Theory was constructed and put forward in the 1940's. According to this theory, the entire Universe was created instantaneously *ex nihilo* out of nothing 15 billion years ago, in just one go. Before that, space and time did not exist.

Another basis for the Big Bang was the assumption that when this explosion happened, it left a *trace* known as the cosmic microwave background radiation, billions of years old and with a *specific* temperature. If it was ever found, that would prove the theory is "correct". The Big Bang became entrenched orthodoxy, to which a *confession of faith* was demanded of those aspiring to grants, tenure etc. in academic physics departments.

However, if the mechanics of these redshifts which gives the foundation for the Big Bang, is *misunderstood*, then distances can be wrong by factors of hundreds, and luminosities and masses will be wrong by factors of thousands. We would have a totally erroneous picture of extragalactic space, and be faced with one of the most embarrassing boondoggles of our intellectual history.

Because objects in motion in the laboratory, or orbiting double stars, or rotating galaxies all show Döppler redshifts to longer wavelengths when they are receding, it has been *assumed* throughout astronomy that redshifts always and *only* mean recession velocity. No direct verification of this *assumption* is possible, and through the years many *contradictions* have arisen and been ignored.

Then in the 1960s, came the *fatal blow* given by Halton Arp through his work. For more than a decade, Arp has compiled through extensive *observational astronomy* using some of the world's finest telescopes a sizable database of "peculiar" galaxies and redshift "anomalies." These peculiarities and anomalies hardly are insignificant or few in number.

Quite the opposite Arp has produced an entire *Atlas of Peculiar Galaxies,* a remarkable and valuable work (currently administered by the California Institute of Technology, see Arp 1966). The images that he has produced, and the implications stemming from them, have *struck* at the very heart of current cosmological theory. A key concept to understand Arp's work is *quasar.*

Quasar means "quasi-stellar radio source". A compact, with starlike visual appearance, celestial body with a power output greater (high luminosity) than our entire galaxy, discovered in 1963 because of their radio emission. Quasars radiate almost over the entire electromagnetic spectrum with a typical power of $10^{46\text{-}47}$ erg sec^{-1}. The Sun, for example emits $3.9\text{x}10^{33}$ erg sec^{-1}.

Believed to be the oldest and most *distant* objects ever detected, quasars are billions of light-years from Earth, that's why they are only visible in big telescopes. They are now known to be the cores of very energetic galaxies, technically known as *active galactic nuclei* or AGN. What makes them so luminous is due to collective *plasma processes* which generate electromagnetic radiation and accelerate particles (dust, ions, electrons, etc.) in astrophysical objects such as galactic nuclei, quasars, or pulsars. Quasars are the most luminous objects in the Universe.

Arp was Edwin Hubble's assistant. Working at the Mt. Palomar and Mt. Wilson observatories in the US, he discovered that many pairs of quasars (quasi-stellar objects) which have extremely high redshift z values (and are therefore thought to be receding from us very rapidly and thus located at a great distance from us) are physically *associated* with galaxies that have low redshift and are known to be relatively *close by*.

How it can be? It is like an object is hot and cold at the same time. That's impossible. Something was wrong, quite wrong. If the lines in the spectrum of the light from a star or galaxy appear at a lower frequency (shifted toward the red) than where they are observed in the spectrum of the Sun, we say this object exhibits 'positive redshift'.

The accepted explanation for this effect is that the object must be moving away from us. This interpretation is drawn by *analogy* with the downward shift in the pitch of a train whistle as it passes through a railroad crossing and then speeds away from us.

The question is: Is recessional velocity the *only* thing that can produce a redshift, as modern astrophysicists presume? The reader should recall that Hubble was pointing to the same question from the very beginning. A high redshift value *does not* necessarily mean the object is far away.

There is *another*, more important cause of high redshift values. Also, the reader should recall the fallacy discussed before: what the A that the scientists thought was being caused by the B was really being caused by X or Y or Z

Experimental evidence shows that large redshifts differences are observed between whole extragalactic objects which are at the *same* distance. To *explain* this conundrum, Arp argues that the observed redshift value of any object is indeed made up of *two* components: the inherent component and the velocity component.

The velocity component is the *only* one recognized by mainstream astronomers. Therefore, *intrinsic* redshifts are required to explain the discrepancies observed. The inherent redshift is a property of the matter in the object, which will emit a specific wavelength of light. This matter would *change* over time in discrete steps.

In other words, the matter involved in galactic interactions, has *no* constant mass, which is exactly the opposite of what is thought in mainstream science. Cosmologically, the physics that assumes particle masses constant with time is *not* valid.

But now what is the consequence of having low mass, the so- called fundamental particles? It is simply that low mass electrons transitioning between atomic orbits will emit and absorb lower energy photons, so they will appear redshifted compared. to atoms with heavier particles which are less redshifted or tend to be blue shifted.

Arp explains that redshift is primarily a function of *age*. He presents abundant *observational evidence* to show that low-redshift galaxies sometimes eject high-redshift quasars in opposite directions, which then evolve into progressively lower-redshift objects and finally into normal galaxies. Ejected galaxies can, in turn, eject or fission into smaller objects, in a cascading process. Within galaxies, the youngest, brightest stars also have excess redshifts.

The reason all distant galaxies are redshifted is because we see them as they were when light left them, i.e. when they were much younger. About seven local galaxies are *blue shifted.* The orthodox view is that they must be moving towards us even faster than the Universe is expanding, but in Arp's theory, they are simply *older* than our own galaxy as we see them.

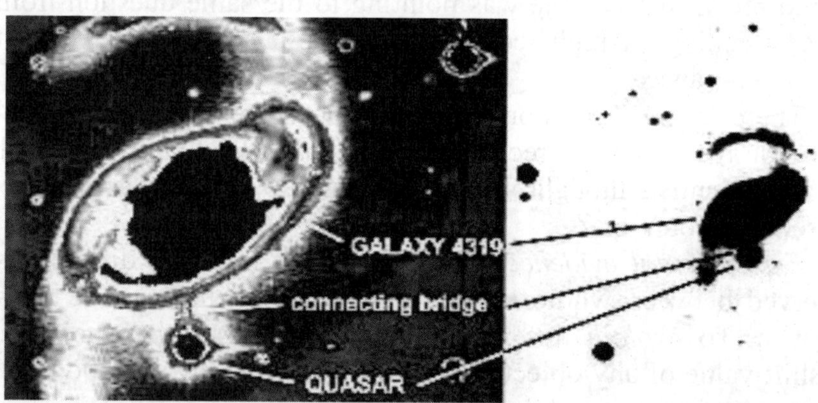

Fig. 14- This isophote image of the galaxy NGC 4319 (above) and the quasar Markarian 205 (below), made by superimposing a number of photographic plates taken by Halton Arp using the 200-inch Palomar telescope, clearly shows the luminous bridge connecting the two objects (north is up, east is left). This photo appears in Arp, Ref. 2 and Arp et al., Ref. 3. According to the Hubble law, this galaxy is 107 million light years away and the quasar is 12 times further away at 1.2 billion light years! Obviously, this simply cannot be, because the galaxy and the quasar are clearly connected together by a 'bridge'.

Summarizing, Arp contends that redshift is caused *mainly* by an object's being young, and only secondarily because of its velocity. Therefore, quasars are *not* the brightest, most distant and rapidly moving things in the observed Universe but they are among the youngest.

Arp has photographs of many pairs of high redshift quasars that are symmetrically located on either side of what he suggests are their parent, low redshift galaxies. These pairings occur much more *often* than the probabilities of random placement would allow

Arp has photographs of many pairs of high redshift quasars that are symmetrically located on either side of what he suggests are their parent, low redshift galaxies. These pairings occur much more *often* than the probabilities of random placement would allow.

Mainstream astrophysicists try to explain away Arp's observations of connected galaxies and quasars as being "illusions" or "coincidences of apparent location". But, the large number of physically associated quasars and low red shift galaxies that he has photographed and catalogued *defies* that evasion. It simply happens too often.

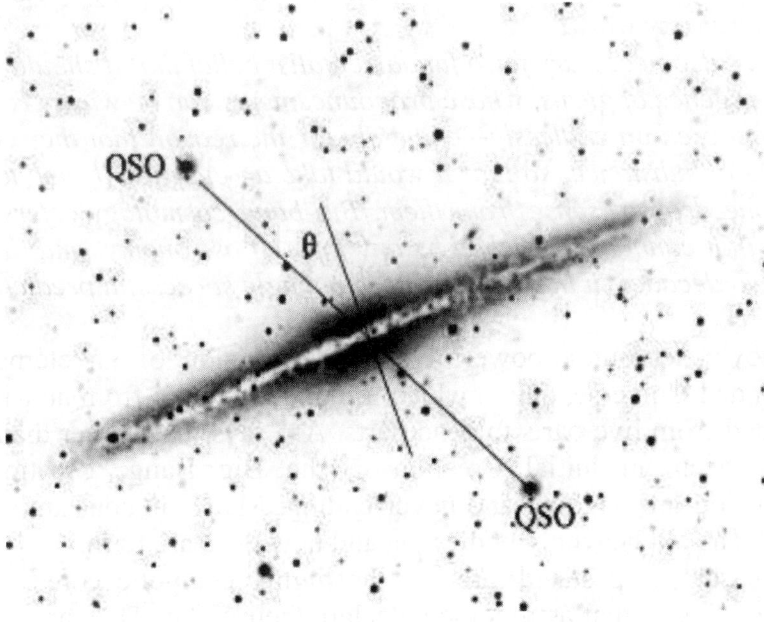

Fig. 15- Schematic drawing of commonly observed pairing of ejected quasars (QSO) from an active galactic nucleus. The objects at the ends of the line are quasars, which are usually measured with about the same redshift and fall within θ = ± 20° of the minor axis, the line drawn perpendicular to the plane of the galaxy.

Because of Arp's photos, the assumption that high red shift objects must be very far away on which the "Big Bang" theory and all "accepted cosmology" is based has been proven to be *false*. He wrote a book about it called *Seeing Red: Redshifts, Cosmology and Academic Science,* which has been so far, ignored by mainstream science.

However, the observational evidence has become overwhelming, and the Big Bang theory *cannot* account for it. Arp has not worked alone. For decades, he was a colleague of English astronomer Sir Fred Hoyle (1915-2001), a leading scientist now in oblivion, who had a far greater scientific calibre, not matched by none of today's cosmologists, astrophysicists and "professors". Here are his wise words in regards to the Big Bang:

"Big-bang cosmology is a form of religious fundamentalism, as is the furore over black holes, and this is why these peculiar states of mind have flourished so strongly over the past quarter century. It is in the nature of fundamentalism that it should contain a powerful streak of irrationality and that it should not relate, in a verifiable, practical way, to the everyday world.

It is also necessary for a fundamentalist belief that it should permit the emergence of gurus, whose pronouncements can be widely reported and pondered on endlessly—endlessly for the reason that they contain nothing of substance, so that it would take an eternity of time to distil even one drop of sense from them. Big-bang cosmology refers to an epoch that cannot be reached by any form of astronomy, and, in more than two decades, it has not produced a single successful prediction"

Hoyle set out a powerful *heterodox* vision of an eternal and unbounded Universe, one in which life does not arise from non-life but is seeded from live parts to dead parts. Arp says that, rather than there having been an initial Creation in the Big Bang. Creation and Destruction are ongoing and never-ending. Matter is constantly being *recycled* as old heavenly bodies die and new ones are born.

Quasars, Arp stated, are *not* the highly energetic cores of very *distant* galaxies that astronomers declare them to be. They are actually fairly *nearby*, and ejected from active galactic nuclei. The Big Bang? Never happened. The Universe, as his colleagues' astronomers Geoffrey Burbidge and Fred Hoyle also contended when they were alive, has always been here in a *steady state* of existence and always will be.

The cosmic microwave background radiation (CMB)? That's *not* the afterglow of the Big Bang fireball, it's the limiting temperature of space heated by all the stars in all the galaxies in the Universe.

Without he hypothetical "inflation field," when a very early Universe expanded faster than the speed of light (c), the Big Bang *does not* predict the smooth, *isotropic* cosmic background radiation that is *observed*, because there would be no way for parts of the Universe that are now more than a few degrees away in the sky to come to the same temperature and thus emit the same amount of microwave radiation.

This misunderstanding of the CMB is because most astrophysicists are ignorant of the *real* electromagnetic activity of space *plasma,* which is far more complex than the 'magnetized hydrodynamic fluid' (MHD) theory of plasma behaviour they are trained to employ.

Besides, when this alleged microwave background radiation was "discovered" in 1965, it had far *lower* temperature than *predicted*. Of course, this fact was quickly swept under the carpet by the scientific orthodoxy.

In the standard Big Bang model, the very large redshifts of quasars are interpreted according to the Hubble Law, to mean they are the most distant sources in the Universe. According to Arp's alternative model, *evidence* strongly suggests that they are *associated* with relatively *nearby* active galaxies and that they have been ejected from those parent galaxies.

One extremely good example of this was reported in the *Astrophysical Journal* in 2004 where a quasar was found embedded in the galaxy NGC 7319 only 8 arc minutes from its centre. This finding was presented by Margaret Burbidge in January 2004, at the American Astronomical Society meeting in Atlanta.

However, the response according to Halton Arp, was "overwhelming silence." Instead of nominating Halton Arp for a prize (and simultaneously re-examining their assumption that "redshift equals velocity"), he was systematically *denied* publication of his results and *refused* telescope time.

One would at least expect the "powers that be" to immediately turn the Chandra X-ray orbiting telescope, the Hubble space telescope, and all the big land-based telescopes toward Arp's exciting discoveries in order to either confirm or disprove them once and for all.

Instead, these objects have been completely *excluded* from examination. Those familiar with the Galileo story will remember the priests who refused to look through his telescope. Official photographs are routinely *cropped* to exclude them and the public is being told a lie.

Galileo introduced a simple new concept that changed the Universe as it was known then. Arp introduces a simple new concept that will change the Universe as we know it now. Seventeenth century educators taught that the Earth was the centre of the Universe.

The Sun, the moon, the planets and the stars revolved around it. Galileo confronted his contemporaries with a known Universe centred around the Sun. Today's educators teach that the Universe started from a Big Bang 15 billion years ago, and has been expanding ever since.

Galaxies and quasars are scattered according to their redshift. Arp confronts us with a Universe of *ejected* galactic families. In general, anybody who comes up with arguments or proofs, that challenge a prevalent scientific paradigm; is likely to be rejected, ridiculed or branded as "conspiracy theorist".

Challenging the paradigm of a solid, massif Earth that is currently taught; would be destined to a similar fate. It is noteworthy to mention that science after all, is a human activity performed by human beings, regardless of the accuracy or sophistication of the equipment and methods used. Therefore, scientific knowledge is not a 100% correct and infallible. There is always room for some *errors*, misconceptions, misinterpretations and mostly bias and *dogmas*. That's part of the human nature also an unnecessary *hindrance* for the advancement of science.

An outdated and obsolete paradigm is actually being used in modern science

Believe it or not in this 21st century, much of scientists, regardless of their field of research; are still anchored in the old paradigm born in the seventeenth century, during the Scientific Revolution: *materialism*. A close relative of it is *reductionism.* Both gave rise to the *mechanistic* paradigm of modern science.

Despite notable Quantum Theory devised in the 1920's and 1930's, which demolished the basic tenets of materialism and emphasizes the importance of fields in physics, coupled with the Theory of Gestalt (*Gestalttheorie*) developed in the beginning of the 20th century which makes materialistic reductionism unviable as a "fundamental" principle of Nature; the fact is that currently, materialism is still hold as the *only* valid approach of the reality of Nature in science.

In science, *materialism* is a paradigm which believes that the world ultimately consists entirely of hard, massy material objects, which, though perhaps imperceptibly small, are otherwise like such things as stones. (a slight modification is to allow the void or empty space to exist also in its own right).

These objects interact in the sort of way that stones do: by impact and possibly also by gravitational attraction. The theory denies that immaterial or apparently immaterial things (such as minds) exist or else explains them away as being material things or motions or by-products of material things. The physical concept of *field* doesn't fit very well within this paradigm.

Materialism states that ultimately everything is composed of material particles (or physical entities generally) but also holds that there are special laws applying to complexes of physical entities, such as living cells or brains, that are not reducible to the laws that apply to the fundamental physical entities.

Materialism states in fact that matter (protons, electrons, quarks or anything that has a mass) is the *only* reality. Also, it believes that these material particles, colliding and bouncing each other, driven by randomness will "eventually" form complex structures, ranging from cells to galaxies. However, materialism in itself, is *not* a scientific concept but a philosophical belief or *doctrine*. Also, materialism is very akin to the philosophical doctrine of *atomism* created by Democritus.

According to this doctrine, the world consists of nothing but atoms (indivisible chunks of matter) in empty space (which he seems to have thought of as an entity in its own right).

These atoms can be imperceptibly small, and they interact either by impact or by hooking together, depending on their shapes. The great charm of atomism was its ability to explain the changes in things as due to changes in the configurations of unchanging atoms.

This doctrine proved to be a workable hypothesis, in the beginning of the 19th century to initially, explain the chemical combinations of elements into compounds, in proportional parts to integer numbers. Later it developed into the Atomic Theory of modern chemistry.

On the other hand, *reductionism* also known as *elementalism*; is a view that asserts that entities of a given kind are collections or combinations of entities of a *simpler* or more basic kind or that expressions denoting such entities are definable in terms of expressions denoting the more basic entities. Thus, the ideas that physical bodies are collections of atoms or that thoughts are combinations of sense impressions are forms of reductionism.

Bear in mind that reductionism is akin to atomism and shares the notion that examining *parts* of things and *believing* that these parts could then be put back together to make *wholes*.

Atomists believed the nature of things made by atoms, to be absolute and not dependent on *context.* Reductionists are those who take one theory or phenomenon to be *reducible* to some other theory or phenomenon.

For example, a reductionist regarding physics might take any given mass of gas to be reducible to tiny particles in random, straight line motion, they move rapidly and continuously and make collisions with each other and the walls.

Or a reductionist about biological entities like cells, might take such entities to be reducible to collections of physical-chemical entities like atoms and molecules.

The aim is to explain or derive a phenomenon from its "constituents" parts. To explain the macroscopic observables from its microscopic "blocks". Reductionism has been proven successful to some extent in science and technology. However, this approach just doesn't work in explaining certain phenomena.

For instance, everybody knows water is a *liquid* made of two parts of hydrogen and one of gas. This is quantitatively valid. However *qualitatively*, according to reductionism, water should be a gas because its parts or blocks are made of *gases* hydrogen and oxygen, which is not the case. Edible table salt has very different qualities from its constituent "blocks" or elements: sodium and chloride.

Reductionism and atomism have much in common and still are *entrenched* in most of today's science. This analysis or reductionism believes that to understand any complex phenomenon, you need to take it apart, i.e. reduce it to its individual components.

If these are still complex, you need to take your analysis one step further, and look at their components. If you continue this subdivision long enough, you will end up with the smallest possible parts, the atoms (in the original meaning of "indivisibles"), or what we would now call "elementary particles".

Particles can be seen as separate pieces of the same hard, permanent substance that is called matter. Reductionism like materialism, is not a scientific concept but a philosophical one. In science, these concepts are *assumptions* that gave origin to scientific materialism and scientific reductionism.

This combination of materialism, atomism and reductionism; which started in the seventeenth century alongside modern science, led to the scientific approach called *mechanicism*: interpreting and explaining the phenomena of the Universe by referring to causally determined material forces. It sought to explain all-natural phenomena in terms of *matter* and *motion* without recourse to any kind of action at a distance (cause and effect without any physical contact).

Mechanicism attempted to explain all-natural phenomena in terms of configurations, motions, and collisions of small, unobservable particles of matter or bigger. *Isolation* and *randomness* are key features. This approach is still practiced and advocated today by most scientists such as English biologist Richard Dawkins. However, it is a half-truth so to speak and has its limitations, leading sometimes to a dead end as it will be discussed in chapter three.

This materialistic- reductionist- mechanistic approach, proved to be formidable in the advancement of science since the 17th century onwards, specially within physics; increasing to some extent our understanding of Nature and bringing greater control and freedom through advances in technology.

This paradigm is known as the Newtonian material Universe of the Classical Physics. However, at the end of 19th century and early 20th century, scientists researching the relationship between energy and the structure of matter, found several empirical phenomena that couldn't be explained by this paradigm. Other branches of science such as Psychology, faced similar issues.

This led to the development, during the 1920's and early 1930's, of a revolutionary new branch of physics called Quantum Mechanics (QM). It has questioned the material foundations of the world by showing that atoms and subatomic particles are not really *solid objects*, they do not exist with certainty at definite spatial locations and definite times.

Also, subatomic particles behaviour shows *coded* information. Most importantly, QM explicitly introduced the *mind* into its basic conceptual structure since it was found that particles being observed and the observer the physicist and the method used for observation are *linked*. According to one interpretation of QM, this phenomenon implies that the consciousness of the observer is vital to the existence of the physical events being observed and that mental events can affect the physical world.

In fact, US scientist Edward A. Tiller presents conclusive experimental data showing that in the same cognitive space, basic chemical reactions and basic material properties can be strongly *altered* by human intentions. This suggests that the physical (material) world is no longer the primary or sole component of reality. Scientists began to recognize that everything in the Universe is made out of energy or *vibrations* coupled with *fields* and *information.* This cannot be fully understood without referring to the mind or consciousness, all of this imbibed in a wholeness of interactions characterized by *coherence*.

It is important to recall that materialism and reductionism, are *not* scientific facts borne from the application of the scientific method but philosophical constructs or opinions.

In other words, what we are dealing in here are assumptions instead of facts. Indeed, materialism is an *indoctrination*. One of the firsts scientific facts, which materialism and its associates couldn't account for, was the concept in physics known as *fields*, introduced in the 19th century. Materialism already was having some troubles with the phenomena known as waves. But its main concern was *action at distance*.

In fact, scientists since the very beginning and following the materialistic-mechanistic paradigm, where unable to explain how in the world the attraction could take place when the objects were very far apart; for example, the Sun and Earth are separated by about 150 million kilometres, but nevertheless there is an attractive gravitational force between them.

How can this be? One can understand that two colliding objects will exert a force on each other, because they come into contact. How can two distant objects exert a force on each other when they are so very far from touching?

Here's when the genius of English *experimental* physicist-chemist Michael Faraday (1791-1867) comes up with the right solution, when studying electrical and magnetic phenomena, among others. Faraday was motivated by the way iron filings align themselves near a magnet: He imagined that the magnet extended some sort of "tentacles" of force throughout the space around it, exerting some sort of tension.

This "tension" in the space around the magnet is what causes the iron filings to align. Might not gravity act in the same sort of way? And electrical forces too? Perhaps all forces arise in the same way: whatever the source of the force is (mass in the case of gravity, electric charge in the case of electric forces, and electric currents in the case of magnetism), the source creates a "tension" of some sort in space.

Then an object that can respond to the force simply "feels" the "tension" in the "tentacles." Later, Faraday concluded the location of electromagnetic phenomena, as occurring in a *medium* or field, not in objects.

In the study of the newly discovered area of fields, the important step was taken by Michael Faraday and completed by remarkable Scottish physicist James Clerk Maxwell (1831-1879). Faraday and Maxwell not only studied the effects of the electric and magnetic forces, but made the forces themselves the *primary object* of their investigation.

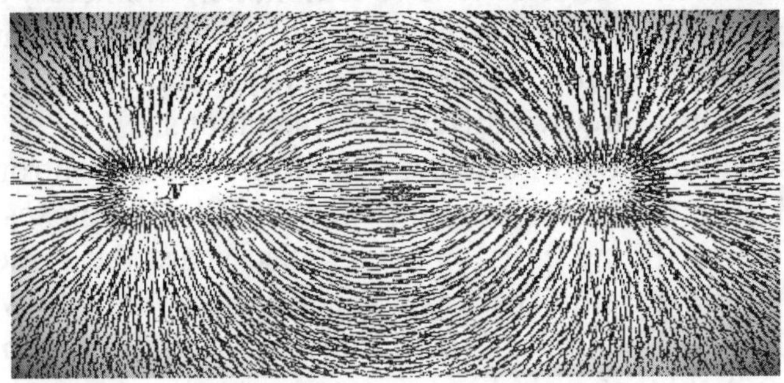

Fig. 16- The spatial organization and shaping of iron filings due to an unseen entity, apparently emanating from a magnetic bar, inspired Faraday to introduce the concept of *fields* in physics. The field is a property of space itself. It, not atoms ("centers of force"), is the field's home. In this case, the magnetic field still exists, even when the iron fillings are absent. It exists as a spatial structure (potential) even though the magnetic field has not been yet actualized in the final form of the system.

By replacing the concept of a force with the much subtler concept of a *force field* they were the first to go *beyond* Newtonian physics, showing that the fields had their *own* reality and could be studied without any reference to material bodies. Fields are not matter. That is probably one greatest sources of confusion regarding what fields are. In fact, fields govern matter and it is subdued to these. Fields are more *fundamental* than matter.

Today, fields are a way of describing physical effects that occur at a distance, by giving a quantity (scalar or vector) at every point in space and time. Examples include gravitational fields and gravitational potentials which relate to the gravitational attraction of a mass, or electric fields and electric potentials which relate to the attraction or repulsion of charged objects due to "electrical charges".

In general, fields involve the region of influence of some physical phenomena. Bear in mind that fields are *invisible* and *immaterial* "things" that give shape and organize *material* objects, whatever they might be. Contemporary science would be unthinkable without the use of fields, though it uses a mediocre and distorted version of the *original* concept of force fields, proposed by Faraday and Maxwell.

Most important, fields are scientific concepts based on *empirical* scientific findings not philosophy. Fields are the first conceptual step that washes away, the isolation and randomness of phenomena, preached by the materialistic-mechanistic paradigm.

In fact, the new physics tells us that matter, ultimately at its core is nothing more than a series of patterns out of focus and that subatomic "particles" aren't really made of energy, but simply are energy (vibrations). These are two different explanations.

The subatomic world of electrons, protons, and neutrons may thus be viewed as patterns of vibration within a field. Complexes of physical entities, such as living cells or beings also share the same pattern of vibration within a field. There is a specific field, the one what English biologist Rupert Sheldrake calls a *morphogenetic field*, an organizing and shaping field that underlies a system's structure. There's simply, no room for *isolated* objects acquiring a determinate structure by "chance."

Morphogenetic fields play a causal role in the development and maintenance of the forms of systems at ALL levels of complexity. The word "form" includes the *shape* and its internal *structure*. It is a *formative* causation in contrast to the energetic type of causation which conventional physics, already deals so thoroughly. For although morphogenetic fields can manifest in conjunction with energetic processes, they are not in themselves energetic.

US-British physicist David Bohm, warned against this outdated materialistic-mechanistic-random approach of contemporary science, in his book *Wholeness and Implicate Order* (1980). An unseen and active entity pervading all space (implicate order), *organizes* simple and complex material things, forming stable structures (explicate order).

For example, conventional physics is *unable* to explain the unique structure of complex molecules and crystals, it permits a range of minimum-energy (stable) structures but there is no evidence that conventional physics, can explain that only *one* of the possible stable structures, is realized. Some other factor (implicate order) than energy *unrecognized* by ordinary physics, unequivocally "selects" between these possibilities, determining the specific structure of the system.

Subatomic particles are no longer considered "material". Sometimes they behave like waves of light. The current consensus is that they are both wave-like and particle-like, as is all matter. In fact, in the world of quantum physics, it seems these elementary, *individual* "particles" (including electrons) don't really exist at all. What does exist are relationships, correlations, tendencies to actualize from a multifaceted set of potentials.

What the scientific community once thought was there in the sub-atomic realm and what the educated world was taught to perceive as real, simply does not exist. How this can be? As we should know, all true science is based on and only *experience*. When natural scientists observe and record natural events, they do so through their sensory experiences or *perceptions*.

And here's the catch: Most scientists and laymen alike, are still trapped in what is known as *realism* which in itself, is another philosophical concept or *opinion*. Realism states that the "reality" of material objects, and possibly of abstract concepts, exists in an external world *independently* of our minds and perceptions.

However, the abstract statement of realism or the independent objective physical reality, from the personal subjective world is *false*. In physical reality, *nothing* exists in isolation.

What can we know? Scientific knowledge is constrained to the natural, empirical world and it is shaped by our minds

We perceive because we are conscious of the perceptions or sensations, conveyed in our brain trough sensory organs. Those are organized in consciousness (mind) to create qualities of the form that may be novel. A table has a "form quality" that persists even when the sensations change. We look at a table from the side or top, but we still see it as a table.

Similarly, shapes may be reduced in size or enlarged. shown in one part of field or another, their colour changed, and they are still perceived as the same shape. Therefore, what we think as something being of whatever shape, colour or consistency is due to the *raw* energetic data that reaches our sensory organs, is conveyed to the brain and it filters, decodes and structures this information in colours, sounds or anything that we usually "perceive" only in our *consciousness*, nowhere else.

Moreover, animals such as dogs, cats and others; have their sensory organs tuned in such way that they hear and see things humans can't.

Consciousness is everything and everything is in the mind. Is the *same* individual conscious mind that is focusing either inside (subjectivity) or outside (objectivity), they are *not* separated. Therefore, this statement of "independent" objective physical reality is arbitrary and thus, has no scientific basis.

There is no "out there" as quantum physics demonstrates thought the notable double slit experiment, as it blurs the line between the observer and the system being observed. Recent neuroscience research support this "merge". The human world of experience is the *only* immediately given reality, the three-dimensional world we live in and visualize things or *observables*.

By method, *real science* is anchored in this 3-D realm. This is key to understand. Because the alleged mathematical constructs of 4-D or N-D universes, can *never* be tested and therefore, belong to the realm of philosophy/pseudoscience.

Science deals with specific and methodological knowledge. It is important to briefly examine the following questions: how do we gain knowledge of objects? To what extent we can know an object? How do we make sense of our experiences?

In general, how do we get to know anything at all? At glance, it might seem that a scientist has a *passive* stance when doing any scientific work. Scientists are just reading the book of Nature and obtaining knowledge.

Most people think science is basically collecting data about a certain phenomenon trough experiments or fieldwork and with some logical reasoning, laws and theories will come out. The scientist' inquiring mind conforms or adjusts to what is studying. Indeed, it is far more than that. Human mind is the *decisive* factor in elaborating scientific knowledge and it relies on raw data from *experience.*

German scientist and philosopher Immanuel Kant (1724-1804), after a meticulous study of the Newton physics' success, called this discovery his 'Copernican Revolution'. In the past, according to traditional psychological and philosophical thinking, knowledge is a simple reflection or duplication of the external world in the mind. In his masterpiece *Kritik der reinen Vernunft* (1781) or Critique of the Pure Reason, Kant criticised this view, arguing that the mind is *not* a passive reservoir or mirror for sensory input. Instead, he demonstrated that any knowledge we obtain, is necessarily *shaped* and *structured* by our cognitive faculties.

It is no longer our perceptions that give us a sense of objective reality and the outside world. The human mind sets the *conditions* of possibility of knowing and thus actively *constructs* our scientific knowledge. Nothing in our experience is just "given" to us in a pure form *unadulterated* by the way we think. Our cognitive apparatus is always both receptive and active.

To flesh out his revolution, Kant constructed an incredibly detailed account of how the *chaotic input* of the senses is *structured* by our cognitive faculties. In the first stage the raw sensory input must pass through what Kant terms intuition, the mental faculty that orders things according to the dimensions of *space* and *time*.

Kant asserts that space and time are *not* objective, self-subsisting realities, but *subjective* requirements of our human sensory-cognitive faculties to which all things must conform.

Space and time serve as indispensable tools that arrange and systemize the images of the objects imported by our sensory organs. The raw data supplied by our eyes and ears would be useless if our minds didn't have space and time to make sense of it all.

Kant further argues that this sensory-spatiotemporal process requires a supreme mediator or the human *reason* that will *synthesize* the sensory input within our cognition so as to turn it into meaningful knowledge.

Second, after receiving the basic structures of time and space, the sensory input must be further synthesised in our understanding. Kant puts forward a set of basic concepts such as *causal* and *logical* relations in our experiences, that form the basic precondition for us to receive and understand the temporally and spatially informed data from the intuition.

They allow us, for instance, to make connections between separate events as a causal sequence. Thus, our experience is *shaped* both spatial-temporally and logically by our *cognitive* faculties that spring from a *conscious mind.* But receiving and structuring information from the outside world isn't all our minds can do. Adding to his complex descriptions of intuition and understanding,

Kant describes some other mental faculties that contribute to forming our experience. For instance, *judgement* acts as a guide to the faculty of understanding, and the *imagination* provides information to the understanding in the absence of immediate sensation from the external world. After more than two centuries, Kant's ideas are still valid.

Given all the details mentioned before, *scientific knowledge* cannot be produced or processed through our minds, *without* any input of *empirical* raw data from the external (3-D) world. It is like a mill which needs wheat to produce flour.

If there is *no* raw empirical data, the mind will produce what Kant would say: "metaphysical flour". For this reason, it is obvious that speculations of theoretical physicists, are *not* a reliable source for learning the nature of the Universe. These are in fact, pseudoscience.

A new paradigm: Wholeness and Gestalttheorie

The reductionist-mechanistic paradigm is *outdated*. We need to learn to perceive holistically because our world and the entire Universe is actually *interconnected*.

It is erroneous to continue to perceive our world as a conglomeration of separate, unrelated parts, in an "empty" space (which cutting-edge scientific research proves otherwise), assuming happenstance or stochastic as the main driven force in the Universe and ignoring *organizational patterns* originated by morphic fields, observed in Nature such as fractals and vortices.

Considering emergent scientific principles, the reductionist-mechanistic world view is decidedly misleading. For instance, it has caused serious issues within astronomy, when trying to explain the origin of the Solar System and Earth as we will see in chapter three.

In Nature, is the *wholeness* which organizes and makes the components work together, with *coherence*. Only a manmade thing like a machine would work from its parts and therefore, be fully explained (mechanically) from these parts.

The basis of any science is grounded on empirical observations (perceptions), which indeed are nothing more and nothing less than scientist's direct *psychological experiences*. It is worthy to examine this closely. In the beginning of the 20th century, a branch of science known as *Gestalt* psychology developed in Germany, made a research of *human perception.* The founders of this pioneering work were Kurt Koffka (1886-1941), Wolfgang Köhler (1887-1967) and Max Wertheimer (1880-1943). Gestalt is a German word which could be translated as "shape" or "structure".

When trying to make sense of the world around us, *Gestalt* psychology suggests that, contrary to the reductionist-mechanistic paradigm; we do not simply focus on every small component. Instead, our minds tend to perceive objects as part of a greater *whole* and as parts of more complex systems. This ability is often truncated by the formal academic education.

Gestalt experiments show that the mind does not act like a sponge (as a passive receiver of information) but actively filters, structures, and matches all incoming information against known patterns to make sense of it. *Gestalt* does not question the contribution of experience acquired in the past but asserts that thinking involves *more* than recall. Current "education" stresses *only* on recall.

Gestalt psychology proposes that what is 'seen' is what appears to the seer and not what may 'actually be there,' and that the nature of a unified whole, whatever it might be; is *not* understood by analysing its parts. It has the sense that meaning *cannot* be found from breaking things down into parts but rather from appreciation of the *whole*, which is an entire *field* with differentiated parts.

Gestalt is key in learning and problem-solving situations like those faced in any scientific research. It stresses the importance of seeing the *ganze Gliederung* or whole structure of any given problem. This leads to the *Einsicht* (insight) experience where a consistent solution of the problem is found.

In fact, one is able to see the overall structure of the problem: A certain region in the field becomes crucial, is focused; but it does not become isolated. A new, *deeper* structural view of the situation develops, involving changes in functional meaning, the grouping, etc. of the items.

Directed by what is required by the structure of a situation for a crucial region, one is led to a reasonable prediction, which like the other parts of the structure, calls for verification, direct or indirect. Two directions are involved: getting a whole consistent picture, and seeing what the structure of the whole, requires for the parts.

This approach is advantageous in any scientific endeavour because it gives an *Übersicht* or a wide, clear and meaningful perspective in any given situation. However, the current scientific paradigm still is anchored in the old reductionist-mechanistic tradition, which in isolating or stressing the attention on a few elements of a system and *ignoring* its context or environment, only solves or explains partially the situation. Sometimes contradictions and inconsistencies do arise, which is often the case.

Moreover, this reductionist-mechanistic mindset can be linked with what is known as "compartmental thinking" which is very common in the academic community. Due to highly specialized branches of current science, each branch develops in isolation compared with the rest or what they call, *fields of specialization*.

This "atomization" leads to problems and inconsistencies when a specific branch of science is trying to "understand" our *complex* Universe. This excessive specialization makes scientists to be *unable* to see the Universe in a holistic, cyclical and interconnected manner, which is the way it actually operates.

According to the *Gestalt* school of thought, humans are naturally capable of perceiving objects as orderly and organized forms and *patterns*. We prefer things that are simple, clear and ordered. Instinctually these things are safer.

They take less time for us to process and present less dangerous surprises. Also, being able to recognize patterns is key in science.

This refers to *prägnanz*, a German word that means "pithiness" or precisely meaningful. Rather, wholes are entirely *different* from any summation; they have their own inherent properties that in turn determine the nature of the parts. There is no need to refer to "elements."

The *Gestalt* motto says:

"The whole is more and before the parts. The whole is more than the sum of the parts"

In technical jargon this means: *non-linearity* or that a system has non-linear properties. Also, it means that physical reality in any given (local) system is the mutual interplay of three factors: matter and energy within a specific, coded environment or context. In conventional science, the context is often ignored.

This context refers to the *spatial entity* (morphic field) capable of ordering physical changes and like electromagnetic and gravitational fields, it is invisible and intangible. It is endowed with the property of traversing "empty space". In fact, it IS space. However, morphic fields *differ* radically from electromagnetic and gravitational fields, because these depend on the *actual* state of the system, on the distribution and movement of "charged particles," or masses, whereas morphic fields correspond to the *potential* state of a developing system and are already present, *befor*e it takes up its final form.

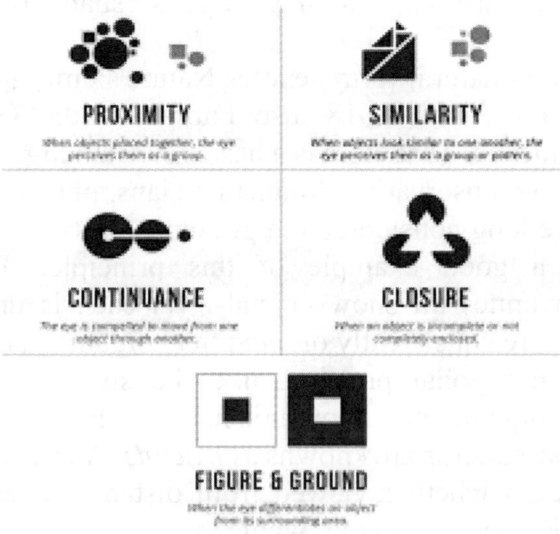

Fig. 17- Theory of Gestalt aims to explain how the world as a whole is structured and perceived by individuals. It is based on five principles. Pattern perception, or our ability to discriminate among different figures and shapes, occurs by following the principles described above. The whole of an image or structure is more important to our knowledge, than the individual components.

This spatial entity is not in itself energetic nor originates from matter; but nevertheless, it plays a *causal role* in determining the forms of the systems (*Gestalt*), which it is associated.

Also, this spatial entity in contrast to ordinary physical quantities such as mass, momentum, energy, temperature, pressure, electric charge, etc. is *not conserved*. If a bunch of flowers is thrown into a furnace and reduced to ashes, the total amount of matter and energy remains the same, but the *form* of the flower simply disappears.

Gestalttheorie is more closely linked to the Nature's unity, order, and *Veranstaltung* or organization than the isolated and fragmentary approach of the reductionist-mechanistic paradigm, where *randomness* is the driven force. We are part of Nature's ordered complex, and so have an innately ordered sensibility.

Nothing exists in isolation nor is driven by chance. It is erroneous to continue to perceive our world as a conglomeration of *separate*, unrelated parts. For example, the various parts of the plant: roots, stem, leaves, petals, sexual organs, (etc); all must to be understood as transformations of an underlying *Bauplan* or pattern, already implicit in the seed. The same applies in general to shape planets, stars and galaxies.

Research on natural form denotes Nature's unity in the remarkable irrational number, Phi 1.618 also known as the Golden Ratio or Fibonacci sequence. That there is a basic *unity* among Nature's diversity is a centuries old observation. Mathematicians, philosophers, and some scientists have long considered the principle 'unity in diversity.' Snow crystals are a good example of this principle. Their hexagonal configuration unites all snow crystals, yet each is unique. Inorganic patterns are more consistently ordered than organic patterns.

Further, hexagonal patterns, like the snow crystals, are more common in inorganic than in organic nature, which favours pentagonal patterns. These patterns are knowns as *fractals*. A fractal is a self-similar repeating pattern whether viewed from distance or close-up. So, the 'part' is almost identical to the 'whole'.

The spiral is another recurring configuration throughout organic nature. Botanists call the spiral growth configuration in plants *phyllotaxis*, and it has the 1.618 proportion. The 1.618 proportion or Phi vortex appears in the curve-cross configuration of the pineapple, pinecone, and sunflower, and in the animal world for example, in the spiral of snails and the chambered nautilus, and in the proportions of the human body.

From water flows, to kelp patterns, to shell architectures, Nature repeatedly utilizes 3-dimensional centripetal spirals oriented toward the center of curvature for liquid flows. Experimental evidence shows that the most *efficient* way to move matter and energy seems not to be a straight line, but a curve known as Phi.

Other examples are the vortex configuration of *galaxies* and hurricanes. It doesn't matter if one is a flower, fish, alien, tree, human, bird or whatever one has been created, comes from the same geometrical pattern, appearing when self-organization processes are at play. In the past, Nature has long guided architects, artisans, artists, scribes, and typographers in the shaping of our built world. Certain proportions recur in their work, and comprise simple geometric figures: equilateral triangle, square, regular pentagon, hexagon, octagon, and others. Often, builders-based measurements and proportions on the human body (anthropomorphism). Moreover, recent research shows this Golden Ratio, prominent in Nature and art, has also its presence in *Fluid Dynamics*. The first example draws from the investigation of the resonance in wind tunnels with ventilated walls.

Using *Acoustic Wave Theory*, the reciprocal Golden Ratio is shown to *determine* the critical Mach number below which refraction is possible and above which total reflection takes place. The second example concerns the vortex merger, such as observed in aircraft turbulence and large-scale atmospheric or oceanic flows.

Based on a numerical simulation and available experimental data, a conjecture is made that the distance below which two identical Rankine vortices merge and above which they do not, is the *product* of the vortex diameter and Golden Ratio. Even in Plasma Physics, scientist have *discovered* microscopic Fibonacci spirals.

From the biggest to the smallest, really everything that *works fine* shows these patterns that the Fibonacci sequence delivers. In almost all flowering plants, the number of petals on the flower is a Fibonacci number.

The Golden Ratio is seen in these flowers in terms of petal arrangement. All the petals exhibit a twisting of about 1.618034°, in order to *optimize* exposure to sunlight. Thus, the Golden Ratio determines *efficiency* in any natural system.

The natural design that connects vision and cognition is a theory that flowing systems, from airways in the lungs to the formation of river deltas, evolve in time so that they flow more and more easily

The evidence presented here for the pervasiveness of the Golden Ratio along *vortices*, throughout the sciences merely scratches the surface and one could easily cite more instances of its occurrence. The potential applications for the Golden Ratio stretch from the macrocosmic down to the microcosmic scale, appearing to have scale independent relevance to everything in between.

From this perspective, the whole system of our Universe is essentially governed by *flow dynamics* or the physics of turbulence. Turbulent systems are characterised by the eddies, currents and vortices of “flow”. Vortices are self-sustaining *Gliederung Energies* (energy structures) that occur everywhere naturally, expressing minimum energy and stable configurations.

Nature uses centripetal, vorticial form of motion, moving from the outside to the inside with increasing velocity, which acts to cool, to condense, to structure, assisting the *emergence* of higher quality and more *complex* systems with coherence and efficiency. Vortices manage energy gathering and dispersing it, keeping the entire Universe organized and alive. A vortex is the epitome of Nature's use of opposites in the maintenance of balance. In fact, vorticial movement is a perfect example of balance through polarity.

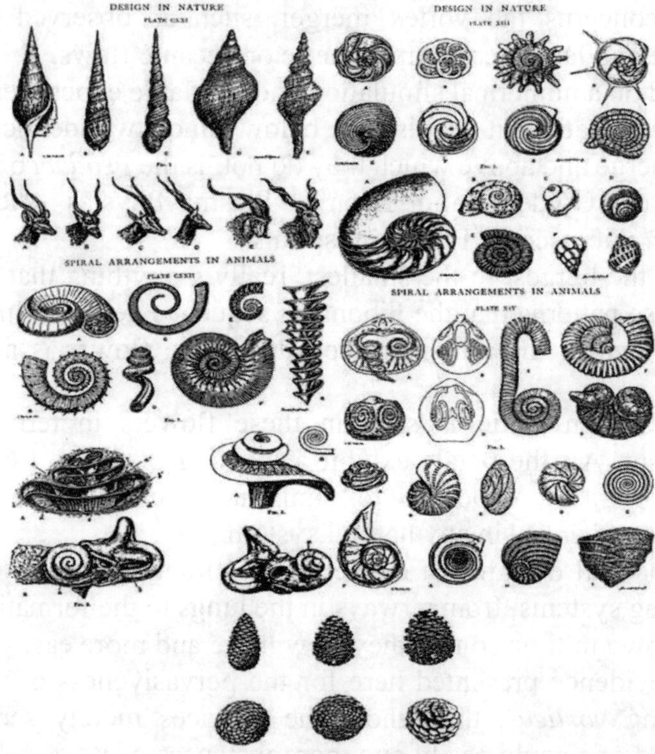

Fig. 18- Nature is filled with spiral designs expressing the Golden Ratio. Here are shown some examples. In fact, tons of lifeless and alive things have spiral designs, and they are not random spirals. Golden Ratio can be seen all over the place in nature. They are, found across the cosmos. From the human ear's curve to the human bone structure, to the growth pattern of leaves and flowers, to the spirals of a nautilus seashell or the spirals of galaxies; spirals can be seen everywhere. The Golden Ratio's generality has led to be recognized as the geometrical blueprint for life. It was even named "The Key to the Physics of the Cosmos."

It is also Nature's finest creative/organizing force. This concept will be extremely helpful in astronomy as we shall see in chapter three, in regards of the origin of the Solar System and thus planet Earth.

A vortex such as flowing water is the least complex of the physical structures, which have been labeled *dissipative structures*. By definition, dissipative structures are *open energy systems* that constantly incorporate energy and matter as inputs and dissipate them as output. The form they take is dissipation. Dissipative structures theory is ideal for studying the generative emergence of molecules, *planets* or galaxies. Firstly, once initiated, the system can move into a far-from-equilibrium state. Secondly, nearing a threshold, fluctuations (turbulence) arises throughout the system.

Thirdly, at the threshold, the system exhibits nonlinearity as well as bursts of amplification. Fourthly, the emergent order that happens is a recombination of existing elements in the system. Lastly, emergent order remains *stable*, even when perturbed. In terms of outcomes, (a) Emergent order increases the capacity of the system to a large degree, and (b) the emergent order transcends but includes its components.

This sort of formation/shaping or *Gestaltung* is key and determinant in Nature and it is neither a theory nor hypothesis. Let alone a mathematical construct. It is observable and testable. *Gestaltung* is what organizes and shapes material things and phenomena, within a context or environment, in *stable* and *efficient* systems.

Gestaltung exhibits none of the elements of the straight line, circle and point, the basis of modern mechanical and technological constructs. It can be interchangeable with *morphic fields*.

Reflecting Nature's principal constant, namely that of continuous change and transformation, one of the most common *Gestaltung*, the vortex epitomizes this form of open, fluid and flexible motion.

Nature's energies spin clockwise and anti-clockwise, they expel centrifugally then pull back in centripetally. It excludes the traditional reductionist-materialistic-mechanistic paradigm (indoctrination) of how Nature allegedly works, which is the mindset of *mainstream science*.

In chemistry, is routine to break down chemical components into its elements (analysis) and combine elements to produce compounds (synthesis). Through this way, millions of tons of chemical compounds are produced annually.

In the past, it was believed that chemicals compounds, produced by living forms and called organic compounds, could not be synthetized in laboratory, using mineral inorganic substances or elements. Then in 1828, German chemist Friedrich Wöhler, demonstrated that it can be done by combining cyanic acid and ammonium in vitro.

In these experiments, the synthesis of an organic compound from two inorganic molecules was achieved for the first time. Apparently, there was *no difference* whether a specific chemical compound was made by a living form or in the laboratory. Being made of the same elements or blocks, it is just matter of combining them to fabricate the known compound according to the reductionism-materialistic paradigm. Nothing else matters. If that were true, then the chemical compound, besides its chemical properties, would have the same *physical properties*; regardless how it originated. However, this is *not* the case.

Louis Pasteur the famous 19th-century French chemist, was the first to describe chemical handedness, or "chirality." He was puzzled by the fact that crystals derived from the dregs of wine twisted light in a specific direction, but the same crystal synthesized in the laboratory *did not*. Examining the crystals under a microscope, he discovered that the synthetic chemical came in two mirror-image forms, which cancelled out the polarizing effect. The crystal derived from wine had only one.

Whereas both compounds are made of the *same* building blocks of matter, their context or *environment* where they have been assembled, was totally *different.* Therefore, they have different physical properties though being chemically, the same substance. What makes them *physically* different? Why the synthetized compound *never* matches the physical characteristic of the natural one?

Scientists later discovered that this phenomenon of chirality encompasses the *entire living world*. In other words, chirality is a *Gestaltung* working feature of Nature. Synthetic or artificial chemical processes will generate both left-and right-handed molecules; technically known as enantiomers.

But when Nature makes a molecule, the product is either left-or right-handed. For example, *all* amino acids that are used to make proteins twist light to the left. Indeed, chirality is an *essential* component of biochemistry. It provides a form of molecular "recognition".

The chirality of a molecule *affects* how it interacts with other components of the cell. Molecular locks can only be opened with a key of the *correct* handedness. The origin of this handedness is a complete *mystery* to evolutionists. There is absolutely no room for "chance" or randomness. Chirality is an *asymmetric* phenomenon.

Whereas Nature produces *only* one type of chirality, either left or right handed; laboratory synthesis always produces both. This is technically known as racemic mixtures and when it is used for human consumption, racemic mixtures can lead to serious consequences, to say the least.

A tragic reminder of the importance of chirality is thalidomide. In the late 1950's and early 1960's, this drug was prescribed to pregnant women suffering from morning sickness. However, while the left-handed form is a powerful tranquilliser, the right-handed form can disrupt foetal development, resulting in severe birth defects. Unfortunately, the synthesis of the drug produced a racemic mixture as would be expected and the wrong enantiomer was not removed, before the drug was marketed.

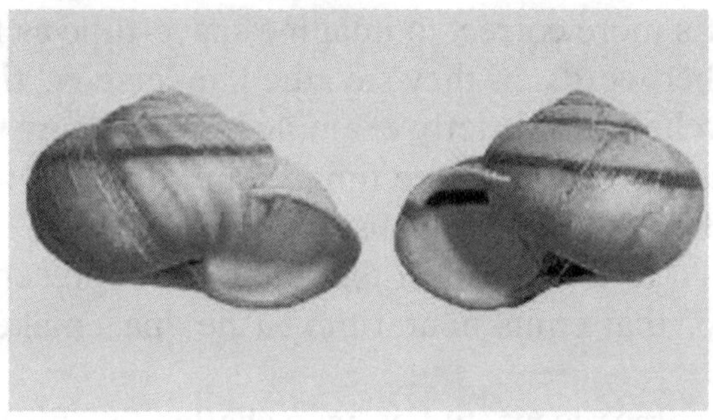

Fig. 19- Mirror image, a left-handed and right-handed snail shell. In the biological world, we see inorganic minerals being shaped with remarkable control. The organic molecules are acting as a scaffold, dictating where the minerals are placed and how they are linked together, a bit like building blocks. And as they do this, the biomolecules transfer their left or right-handed nature, or chirality, to the crystal structure.

On the other hand, another idea of the old paradigm that is being washed away, is the concept of *time*. Some researchers are considering that this Newtonian idea of time as an absolute quantity that flows on its own, along with the idea that time is the fourth dimension of space-time, are fundamentally *incorrect*. In two recent papers (one published and one to be published) in *Physics Essays* (2010), Amrit Sorli, Davide Fiscaletti, and Dusan Klinar from the Scientific Research Centre Bistra in Ptuj, Slovenia; have described in more detail what this means.

They propose to replace these concepts of time with a view that corresponds more accurately to the *physical world*: time as a measure of the numerical order of change. They begin by explaining how we usually assume that time is an absolute physical quantity that plays the role of the independent variable (time *t*, is often the x-axis on graphs that show the evolution of a physical system). But, as they note, we *never* really measure *t*.

What we do measure is an object's frequency, speed, etc. In other words, what *experimentally* exists are the *motion* of an object and the tick of a clock, and we *compare* the object's motion to the tick of a clock to measure the object's frequency, speed, etc. By itself, time (*t*) has only a mathematical value, and *no* primary physical existence, whatsoever.

This view doesn't mean that time does not exist, but that time has more to do with *space* than with the idea of an either absolute or relative time. So, while 4D space-time is usually considered to consist of three dimensions of space and one dimension of time, the researchers' view suggests that it's more correct to imagine space-time as four dimensions of space. In other words, as they say, the Universe is "timeless."

The researchers also briefly examine how this new view of time fits with how we *intuitively* perceive time. Many neurological studies have confirmed that we do have a sense of past, present, and future. This evidence has led to the proposal that the brain represents time with an internal "clock" that emits neural ticks (the "pacemaker-accumulator" model).

However, some recent studies have challenged this traditional view, and suggest that the brain represents time in a *spatially* distributed way, by detecting the activation of different neural populations.

Although we perceive events as occurring in the past, present, or future, these concepts might just be part of a *psychological frame* in which we experience material changes in space. Exactly as proposed by Kant, more than two centuries ago.

The two sides of the current scientific endeavour

Since the Scientific Revolution was carried out in the seventeenth century until beginning of the twentieth century, the scientific activity was unilateral or one-sided. Scientists carried out their work and their results were openly and widely published. Anyone could have accessed it and gain knowledge.

There was no such thing as information being *classified*. Scientific societies flourished everywhere. The Industrial Revolution took place and made a lasting impact worldwide. During the nineteenth century, huge scientific and technical advances indeed, improved the material living standard of mankind.

Railways, electricity, steamboats, photography, advances in medicine and physiology are just a few examples. The geographical discoveries of explorers in remote areas of Africa, Asia and the polar regions; where followed by the press and public was kept well informed.

Scientists pursued their genuine interests without any pressure from outside. This allowed the appearance of remarkable scientists, true geniuses such as Leonhard Euler, Michael Faraday or Nikola Tesla, among others. They were truly creative scientists by their own, not the manufactured products propped up by the media as we see today. Attaining scientific knowledge, spreading it to everyone and applying it to benefit mankind where part of the same equation.

Science was a domain that belonged to universities and colleges (which also provided advice and human resources for the industry), to scientific societies and finally to some individuals carrying out their scientific work by their own. That was the "heroic age of science". Everything was peachy but not for long time. The reason was that, like bystanders, government and corporations were not involved, yet.

This picture began to change since around 1900 onwards. It started with electrical companies seeking more profits in alliance with bankers and investors. Later this included the steel and oil industry among chemical companies. This trend kept growing worldwide and originated the big corporations we see today, which have significant influence over a nation's high finances and political members of the government.

Since then, part of the scientific research was meant to increase corporation's *profits*. Scientists had to follow the corporation's guidelines. Also, several corporations do fund grants for universities and colleges which in a sense, is a subtle *subordination* to corporations.

There is little room for *independent* scientific research, unless it is profitable. Any scientist pursuing genuine interests or offering alternate options, that disagree with the corporation's agenda (profits) is not welcome.

In fact, Serbian-US engineer and physicist Nikola Tesla (1856-1943) the discoverer of the rotating magnetic field and its application to the generation, transmission and utilization of polyphase alternating currents (AC) for power and light, that brought electrification to the world; is a fine example of this attitude.

At age 38 Tesla was just beginning the most fruitful part of his career in applying breakthroughs in science, including some of his own discoveries, to the development of paradigm-shifting technologies. One of such tantalizing discoveries were scalar waves also referred to as Tesla Waves or Longitudinal Waves (non-Hertzian waves) which can penetrate any solid object including Faraday Cages.

Tesla was a pioneer in many fields. The Tesla coil, which he invented in 1891, is widely used today in radio and television sets and other electronic equipment.

His alternating current induction motor is considered one of the ten *greatest discoveries* of all time. Among his discoveries are the fluorescent light, laser beam, wireless communications, wireless transmission of electrical energy, remote control, robotics, Tesla's turbines and vertical take-off aircraft.

Tesla is the father of the radio and the modern electrical transmissions systems. He registered over 700 patents worldwide. His vision included exploration of solar energy and the power of the sea. He foresaw interplanetary communications and satellites.

However, a prominent banker who financed Teslas' research could easily see that the scientist's main motivation was *not* to make money. Rather it was to emancipate humanity with new forms of technology that would liberate people from darkness, drudgery, and various forms of top-down oppression (control).

Neither the bankers nor the class they represented, shared Tesla's goal of conducting research and development to make life easier and better in ways that operate within, rather than against, prevailing patterns of Nature, including what is best in human nature.

Tesla suffered a nervous breakdown after 1904 when his principal financer betrayed him. Tesla's famous tower and laboratory at Wardenclyffe, were subsequently destroyed and obscure interests placed many obstacles in the way of the inventor's ability to gain backing from other financiers. He continued his work in obscurity, making scientific breakthroughs with unprecedented applications.

Indeed, the FBI intervened to seize his papers and prototypes when Tesla died penniless in New York in 1943. Tesla's name was practically ignored (and still is) from science textbooks.

The *only* official reference of him is the unit of magnetic induction or magnetic flux density (Tesla), in the metre–kilogram–second system (SI) of physical units which is named after him and is used in all work involving strong magnetic fields, while the Gauss is more useful with small magnets.

There were other maverick scientists and inventors who suffered similar fate and for the same reason. Broadly speaking, the corporative thought is that science is meant for profit and control, *not* to really benefit mankind.

On the other hand, governments started to get directly involved in scientific research, to develop *non-conventional weapons*, since the world wars. During the First World War, chemicals weapons such as phosgene and mustard gas and the invention of the tank, were made and publicly known. But is during Second World War onwards, that the alliance between scientific research and government is completed and fully developed, very often within *secrecy*.

For example, The *Manhattan Project* that officially, developed the atomic bomb. It resulted in the creation of multiple production and research sites that operated in secret. Project research took place at over thirty sites across the United States, Canada, and the United Kingdom.

The three-primary research and production sites of the project were the plutonium-production facility at what is now the Hanford Site, the uranium-enrichment facilities at Oak Ridge, Tennessee, and the weapons research and design laboratory now known as Los Alamos National Laboratory.

Entire towns were built for short periods of time, employing people, all under secrecy and top national secrecy at that. The government *never* admitted to it, the media never reported on it, and people had no idea for years.

Also, Germany during WWII had its own military scientific research projects dealing with atomic weapons, rocketry and other cutting-edge, projects that included the world's *first* operational jet fighter the Me-262.

Those technologically advanced systems proved to be devastating and went under the name of *Wunderwaffe* (miracle weapons) but came too late to change the course of the war. At the end of hostilities, most of this advanced scientific and technological knowledge was "shared" by the Allies as booty of war and paved the way for the US vs USSR Space Race, among other things.

Currently there are dozens of governmental scientific organizations, that are widely known. For instance, NASA an organization for space research. However, since WWII, a *split* in the scientific endeavour did occur. The results of scientific research carried out by some governmental bodies are *not* widely publicized. Material considered *classified* and *top secret* started to appear and still it continues to do so, being kept away from the public. Several files of this material have been declassified and made available but still, there is a lot to be done.

From this perspective, there are *two versions* of this sort of scientific research and its findings: the one kept for the governments and the other filtered or deemed suitable for the public. This is key to understand because the scientific knowledge, coming from this source may not always be *reliable*. Moreover, it can be manipulated by those government organizations and its allies, to gives us an *official* reality of what we are supposed to believe and most of the gullible public, including many scientists, would buy it.

Based in anomalies, inconsistencies and other facts, those who dare to question this official reality (and have the right to do so) or propose an alternate one, are branded either a "conspiracy theorist" or a pseudoscientist. Besides, there always will be the so-called sceptics, who will "debunk" any alternate version of the official narrative to "convince" the public, otherwise. The reader should bear this in mind through this book.

For example, since its inception, anthropologists and sociologists had been warning NASA of the potentially social instability consequences, if intelligent life elsewhere in the Universe or their ruins would be ever found. Then, in 1976, NASA sent a pair of space probes, known as Viking 1 and Viking 2, to Mars. Viking 1 was launched on August 20th, 1975, and Viking 2 was launched in September of the same year. Both probes photographed the surface of Mars from orbit, and one studied the planet from the surface.

The first one touched down on the surface of Mars on July 20th, 1976 and the second a couple of months later. The main objective of the mission was to obtain high-resolution images of Mars, to look for any evidence of life, and to learn more about the structure and composition of the atmosphere.

As Viking 1 spacecraft was circling the planet, it spotted the shadowy likeness of a human face. An enormous head nearly two miles from end to end seemed to be staring back at the cameras from a region of the Red Planet called Cydonia. A pyramid structure was also seen. NASA has attempted to downplay this image from the start. Viking leadings project scientists tried to explain it away by saying that a picture transmitted a few hours later, of the same region, revealed that the image in the first photo "…was just a trick, just the way the light fell on it."

Controversy followed and some scientists analysing thoroughly the data received by Viking 1, reached very different conclusions. What's more interesting are the results of the experiments carried out by both space probes, concerning life on Mars.

One experiment, the Labelled Release (LR) experiment, showed that the Martian soil tested *positive* for metabolism a sign that, on Earth, would almost certainly suggest the presence of life. Both the Viking 1 and Viking 2 landers collected samples of Martian soil, injected them with a drop of dilute nutrient solution, and then monitored the air above the soil for signs of metabolic by-products.

Since the nutrients were tagged with radioactive carbon-14, if microorganisms in the soil metabolized the nutrients, they would be expected to produce radioactive by-products, such as radioactive carbon dioxide or methane. They did. US scientist Dr Gilbert V. Levin, the mastermind behind the experiment was given congratulations for the first discovery of life beyond Earth.

However, days later he was told that it had all been a mistake. Indeed, NASA scientists *changed* their minds saying that a related experiment allegedly found no trace of organic material, suggesting the absence of life. Bear in mind this experiment was designed and tested precisely, to *avoid* any false conclusions.

Before launching the Viking spacecraft, the researchers tested the experimental protocol on a wide variety of terrestrial soils from harsh environments, from Death Valley to Antarctica. In each case, the experiments tested positive for life. Then as a *control*, the researchers heated the samples to 160 °C to kill all lifeforms, and then retested. In each case, the experiments now tested negative.

To further confirm that the experimental procedure would not produce false positives, the researchers tested it on soils known to be sterile, such as those from the Moon and the Surtsey volcanic island near Iceland, which produced negative results as expected. This control experiment was carried out on Mars with the same result.

To rule out the possibility that the strong ultraviolet radiation on Mars might be causing the positive results, the landers collected soil buried underneath a rock, which again tested positive. The control tests also worked, with the 160 °C sterilization control yielding negative results.

In addition, it seemed that whatever was doing the metabolizing was relatively fragile, since metabolic activity was significantly reduced when heating the sample to 50 °C, and completely absent when storing the soil in the dark for two months at 10 °C. Gilbert Levin and others are sure that these results provide some of the strongest evidence that the soil contained Martian life.

It is noteworthy to mention that, for years we've been made to believe that Mars is a dry and arid planet, completely *devoid* of life, but that's not the case. Mars actually used to be an Earth-like planet, with giant oceans and extensive greenery.

The soil is moist and wet, and there is a very high likelihood that some type of life exists within the interior of Mars today. And here's the *contradiction* with NASA, who tried really hard to get the world to believe ancient bacteria on a meteorite came from Mars, would put so much energy into the notion that intelligent or basic life *couldn't* or didn't exist on Mars.

The alleged reason against life in Mars is the harshness of the environment but there are *extremophile* bacteria on Earth, microbes that have been discovered thriving in some of the most inhospitable places on our planet. It just doesn't make any sense at all.

Besides, critics had come forward to public attention, exposing that several photographic materials from missions send to the Moon and planets by NASA, have been *deliberately* doctored or pixelated to hide some key information. This leaves an open question about the reliability of the information given to the public, by NASA.

The same applies to Google Maps. It is widely known that there are places on Earth that are withheld or censored from public view. Images are deliberately either blurred or pixelated in some places. The explanation of hiding military or energy installations, governmental buildings, faulty imagery and the like are not enough to cover *all* these anomalies.

On the other hand, there are within the governments, special programs of scientific research linked mainly with the military. These programs are *extremely* classified dealing with technology, information and more. No expenditure is questioned. The main purpose is to develop *exotic* weapons systems.

Obviously, their findings never will appear in scientific journals nor conferences. Only *after* decades, some information is released to the public, if so. Basically, as far as technology is concerned, for every calendar year that elapses, *military technology* increases about 10-20 years (compared with the increase rate of 'conventional' technology).

One must understand this fact, because there are projects dealing with subjects beyond the belief or experience, beyond the imagination of many public domain scientists and engineers. Scientific military research is far more *advanced* than ordinary research, carried out in universities and industries, pretty much like in the days of the Manhattan Project.

They are the smartest people on the planet, but even so remain deeply puzzled by much of what they have learned. The advantage is that they have no *limiting beliefs* about reality as many conventional scientists do. Indeed, these special programs do test the boundaries of reality to the extreme, squeezing the *scientific method*. There is no room for theorists and their untestable hypothesis (pseudoscience).

Those leading the special programs of scientific research or clique, tend to regard the public with disdain, like undisciplined and unruly children incapable of handling information of extraordinary complexity.

This information deals with the fact that reality is more complex than our *conventional* scientific notions of space, time and matter, because conventional science has locked up itself into such a materialist-reductionist corner that it could not, at present, cope with this sort of knowledge.

For instance, they know that *high-energy* processes indeed do create a rip in the so-called fabric of space as Tesla proved through his experiments with strong *electromagnetic fields*; not strong gravitational fields as it is commonly believed. Therefore, many conventional scientists with several PhD's think they are smart but they aren't. Also, this clique knows indeed what's *really* out there in space, planets and moons.

We are only given leftovers, so to speak. This includes what's really going on in the inaccessible areas of Earth such as the *polar regions,* which will be dealt in the chapter four of this book.

This clique believes, has earned the right to know and to make decisions in the best interest of civilization, to which the ordinary person, being lazy and easily distracted, is not motivated or qualified to contribute anyway.

Some declassified files show that microwaves are key in the evolution of a new series of non-lethal weapons, that the military has been developing. This is not to be taken lightly, this kind of waves do modify/alter physical and biological systems *from distance,* as dozens of experiments carried out, successfully proved. By now, the classified world has moved far beyond the reach of the public world, and far beyond in its power and capabilities.

For example, an underground base is a subterranean facility used for military or scientific purposes. Some of them are known to the public other are classified (off-limits zone). To build these subterranean facilities, covert techniques are used to avoid detection. The techniques used for covert tunnel construction are very *different* than ordinary civilian tunnel construction. Covert tunnel construction makes use of a boring machine that actually *melts* earth and rock (by chemical valence disruption), thereby forming a glass like tunnel wall. This has several advantages over civilian tunnel boring methods.

This is known as *nuclear tunnel boring*. The headline of the patent says: "a machine and method for drilling bore holes and tunnels by melting in which a housing is provided for supporting a heat source and a heated end portion and in which the necessary melting heat, is delivered to the walls of the end portion at a rate sufficient to melt rock and during operation of which the molten material may be disposed, adjacent the boring zone in cracks in the rock and as a vitreous wall lining of the tunnel so formed. The heat source can be electrical or nuclear but for deep drilling is preferably a nuclear reactor."

This sounds very interesting; the melted rock is forced into cracks wherein heat is given up to the crack surfaces and freezes as a glass at some distance from the penetrator. This amazing boring device is capable of drilling at depths totally inaccessible with previous *conventional* drilling techniques, even, according to the patent claims, down to 30,000 meters.

Normal rate of speed is approximately between one and six miles per hour, depending on type of rock, sand etc. Truly amazing. The nuclear tunnel boring machines were invented by scientists and engineers at Los Alamos National Laboratory during the 1970's.

They called their new machine, the "Subterrene" and it can be found online. From the perspective of secrecy, the biggest advantage is that little or no waste material (rock, dirt, etc.) is produced by the boring process, thereby alleviating the need for above ground disposal sites, very common in civil construction sites.

Another advantage is that tunnels can be bored through lose rock, sand, etc. or other locations that would be unsuitable for conventional civilian boring methods (for instance, a river bed). Large underground cavities are also constructed using this technique.

Usually the governmental special (classified) programs of scientific research, are focused to develop weapons and they aim at *anything* that could potentially become a weapon.

In fact, they have produced weapons and other experimental technologies that can be used in military operations or *civilian control* activities. Weather is an example. In the last decades, Meteorology has increased its knowledge of weather and *atmospheric patterns* trough meteorological balloons, radars and artificial satellites.

The natural causes that govern weather are linked to disparities in temperature of the wind belts and the jet streams girdling the planet, that are steered by three convection cells: the Hadley cell, the Ferrell cell, and the Polar cell and ultimately depend on thermal energy coming from the Sun.

Once the factors or rules of the game are well known, the key is to figure out how and where is feasible to *alter* or modify a specific *pattern* and thus *change* the game to one's favour.

Fig. 20- Several textbooks such as this one (1974) have been written stating that for the purpose at hand, the atmosphere may be viewed as a complex physical system in which ascertainable changes in air motion take place in response to identifiable forces. In principle, by altering these forces, consequent changes in the air motion can be influenced. Thus, in principle, controlling the weather or modifying the climate is scientifically possible. It is not a "conspiracy theory."

The answer is simple: to direct aimed energy in form of electromagnetic or other type of waves, with certain frequency at a specific cell where one intends to modify the weather. Jet streams loaded with humidity can be diverted from a specific region, creating droughts. This is a way of creating atmosphere altering weapons.

Part of this technology was developed using Tesla's *scalar waves* research, which we will see in chapter four. To enhance the transmission of these special electromagnetic waves, or heat up a specific area in the atmosphere, the releasing of *metal oxides* into the air is essential. When an atmospheric weapon is heating those metal oxides, the skies heat to over 38 degrees Celsius, preventing the gathering of water vapour that would otherwise turn into clouds. No clouds mean no rain.

Weather manipulators have looked at flooding. *Silver iodide* is commonly used for cloud-seeding, which forces tiny ice crystals that make up the cloud to fuse together. Consequently, they become too heavy and fall as rain. Its only matter of spraying this chemical substance, wherever the mission needs to be accomplished.

Indeed, some scientists have been learning to *influence* droughts, rain and hurricanes, and turn them into *weapons*. This also includes *climate modification, polar ice cap melting, ozone depletion, earthquakes, and ocean wave control*. Several hurricanes have shown "unnatural" behavioural patterns such as: abrupt 90 degree turn, staying stationary or following straight lines.

Geoengineering is part of this agenda. Flood and droughts can be indeed *engineered*. The same principle is used in producing earthquakes. Extremely low frequency (ELF) waves can trigger catastrophic earthquakes by beaming millions of watts of waves into the ionosphere and reflecting them back down to Earth and the ocean.

If the reflected energy hits an *unstable* fault line, the consequences could be catastrophic. However, conventional or academic scientists *ignore* these facts and blame those "natural disasters", either to climate change or mysterious circumstances that deem further studies as in the following case:

From 1987 through 1992, California experienced its worst-ever drought. Food prices went up as crops and livestock died. Scientists believed the problem was a mysteriously *stalled* high-pressure system 800 miles off the coast of California, which was *blocking* the usual flow of moist Pacific air. Normally, the winds blow in an easterly direction. However, the National Oceanic and Atmospheric Administration (NOAA) showed the winds were blowing in a westerly (opposite) direction.

That meant the jet stream was much weaker. The jet stream only returned to normal in 1995. Scientists and the NOAA *could not* explain this change. But some individuals already knew what was really happening. If a public representative comes up with terms such as “weather manipulation” or “atmospheric weapons”, a deflecting campaign of misinformation follows.

In view of the above, the likelihood exists that more than seven decades later after the Manhattan Project, there is a covert group that possesses: A technology that is vastly *superior* to that of the conventional world. The ability to explore areas of our world and surroundings presently unavailable to the rest of us. Scientific and cosmological understandings that give them greater insights into the nature of our world, including what’s *inside Earth*.

Chapter II

Questioning the validity of the current model of Earth's interior. Here is a need to analyze, how one premature reasoning influences others, how hypotheses become unproven "theories" simply due the fact, that no one objected them for certain time period.

I shall not mingle conjectures with certainties.
Sir Isaac Newton (1642-1727)

Let's face it. The fact is that it is extremely *difficult* to determine the material of the Earth and the physical state along the chemical composition of this material at depths where is no *direct* observation. Rock samples originate from depths not exceeding 12 km since this is the greatest depth attainable by the present drilling techniques.

Though some rock materials find their way to the surface by geological processes, primarily by volcanic eruptions from depths exceeding those of drillings, even these samples provide only *limited* information concerning at most, only some portion of the upper mantle of the Earth.

At present, *no* means are available for the direct observation of deeper regions of Earth. The only indirect way to obtain a model of Earth's interior is through *Seismology* and the most it can achieve is only an *interpretation* of the raw data, regardless of the advances in technology. Most information on properties and processes below the uppermost portion of Earth crust results from the application of theory to phenomena observed at the Earth's surface (extrapolation). Therefore, estimates of the temperature as well as of the composition of the Earth's interior and structure become more and more *uncertain* as the depth *increases*.

Today, ascertaining what happens below Earth's crust relies mostly on computer simulations and modelling, or measurements taken from the surface and then make educated guesses, of what ought to be below it. Furthermore, in investigations of properties and processes at greats depths in Earth, extrapolations of theoretical equations, are frequently required extremely far *beyond* the range of conditions for which they have been developed.

Similarly, numerical values of constants are sometimes used far beyond the range for which laboratory data on the one hand and theoretical information on the other hand exist. Extrapolations are uncertain and usually contain theoretical and as well numerical *assumptions* which may lead to grave errors.

Unfortunately, geoscientists as any specialist, are not often familiar enough with other scientific theories involved in their studies because of their tunnel vision and realize the underlying assumptions. Frequently, the mathematical development of the equations is not the main source of errors, but the insufficiently know or incorrect solution of the *physical* side of the problem.

Several theories of the Earth have been proposed at different times: its central substance has been supposed to be fiery, fluid, solid and gaseous in turn; until geologists have turned in despair from the subject and become inclined to confine their attention to the outermost crust of the Earth, leaving its centre as a playground for mathematicians.

Seismological studies of Earth's interior

Seismology as a science started in the early 20th century. By then and based on the mean Earth's density, average chemical composition of its crust and *supposing* a solid sphere; which originated from the outdated chaotic-accretion hypothesis that formed the planets. The heat energy of this process, melted the entire Earth, which lead to its differentiation into a layered planet as materials migrated toward the centre and lower-density materials "floated" to the surface.

Scientists then, *postulated* the existence of heavier and denser molten inner core made of nickel and iron, that could explain this mean density (5.5 g/cm^3). Here is when comes the Seismology handy in trying to elucidate the Earth's interior and thus, obtain a *model.*

Bear in mind that from the beginning, seismological data was *interpreted* (not observed) to fit the hypothetical molten and differentiated origin of the Earth. In this case, geoscientists committed the deadly sin of *reification:* the mistaking of models for reality.

It leads to strange and untestable creations such as the "geodynamo" hypothesis, a futile attempt to produce Earth's magnetic field consistent with observed data given certain conditions and equations.

The standard definition of Seismology is the study of *mechanical vibrations* or waves traveling within Earth. These vibrations are caused by various events, including earthquakes, extra-terrestrial impacts, explosions, storm waves hitting the shore, and tidal effects.

Of course, seismic techniques have been most widely applied to the detection and study of earthquakes, but there are many other applications and arguably, even with its limitations, seismic waves provide the most important information that we have concerning Earth's interior.

Before going any deeper into Earth, we need to look at the properties of seismic waves. The types of waves that are useful for understanding Earth's interior are called *body waves*, meaning that, unlike the surface waves on the ocean, they are transmitted through Earth's materials. Thus, we have a specific type of *energy* that interacts within the Earth.

The conventional approach is that depending on how these body waves behave inside the Earth, seismologists may be able to figure out how the Earth's interior is like. However, as we will see later, this is a half- truth. Let's know more about these seismic waves. All of Seismology is based on the mechanical behaviour of seismic waves.

Earthquakes and seismic waves

In general, seismic waves are produced when some form of energy stored in Earth's crust is suddenly released. Earthquakes are usually caused when rock underground suddenly *breaks* along a fault. This sudden release of energy causes the seismic waves that make the ground shake and propagate away from the rupture front at a much faster speed than the rupture propagates.

They are the energy that travels through the Earth and is recorded on *seismographs*. When two blocks of rock or two plates are rubbing against each other, they stick a little. They don't just slide smoothly; the rocks catch on each other.

The rocks are still pushing against each other, but not moving. After a while, the rocks break because of all the pressure that's built up. When the rocks break, the earthquake occurs.

During the earthquake and afterward, the plates or blocks of rock start moving and they continue to move until they get stuck again. At some point, the rupture will cease to grow and will slow down and stop.

The spot underground where the rock breaks is called the *focus* of the earthquake. The place right above the focus (on top of the ground) is called the epicentre of the earthquake. The size or magnitude of the earthquake depends upon how much the fault has ruptured (the slip) and the area over which the rupture has occurred. It is estimated that there are 500,000 detectable earthquakes in the world each year. 100,000 of those can be felt, and 100 of them cause damage.

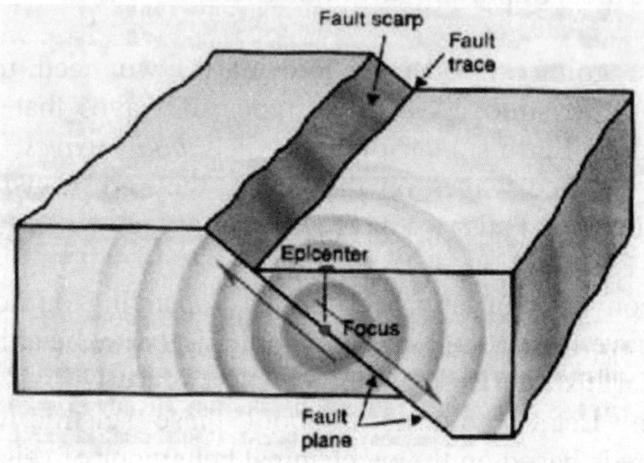

Fig. 21- Diagram showing the general earthquake mechanism due to a rock masses moving along a fault. This mechanism is applicable only to *tectonic* or crustal earthquakes.

Types of Earthquakes

Shallow focus earthquakes are commonly occurring "crustal" earthquakes, caused by faults and movements of the continental plates. These are earthquakes with their focus nearer the surface of the Earth, occur at depths less than 70 km. Shallow focus earthquakes are usually of large spread, causing greater damage at the surface or the Earth's crust.

These occur quite frequently and at random. However, being of smaller magnitudes and at lesser depths, very often they are not even felt. Nevertheless, about 75 % of the world's energy released from earthquakes is from shallow-focus ones.

Deep focus earthquakes or intra plate earthquakes, are believed to occur within the "subducting oceanic plates" as they move beneath the continental plates. Appearing along fault lines, these are earthquakes with focus much deeper within the Earth, at greater focal depths of 300 – 700 km.

These are typical of the "subduction zones" of the Earth which are seismically active zones. Deep focus earthquakes happen as huge quakes with larger magnitudes, as a great deal of energy is released apparently, with the forceful collision of the plates. For a long time, seismologists believed that all earthquakes had a depth less than 60 kilometres from the surface.

This belief was based on our understanding of how materials fail at high pressures. The mechanics of deep focus earthquakes, have *puzzled* geoscientists, as brittle fracture and frictional sliding at depths exceeding 100-200 km and even more would require great rock strengths which does not seem possible because of the high temperatures in such zones, which makes the material *ductile*.

Approximately one in four measurable earthquakes can be classified as a deep earthquake. This might seem like a small number, but if one considers the fact that geoscientists have *no* convincing explanation for this phenomenon, it is certainly a significant number.

Seismic waves P and S

When an earthquake occurs, different spherical *wave fronts* spread out as they travel inside Earth and like all waves, seismic waves they transfer energies from one place to another without moving material. This process is not completely reversible.

There is energy loss due to shear heating at grain boundaries, mineral dislocations etc. Seismic waves are always *attenuated* as they travel through rocks and these may propagate either along or near the Earth's surface (Rayleigh and Love waves) or through the Earth's interior (P and S waves).

Imagine hitting a large block of strong rock (e.g., granite) with a heavy sledgehammer. At the point where the hammer strikes it, a small part of the rock will be compressed by a fraction of a millimetre. That compression will transfer to the neighbouring part of the rock, and so on through to the far side of the rock, from where it will bounce back to the top all in a fraction of a second.

This is known as a compression wave, and it can be illustrated by holding a loose spring hat is attached to something at the other end. If you give it a sharp push so the coils are compressed, the compression propagates (travels) along the length of the spring and back. You can think of a compression wave as a "push" wave; it's called a P-wave (although the "P" stands for "primary" because P-waves arrive first at seismic stations). When we hit a rock with a hammer, we also create a different type of body wave, one that is characterized by back-and-forth vibrations (as opposed to compressions).

Fig. 22- Hitting a large block of rock with a heavy hammer, will create a seismic wave within the rock. It is a mechanical wave of acoustic energy that travels through the Earth or another planetary body. It can result from an earthquake (or generally, a quake), volcanic eruption, magma movement, a large landslide and a large man-made explosion that produces low-frequency acoustic energy.

This is known as a shear wave (S-wave, where the "S" stands for "secondary"), and an analogy would be what happens when you flick a length of rope with an up-and-down motion. A wave will form in the rope, which will travel to the end of the rope and back. A shear wave can be seen as a "shake" wave.

As mentioned before, there are surface waves of a lower frequency (L and R) than P and S waves (body waves), and are easily distinguished on a seismogram, as a result. Though they arrive after body waves, it is surface waves that are almost entirely responsible for the damage and destruction associated with earthquakes.

This damage and the strength of the surface waves, are reduced in deeper earthquakes. These L and R waves bear no importance in the Earth's interior studies. Our concern here are the body waves (P and S) traveling through the interior of the Earth and arrive before the surface waves emitted by an earthquake. These waves are of a higher frequency than surface waves.

P Waves

The first kind of body wave is the P wave or primary wave. This is the fastest kind of seismic wave, and, consequently, the first to 'arrive' at a seismic station. The P wave can move through solid rock and fluids, like water or other liquids layers of the Earth and also trough gases.

It pushes and pulls the rock it moves through just like sound waves push and pull the air. For example, when one hears a big clap of thunder and heard the windows rattle at the same time. The windows rattle because the sound waves were pushing and pulling on the window glass, much like P waves push and pull on rock.

Usually, animals can hear the P waves of an earthquake. Dogs, for instance, commonly begin barking hysterically just before an earthquake 'hits' (or more specifically, before the surface waves arrive).

Normally, people can only feel the bump and rattle of these waves. P waves are also known as compressional waves, because of the pushing and pulling they do. Subjected to a P wave, particles move in the same direction that the wave is moving in, which is the direction that the energy is traveling in.

S Waves

The second type of body wave is the S wave or secondary wave, which is the second wave one feels in an earthquake. An S wave is slower than a P wave and can only move through solid rock, not through any liquid (fluid) or gas medium.

As we will see later, it is this property of S waves that led seismologists to infer that the Earth's outer core is a liquid. S waves move rock particles up and down, or side-to-side perpendicular to the direction that the wave is traveling in (the direction of wave propagation).

Compression waves (P) and shear waves (S) travel very quickly through geological materials. For instance, typical P-wave velocities are between 0.5 km/s and 2.5 km/s in unconsolidated sediments, and between 3.0 km/s and 6.5 km/s in solid crustal rocks. Of the common rocks of the crust, velocities are greatest in basalt and granite.

S-waves are *slower* than P-waves, with velocities between 0.1 km/s and 0.8 km/s in soft sediments, and between 1.5 km/s and 3.8 km/s in solid rocks. This data applies to materials ranging from the surface of Earth to a few kilometres beneath.

The *fundamental* observations used in seismology (the study of earthquakes) are *seismograms* which are a record of the ground motion at a specific location, caused by seismic waves.

An earthquake's magnitude may be considered to vary as a function of the amount of energy released at the rupture point. When an earthquake occurs, *two main types* of vibratory waves move through the body of the Earth from the point of fracture.

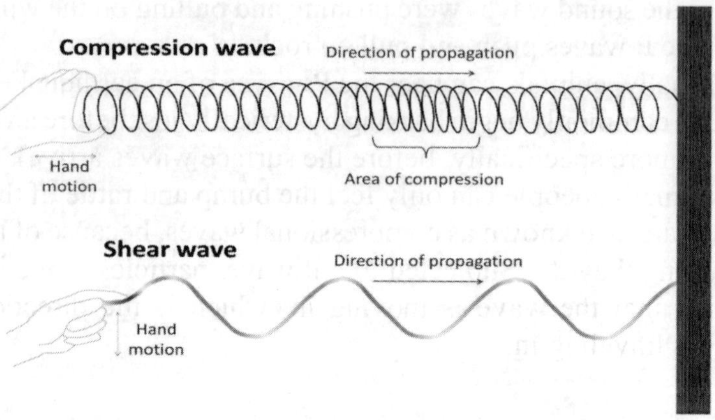

Fig. 23- A compression wave can be illustrated by a spring (like a Slinky), that is given a sharp push at one end. A shear wave can be illustrated by a rope that is given a quick flick.

The primary, or P, waves travel most quickly and are the *first* to be registered by the seismograph. Secondary, or S, waves travel more slowly. As S waves have a greater amplitude than P waves the two groups are easily distinguishable on the seismogram.

By measuring the *time* interval between the arrivals of the P and S wave groups, seismologists can calculate the *distance* between the seismograph and the origin of the earthquake. *Magnitude* is then derived from the amplitude of the waves on the seismogram and the distance of the earthquake from the seismograph.

When P and S waves strike the surface of the Earth they initiate a third kind of wave, the surface waves L and R, which travel over the Earth's surface. These are the slowest waves. On recordings of local earthquakes, the surface waves are small and can seldom be distinguished from the S waves that preceded them.

However, since surface waves attenuate much more slowly than do P or S waves, they are generally the largest waves to appear on long period seismograms of distant earthquakes.

Here's an example of a seismogram:

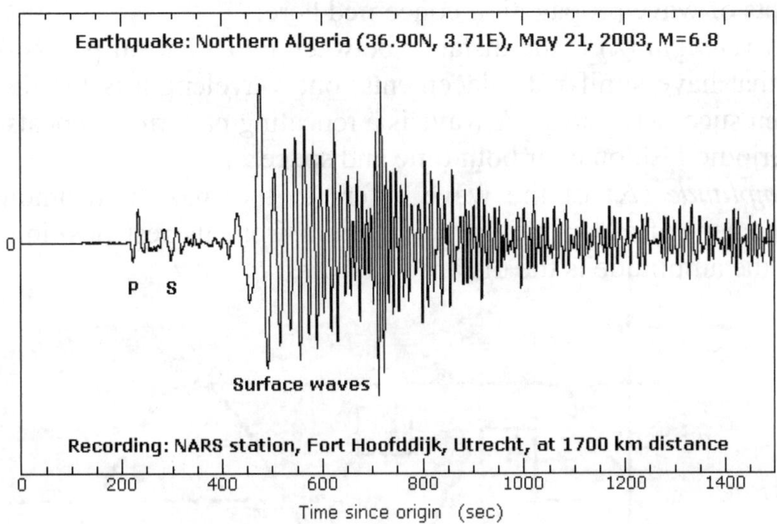

Fig. 24- A seismograph showing the record of seismic waves through time. First come the P waves then the S waves followed by L or surface waves.

A seismogram is the wiggly trace that *records* the vibrations caused by an earthquake at a particular recording station. Seismograms are used to determine the location and magnitude of earthquakes. Modern seismographs record ground motions using electronic motion detectors. The data are then kept digitally on a computer.

When you look at a seismogram, there will be wiggly lines all across it. These are *all* the seismic waves that the seismograph has recorded. Most of these waves were so small that nobody felt them. These tiny *microseisms* can be caused by nearby activities, such as heavy traffic or wind, or by distant sources such as interactions of waves with the ocean floor. They may also be caused by earthquakes that are too small or too far away to be *recognized* as earthquakes.

Bear in mind that the ONLY *direct* and *reliable* information we can obtain from any seismogram is: type of wave and travel times, nothing else. Regarding the magnitude, distance, and speed these can be calculated. Therefore, P and S *travel-time tables* must be constructed. To simplify calculations, P and S velocities are considered (assumed) *constant*.

Seismologists take the Earth's model, *assuming* the likelihood that at most points of its interior, velocity (v) increases with depth. Knowing the basics of seismology, it is also helpful to have a look some basic concepts of wave propagation concerned here:

Wavelength (λ) is the distance between two adjacent points on the wave that have similar displacements, one wavelength is the distance between successive crest. A wave is a repeating pattern. It repeats itself in a periodic fashion over both time and space.

Amplitude (A) of the wave refers to the maximum amount of displacement of a particle on the medium from its rest position. In a sense, the amplitude is the distance *from rest to crest*.

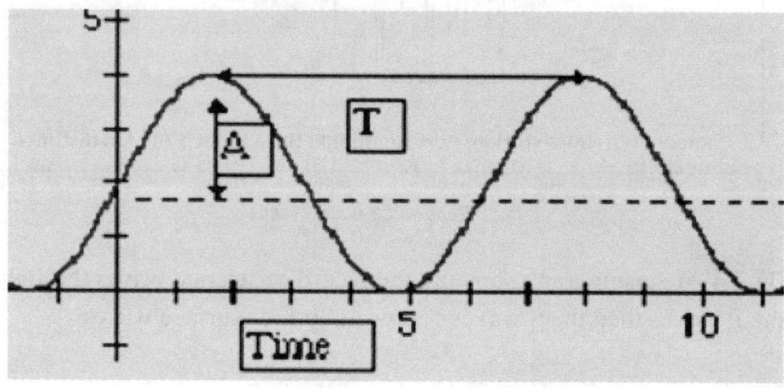

Fig. 25- The main parameters of a wave used in seismology: amplitude (A) and period (T).

Period (T) is the time it takes for two successive waves to pass a reference point or the motion to complete one cycle. This is a *key* parameter in seismology, as it allows to calculate velocities.

Refraction

Snells's Law describes the relationship between the angle of incidence of a seismic wave passing through a boundary between two *different* media. In this model of increasing velocity with depth, the critically refracted seismic rays speed up with depth. Velocity boundaries *redirect* the seismic waves until they reach a critical angle and return to the surface.

As a wave travels through Earth, the path it takes *depends* on the velocity. A principle called *Snell's law*, is the mathematical expression that allows us to determine the path a wave takes as it is transmitted from one rock layer into another. The change in direction depends on the ratio of the wave velocities of the two different rocks. Refraction has an *important* effect on waves that travel through Earth.

In general, the seismic velocity within Earth (according to the accepted model), increases with depth and *refraction* of waves causes the path followed by them, to curve upward or downward.

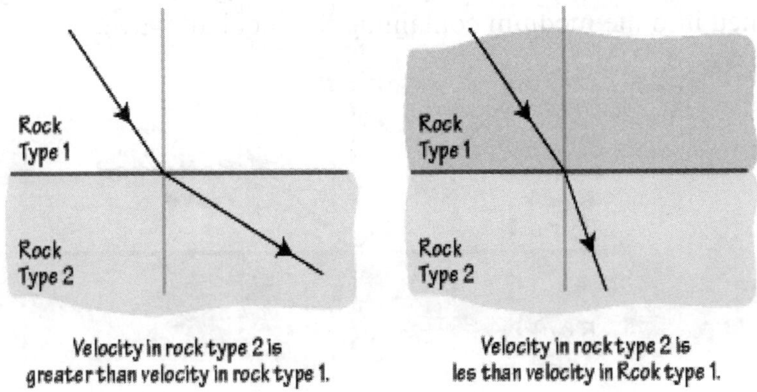

Fig. 26- When seismic waves reach a boundary between different rock types, part of the energy is transmitted across the boundary. The transmitted wave travels in a different way which depends on the ratio of velocities of both rock types. Considering the interior of Earth, the left diagram indicates the path of a wave assuming density increases with depth (rock type 2 > rock type 1). The right diagram indicates the same path assuming density decreases with depth (rock type 2 < rock type 1).

Reflection

The second wave interaction with variations in rock type is *reflection*. Is the same phenomenon as reflected sound waves and we call them echoes.

In seismic prospection, reflections are used to search for petroleum as well as ore deposits and in seismology, to investigate Earth's internal structure. In some instances, reflections from the boundary between the mantle and crust may induce strong shaking that causes damage about 100 km from an earthquake. A seismic reflection occurs when a wave impinges on a change in rock type (which usually is accompanied by a change in seismic wave speed).

Part of the *energy* carried by the incident wave is transmitted through the material (that's the refracted wave described above) and part is reflected back into the medium that contained the incident wave.

The actual interaction between a seismic wave and a contrast in rock properties is more *complicated,* because an incident P wave generates transmitted and reflected P- and S-waves and so five waves are involved.

The amplitude of the reflection depends strongly on the *angle* that the incident wave makes with the boundary and the contrast in material properties across the boundary. For some angles, all the energy can be returned into the medium containing the incident wave.

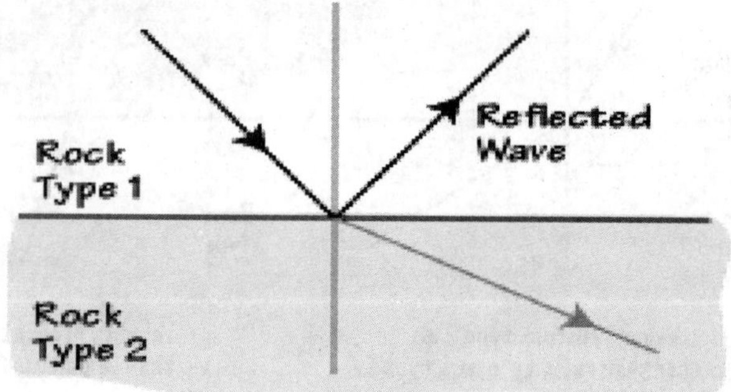

Fig. 27- When a wave encounters a change in material properties (seismic velocities and or density), its energy is split into reflected and refracted waves.

Likewise, when an S-wave interacts with a *boundary* in rock properties, it too generates reflected and refracted P- and S-waves. This leads to a careful and *difficult* analysis of seismograms, to correct these issues.

Dispersion

When a wave propagates through subsurface materials both energy *dissipation* and velocity *dispersion* take place. Energy dissipation is frequency dependent and causes *decreased* resolution of the seismograms when recorded in seismology.

This means that different periods travel at different velocities and the effects of dispersion in seismic waves become more noticeable with increasing distance because the longer travel distance spreads the energy out (it disperses the energy).

Usually, the long periods arrive first since they are sensitive to the speeds deeper in Earth, and the deeper regions are generally faster. Dispersion is caused by material density, the material the waves pass through, will depending on density, pass certain waves and wavelengths of those waves, better than other waves and wavelengths. Additionally, the rock matrix itself has dispersive properties; is a *dispersive medium*.

However, it is curious that so much of classical seismology and wave theory is nondispersive. These studies are based upon *ideal*, not real conditions.

Seismic vs. seismological interpretation

The reader must understand the meaning and differences between *seismic* and *seismological* interpretation; because unlike the former, the accepted seismological interpretation MUST be *validated* by factual *independent* sources, not by circle reasoning.

Whereas both procedures share the same basic principles, is the *scope* of their application which makes the difference and decides the degree of reliability of the conclusions reached by each one.

Seismic interpretation

Is the science (and art) of *inferring* the structural geology at some depth, from the processed seismic record. It is used in the search for commercially economic subsurface deposits of crude oil, natural gas and minerals by the recording, processing and interpretation of artificially induced shock waves in the Earth's crust. Artificial seismic energy is mainly generated on land by vibratory mechanisms mounted on specialized trucks.

Seismic waves reflect and refract off subsurface rock formations and travel back to acoustic receivers called geophones. While modern multichannel data have increased the quantity and quality of interpretable data, proper *interpretation* still requires that the interpreter draw upon his or her geological understanding to *pick* the most likely interpretation from the *many* "valid" interpretations that the data allows.

This is because there are a multitude of *different* possible causes that we can calculate from the observations, and we cannot tell which one *matche*s the reality.

The seismic record contains two basic elements for the interpreter to study. The first is the *time* of arrival of any reflection (or refraction) from a geological surface. The actual depth to this surface is a function of the thickness and velocity of overlying rock layers.

The second is the *shape* of the reflection, which includes how strong the signal is, what frequencies it contains and how the frequencies are distributed over the pulse.

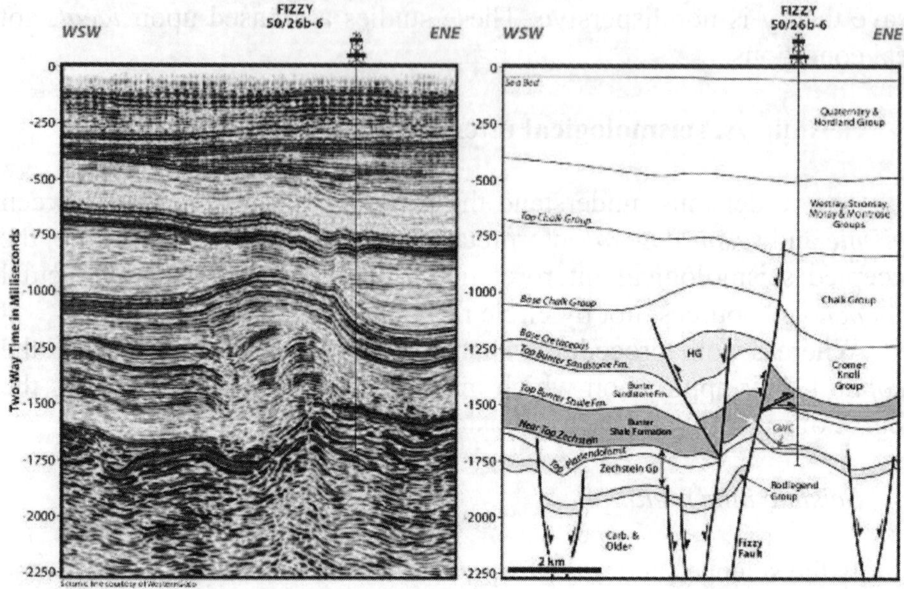

Fig. 28- Uninterpreted (L) and interpreted (R) seismic section. In this context, the interpretation consists in attaining the description of the lithology of the subsurface and structures such as faults or domes. However, seismic interpretation of subsurface faults and other structures, is hampered by factors such as seismic resolution, velocity control for depth conversion and human bias.

This information can often be used to support conclusions about the lithology and fluid content of the seismic geological reflector being evaluated. When the processing and interpretation is complete, a seismic profile is produced. It shows an *interpreted cross section*.

Geoscientists use these profiles, in *addition* to gravimetric, electric and magnetic studies (independent sources) coupled with nearby oil-well information, to *support* their interpretation and produce maps of the rocks buried below.

These maps allow the geoscientists, to accurately position exploration wells in order to find oil and gas deposits. Finally, when *drilling* is completed, and *sampling* and *testing* have been carried out, these will show *how accurate* the interpreted seismic profile or model made by geoscientists, really is.

Seismological interpretation

Now we are in conditions to understand the basis of the of Earth's interior fabricated *model*, according to the seismological interpretation. This involves the following (unproven) assumptions:

- Assuming Earth as a relatively homogenous "solid-plastic" sphere.

- Assuming the existence of a molten nickel-iron core (chaotic-accretion hypothesis).

- Assuming a steady increment of pressure and temperature from the surface through the whole of Earth's radius.

- Assuming the behaviour of waves through unknown conditions would be the *same* as in known conditions.

- Extensive use of *extrapolation* or the action of estimating or concluding something, by believing that existing trends will continue or a current method will remain applicable.

The only source of information are *seismograms* and the *time* it took to record them. The only parameter upon the structure of the whole Earth is studied, relies on the *transmission times* of earthquake waves. Everything else must be calculated and/or modelled. Any *independent* source of information that could *support* the analysis of seismological data is so far, not considered by mainstream scientists.

This is important because velocity of seismic waves travelling inside the Earth (Vp and Vs), cannot be measured it is computed. The Earth's seismological models are based solely on velocity—depth profiles determined from the travel-time—distance curves for seismic waves and on periods of free oscillations.

There is no "direct" method to determine the velocity of seismic waves. Seismologists have measurements of the travel times of earthquake waves at a finite number of places on the Earth's surface and from these they must determine (in theory) both the speed of the waves in the rocks and the *structural changes* within the Earth. For this job, freely available global seismic data continue to increase in volume, with data from thousands of seismographic stations available for any given modern earthquake.

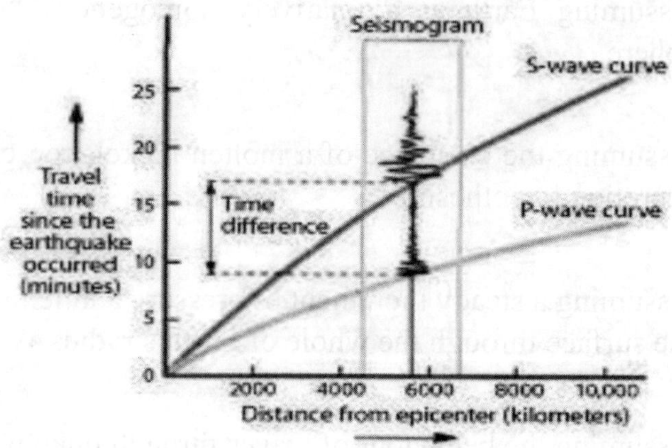

Fig. 29- Travel-time curves for an earthquake thousands of kilometers away. Then amount of time between the P and S wave arrival times is predictable. In this case, the time difference is about 8 seconds.

By making graphs of travel time versus distance between earthquakes and seismograph stations, seismologists found that computed velocity generally increases gradually with depth in Earth, due to increasing pressure and rigidity of the rocks and changes in the curve's slope; showed that there are abrupt velocity changes (discontinuities) at certain depths, indicating layering.

The problem of *inferring* the physical characteristics of the Earth's interior from travel-time curves, belongs to the so-called inverse problems in mathematics. In the present case, this problem was *theoretically* solved in 1907 by Göttingen-based mathematician Gustav Herglotz and then rendered usable in practical application in 1910 by Emil Wiechert.

This is the no less famous "Wiechert-Herglotz method", which was then successfully used by Wiechert and his team to determine the velocity of the seismic waves as a function of depth. The Wiechert-Herglotz (WH) method, which is a classical method of studying the inaccessible deep Earth's structure; *assumes* the medium to be laterally homogeneous, and the travel-time curve of refracted wave to be *continuous* with monotonous derivatives.

However, its application to studies of shallow structures encounters some problems, because the properties of observed travel-time data are usually far from the assumptions of the WH method. It assumes the medium to be vertically inhomogeneous, i.e. the seismic velocity to be dependent on the depth (r) only, v = v(r).

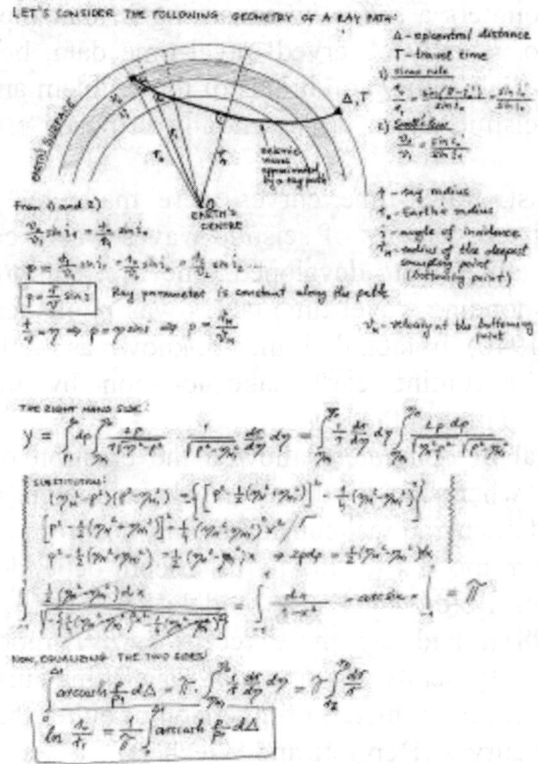

Fig. 30- The velocities of seismic waves P and S are never measured. These must be computed. Here there are some of the mathematical formulations used to derive the integral velocity depth-function, within the Earth, under certain assumptions.

Once a P- or S-wave travel-time curve has been measured, the Herglotz-Wiechert integral (Bullen and Bolt, 1985) can be used to determine the one-dimensional velocity-depth function that matches an observed travel-time curve. This integral relates the velocity-depth function to the slowness (reciprocal apparent velocity) versus distance curve.

All travel-time observations are in some way incomplete or contain scatter due to near-surface effects, lateral velocity variations, and incomplete sampling of the wavefield. For this reason, it is often desirable to know the *permissible bounds* on the velocity-depth structure that a given travel-time curve provides rather than simply deriving one particular velocity-depth model.

However, despite a fairly well-developed theoretical background, practical application of the formulas obtained for the inversion of travel-time data encounters a *difficulty* consisting in that these formulas are inapplicable to actually observed travel-time data, because the latter never meet the solvability conditions of the problem and the kinematic inversion of seismic data is an essentially ill-posed problem (Burmin, 1993).

In the past, travel-time curves were made by trial and error. Calculation of the velocity of seismic waves was a complicated task thus, velocity tables were developed. The first comprehensive model and the corresponding travel time tables was published by Jeffreys & Bullen (1939/1940). In fact. their model, known as the JB model, is still being used for routine earthquake location by the International Seismological Centre in the U.K.

The arrival of computers allowed the creation of seismological Earth models, where the Earth's internal parameters originated from theoretical and mathematical considerations, not from *observation*.

Well known models for the Earth's depth dependent structure are the Preliminary Reference Earth Model (PREM) by Dziewonski & Anderson (1981), and the more recent *iasp91* model (Kennett & Engdahl, 1991). Typically, the fitting of travel-time curves are not done by trial and error but by means of inversion of either the travel times or the travel time curves (Herglotz and Wiechert).

The PREM model was designed to fit a variety of different data sets, including free oscillation center frequency measurements, surface wave dispersion observations, travel time data for a number of body wave phases, and basic astronomical data (Earth's radius, mass, and moment of inertia). In addition to profiling the P and S velocities, PREM specifies density and attenuation as functions of depth. Although these parameters are known *less precisely* than the seismic velocities, including them is important because it makes the model "complete" and suitable for use as a reference to compute seismograms without requiring additional *assumptions* (as if there weren't enough already).

Errors and misinterpretations of seismological data

Once an earthquake has happened, the exact location of seismological stations is known and all the seismograms of the earthquake are gathered; the phase of *interpretation* or elucidation, regarding the Earth's interior, begins. This is *not* an easy task nor it is 100% accurate. The identification of P and S waves depends for example of the magnitude of the earthquake, the magnification and characteristics of the instrument (seismograph), the background noise (microseisms) and on properties of the ground at the station.

Wave dispersion is one of several corrections that must be done. Larger frequencies and therefore, smaller wavelengths are more prone to dispersion inside the Earth. Errors do occur. Interpretation pitfalls fit into one of the following two categories:

1. Pitfalls associated with velocity occur because seismic data are presented in travel time rather than depth.
2. Pitfalls associated with geometry occur because reflections from a three-dimensional space are plotted in a two-dimensional section. However, this issue with the advent of digital technology and computer modelling is practically non-existent.

It happens that *misinterpretation* of waves observed in seismogram or *incorrect* assumptions concerning the path of such waves in the Earth, have resulted in relatively large *errors* concerning the structure of portions inside the Earth, including discontinuities that do not exist. Also, correct data is often misinterpreted.

One important factor, seismologists fail to recognize is that they have no reliable means of determining the *exact* path a given seismic wave or ray, has followed when it reaches a certain point or seismological station. Also, they tend to forget that their interpretation is one out of *several* valid interpretations that the data allows.

For example, recently a team of scientists, from 2120 measurements taken from the ocean floor across the globe, they measured its thickness and density. With this information, scientists observed that in geological terms, the Earth's surface (crust) bobs up and down like a yo-yo. However, they *interpretation* of what causes this movement is attributed to the hypothetical and controversial convection currents in the mantle. But this is one option out of several, that can be chosen based upon available data. An alternative cause can explain the *same* data, even better. Many make the mistake of pointing out that seismological interpretation, is similar to a doctor using an ultrasound device to image a foetus in the womb. It is *not*. Doctors already know what a foetus is and how it looks like, because they have *observed* it beforehand and can correctly, interpret the ultrasound device data.

Whereas scientists never saw what's deep below Earth's crust and thus they never can say *exactly,* what the seismological data is or what represents. We do know heat dilatates bodies, because we *observe* the causal relationship. It is misleading to accept an unobservable *indirect* "cause" without the backup from *independent* sources.

Geological and geodynamic models of the mantle and core, often rely on joint interpretations of published seismic tomography images and petrological/geochemical data. This approach tends to *neglect* the fundamental limitations of, and uncertainties in; seismic tomography results.

These limitations and uncertainties involve theory, correcting for the crust, the *lack of rays* throughout much of the mantle and core, the difficulty in obtaining the *true strength* of anomalies and the choice of what background model, to subtract to reveal anomalies.

This of course, leads to a misinterpreted and unsupported "scientific knowledge" of the Earth's interior, published in peer-reviewed journals that is piling up so fast, one need wings to stay above.

Structure of the interior of the Earth, according to the current paradigm

It begins when an earthquake occurs and seismic waves (P and S waves), spread out in all directions through the Earth's interior. The same applies to nuclear explosions, which by the way are sometimes used to *calibrate* seismographs.

Seismic stations located at increasing distances from the earthquake epicentre, will record seismic waves (seismograms) that have travelled through increasing depths in the Earth. That's all concerned with gathering data or information, regardless of how accurate or sophisticated instruments are. Seismic velocities depend on the *material properties* such as composition, mineral phase and packing structure, temperature and pressure of the media; through which seismic waves pass.

Seismic waves travel more quickly through *denser* materials and therefore generally travel more quickly with depth. Anomalously hot areas slow down seismic waves. Some waves only travel through *solids* or rigid bodies.

If Earth were the same solid composition all the way through its interior, seismic waves would radiate outward from their source (an earthquake) and behave exactly as other waves behave, taking longer to travel further and dying out in velocity and strength with distance, a process called *attenuation*.

Nonetheless, experimentally, things are different. Let's suppose for better understanding, that an earthquake occurs in the North Pole and like any other, it spreads waves P and S through the Earth's interior. Several seismological stations around Earth, record the seismic waves originated.

The experimental evidence shows waves P behaving *differently* from waves S when travelling the Earth's interior. It has to do with the absence of waves P and S at some points on Earth's surface. This absence is also known as *shadow zone*.

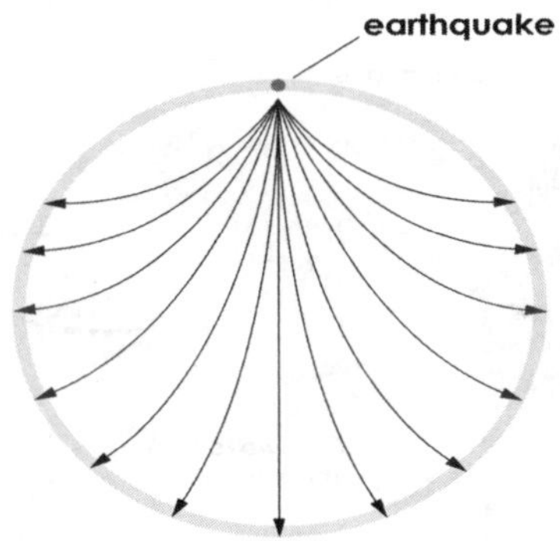

Fig. 31- Supposing Earth as a solid and homogeneous sphere of constant density ρ, this would be the cross section of the propagation of seismic P and S waves through its interior.

P waves

The shadow zone is the area of the Earth from angular distances of 103° to 142° from a given earthquake, that does not receive any direct P waves. In 1906, British geologist R.D. Oldham observed that the travel-times of P-waves recorded at epicentral distances over 100°, were larger than predicted. This meaning that, at somewhere deep in the Earth interior, P-waves were delayed in their passage in a low velocity zone.

From this observation, Oldham inferred the existence of a central "core" in which the P-wave velocity was reduced.

The accepted explanation (interpretation) is shadow zone results from S waves being stopped entirely, by the liquid core and P waves being bent (refracted) by the liquid core. It is assumed a liquid core because of the accretion hypothesis, where Earth is still cooling since its formation, several billion years ago. It is noteworthy to mention that S waves can also be stopped by *gas*.

S waves

Later it was observed that no S-waves arrived beyond an epicentral distance of 103°, and that no P wave emerged in between 103° and 143° (Shadow Zone). In 1913, US-German seismologist Beno Gutenberg, positioned the Mantle-Core Boundary (MCB) at 2900 km. Thus, the basic *interpretation* was a liquid iron core (necessary to account for the Earth's mean density), surrounded by a solid mantle.

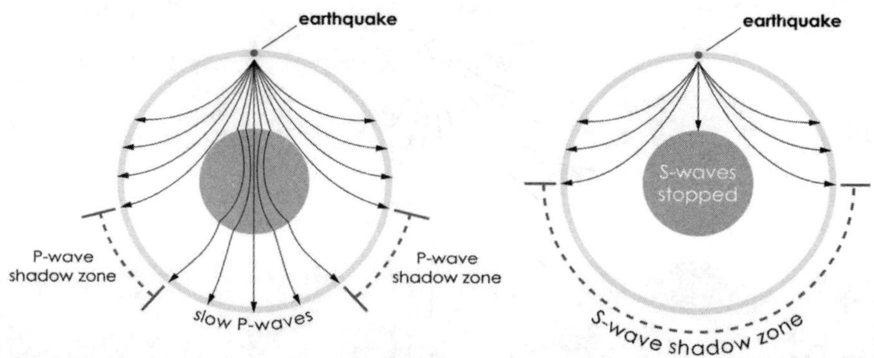

Fig. 32- A structure inside Earth or core, affects the travel paths of the P waves. Due to a double refraction caused by this structure, P waves are "canalized" and aren't recorded by seismographs, between 103° and 142° (shadow zone). Also, this structure blocks the travel paths of the S waves. On both sides of Earth, at 103° these waves aren't recorded by seismographs (shadow zone). Due to the fact S waves, don't travel through fluids; this core is considered a liquid.

The core might be gaseous too, but this interpretation was deemed inviable because a *hypothetical* molten metal was supposed to be the material of Earth's core. The accepted "explanation" is shadow zone results from S waves being stopped entirely by the liquid core, made of molten iron.

Bear in mind that the *only* direct data available is the difference in timing of the arrival of the waves P and S and where they can reach (if possible) on the Earths' surface. Therefore, is valid to make a statement that something, somewhere inside Earth is causing this *variation*. Further research can improve and define more accurately the exact location(s) where the changes do occur (discontinuities).

In 1909, Croatian seismologist Andrija Mohorovičić, discovered by the *refraction* of seismic waves S and P passing from one layer to the other, the discontinuity that bears his name, which is also known as Moho. He deduced from his observations (seismographs and time-travel curves) that there was a sharp change in the structure of the Earth, at the point the earthquake waves were refracted.

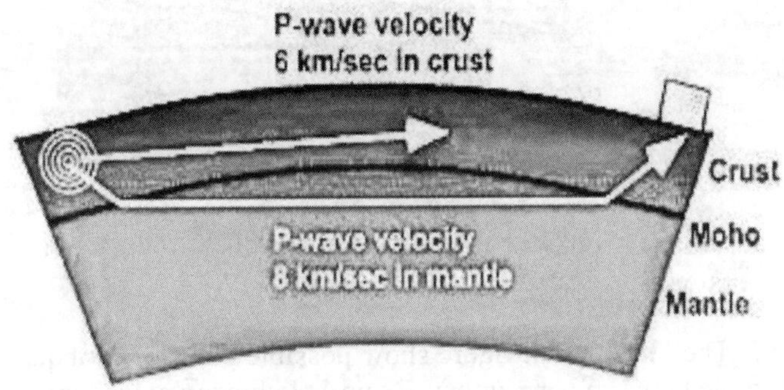

Fig. 33- Mohorovičić discontinuity between the crust and the mantle. Throughout the mantle, seismic waves Pa have a greater velocity than P through the crust.

Because it is a fact that there cannot be two types of longitudinal waves (P and Pa) which would propagate through the Earth at different speeds, Mohorovičić was able to conclude that the P and Pa waves reach the stations via *different* paths.

This is only possible if their speed increases continuously with depth up to a specific depth where there is a sudden (discontinuous) increase in speed. This specific depth at which there is a sharp increase in speed is called the discontinuity surface. Coming to the discontinuity surface the wave rays form smooth curves.

His deductions lead to the idea of the Earth's interior being separated into the crust and the mantle. It marks the level in the Earth at which P-wave velocities change abruptly from 6.7 to 7.2 km/s (in the lower crust) to 7.6 to 8.6 km/s or average 8.1 km/s (at the top of the upper mantle); its depth ranges from about 5 km beneath the ocean floor to about 35 km below the continents. It may reach 60 km or more under some mountain ranges.

In 1929, a large earthquake occurred near New Zealand. Danish seismologist Inge Lehmann, studied the shock waves and was puzzled by what she saw. Waves were discovered in the so-called "shadow zone". A few P-waves with increased speed, which should have been deflected by the core, were in fact recorded at seismic stations.

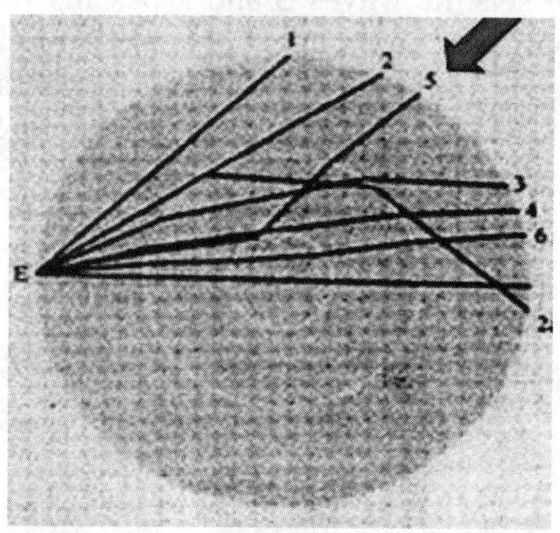

Fig. 34- The lines in that figure show possible paths of earthquake waves. Note the shadow zone between her numbers 2 and 3. Lehmann solved the great mystery by imagining the existence of a solid sphere, now called the inner core, which would reflect earthquake waves into the shadow zone, as shown by the ray highlighted with the red arrow in 5.

Lehmann reasoned that these waves had travelled some distance into the core and then bounced off some sort of boundary, which defines the inner core. Her *interpretation* of this data was the foundation of a 1936 paper in which she hypothesized that Earth's centre consisted of two parts: a solid inner core surrounded by a *liquid* outer core, separated by what has come to be called the Lehmann discontinuity. This simple, elegant approach was accepted and it was quickly adopted by other seismologists over the next few years.

From the above paragraphs, the first thing we can notice is that obviously, the Earth's interior is NOT homogeneous and there is *something* that affects the behaviour of the waves P and S. Thus, is licit to split the Earth's interior in two *distinct* parts at 2900 km depth. The current version or interpretation, proposes a solid mantle made mostly of silicates and a fluid iron core with an inner solid part.

Second, there are only numeric values that gives us the magnitude of the phenomenon but not the exact *nature* of it. Nonetheless we may find a clue regarding in which state of matter, this material is found: whether solid or fluid.

What causes discontinuities?

Discontinuities inside Earth may be caused either by sharp changes in the *chemical composition* of the material inside it or by changes in *phase* of a given material. From these discontinuities, we can *infer* something about the nature of the various layers inside Earth.

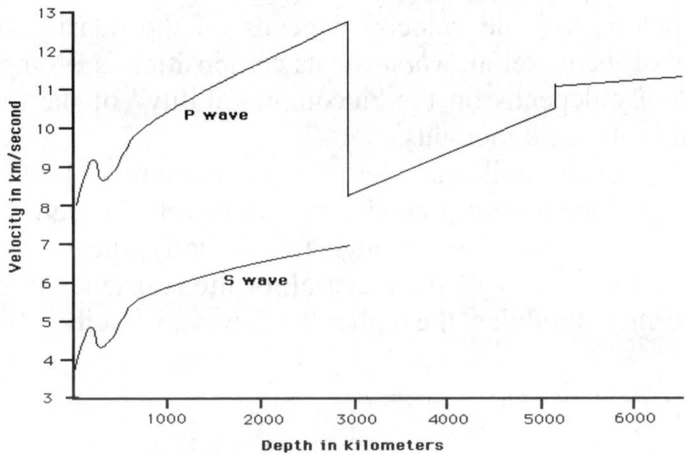

Fig. 35- Velocity (v) of P longitudinal waves and S transverse waves in the interior of Earth. Notice the sharp and dramatic changes at about 3000 km of depth, at the mantle-core boundary. Transverse waves S cannot travel in fluids.

Phase changes may occur between a solid and a liquid phase (change in state) or between two different solid phases. When the changes are progressing slowly or by degrees, this is known as a transition zone. A detailed understanding of the nature of these discontinuities is not quite complete, however this is an area of active geological research.

What controls seismic velocities?

The rate at which a seismic wave travels through a medium, that is distance divided by travel time; can vary vertically, laterally and azimuthally in anisotropic media. Different types of seismic waves travel at different velocities, through any given material. In addition, different materials have different seismic properties, meaning that any *one* wave type can have a wide *range* of velocities, depending on the material properties.

For instance, the P-wave velocity of shale can range from 800-3,700 m/s. Granite can range from 4,800-6,700 m/s, which also falls in the range of limestone. Because of this *variable range*, by themselves, seismic velocities alone are NOT particularly *diagnostic* regarding rock type. Therefore, any statement that says: "by measuring the speed of the waves at different depths, scientists can figure out which types of rocks the waves were passing through" is a *lie*.

Ultimately, seismic velocity depends on the density and elastic properties of the material, *whatever* its composition is. Compressional-wave velocity depends on the "incompressibility" of the material, as embodied in the bulk modulus.

The higher the bulk modulus, the less compressible the material, and the higher the P-wave velocity. Sound travels through water about four times faster than it does through air. Similarly, shear-wave velocity depends on the rigidity of the material, or the resistance to shear. The higher the shear modulus, the higher the S-wave velocity.

Mathematically:

$$Vp = \sqrt{\frac{K + \frac{4\mu}{3}}{\rho}} \qquad Vs = \sqrt{\frac{\mu}{\rho}}$$

Where:
K = Bulk modulus
μ = Elasticity modulus
ρ = Density

We can state that:

• VP is always larger than VS. These are the first to be recorded by a seismograph.

• The stronger material a larger K and μ a larger VP and VS
• VP and VS decrease as density increases
• VP and VS do not depend on frequency (non-dispersive)
• In fluids, μ=0 (and K > 0) and only P-waves travel through fluids (VS = 0). Fluids do not carry shear waves at all. However, there have been several claims of inner core shear S wave observations.

The question that arises is: *how* S waves can travel more than two thousand km, through the liquid outer core and reach the inner core? Geoscientists say that this is possible through the process known as *mode conversion*.

Any time that seismic waves encounter a velocity discontinuity, the energy that transmits through the discontinuity refracts an it is split between transmitted S-waves and transmitted P-waves.

This process of converting P-wave energy to P-wave and S-wave energy occurs at the boundary between the outer core and inner core. This means that P-waves that can travel through the liquid outer core are partially converted to S-waves that travel in the inner core.

By the time the waves encounter so many velocity discontinuities, the converted energy becomes very small, making the detection of S-wave velocities in the inner core, very difficult to observe.

If the wave speed depends *only* on the physical properties of the medium (i.e., the elastic and inertia properties of a mechanical medium) then the wave speed V is a constant, *independent* of frequency. Such a medium is called a *non-dispersive* medium and waves traveling through this medium will maintain a constant shape.

However, there are many examples (including the Earth's interior) of *dispersive media* where, for various reasons, the wave speed depends on the *frequency* of the wave. Most of the seismological work, relies upon the assumption of waves propagating trough *non-dispersive* medium, to simplify calculations.

Thus, the importance of being aware of these common assumptions and simplifications used in synthetic seismogram calculations that are increasingly used nowadays in seismological routine practice. Knowing the theoretical background of *waves propagation*, now comes the application and interpretation of this knowledge to the raw data.

Nonetheless. the 3D information inside the Earth (its structure and properties) is NOT uniquely accessible from the 2D information observed on the Earth's surface.

Therefore, scientists can only construct a *model* of the Earth's interior:

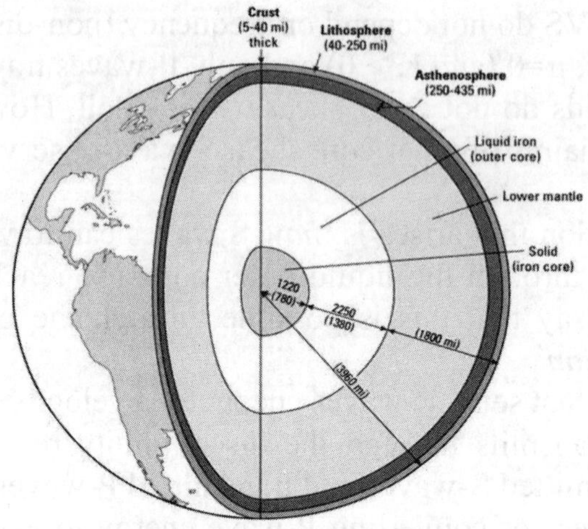

Fig. 36- Model of the structure of the Earth according to the current paradigm. The characteristics of each layer except the crust, are purely *conjectural*. Models are maps of reality: they include only the features one wants to study, and *leave out* everything else.

According to this accepted model:

- Earth has three interior distinctive layers: crust, mantle and core.

- Crust and mantle are rigid/partly-rigid whereas the core is liquid.

- The boundary between crust and mantle is called the Mohorovičić Discontinuity. Ii is marked by a *change* in the velocity of seismic P waves and has a depth of 70 km in the continents and 3 km in the oceans.

- The boundary between mantle and core is the Gutenberg Discontinuity. It is marked by seismic P waves slowed down dramatically and seismic waves S stopped entirely. It lies about 2900 km below Earth's surface.

Posterior studies identified a more accurate seismic waves change, defining the Lehman discontinuity at 5200 km depth, dividing the inner solid core from the outer liquid core.

However, this classical spherically symmetric (or layered) Earth model is undergoing a *revolution*. The Earth has been revealed to be laterally *heterogeneous* everywhere from the crust and mantle to the core, with scales from the size of rocks to the size of continents. The assumed homogeneity and continuity of these layers at global scale is no longer applicable, as we will discuss it shortly.

Earth's inner structure: Crust, mantle and core

The crust

Let's have a briefing of the *only* part or layer of the Earth, that is practically well known and to some extent accessible to *direct* observation. 71% of the crust is covered by oceans. Earth's crust is its outermost layer and represents less than 1% by mass and volume of entire Earth. In fact, this term is *not* quite right and it arose in the past, when scientists thought the Earth had originated from a molten mass and then, cooling gradually, had become covered with a "crust".

The crust is our home and is a *dynamic* place. Volcanic eruptions, earthquakes, erosion, landslides and apparently large-scale *movements* of the continents, rework the surface of our planet constantly. The crust of the Earth is composed of a great variety of igneous, metamorphic and sedimentary *rocks*.

Examples are: granite, gneiss and sandstone. Rocks are made of *minerals* and these are composed by *chemical elements*. Examples are: quartz (SiO_2), pyrite (FeS_2) and calcite ($CaCO_3$). There are roughly 3000 known minerals and these are very important in our civilized world, especially metals.

Prehistory and History have been divided in Stone, Copper, Bronze and Iron Ages. The three most abundant chemical elements (by mass) in the Earth's crust are: Oxygen (O), Silicon (Si) and Aluminium (Al); totalizing 76%.

Crust is classified as oceanic crust or continental crust. *Oceanic crust* is thin (3–4.3 mi [5–7 km]), basaltic (<50% SiO_2) and dense. Compositional chemical studies also claim that oceanic crust is *younger* than the continental crust. However, the recent numerous findings in the

Atlantic, Pacific, and Indian Oceans of rocks far older than 200 million years, many of them continental in nature, provide a strong evidence against the alleged "youth" of the underlying oceanic crust.

Continental crust is thick (18.6–40 mi [30–65km]), granitic (>60% SiO_2), light, and old (250–3,700 million years old). The continental crust mostly composed of less dense rocks than is the oceanic crust. Some of these less dense rocks, such as granite, are common in the continental crust but are apparently, rare to absent in the oceanic crust.

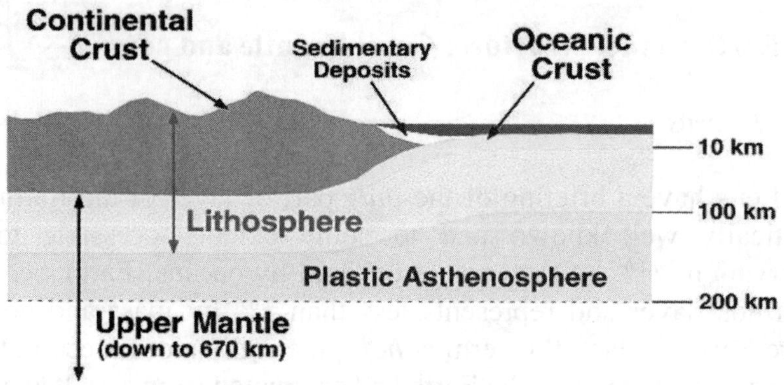

Fig. 37- Subdivision of Earth's crust (lithosphere), in continental and oceanic crusts "floating" on an assumed continuous plastic and hot layer or asthenosphere. However. geophysical data show that, far from the asthenosphere being a continuous layer, there are disconnected lenses (asthenolenses), which are observed only in regions of tectonic activation and high heat flow, nowhere else. Although surface-wave observations suggested that the asthenosphere was universally present beneath the oceans, detailed seismic studies show that here, too, there are only asthenospheric lenses.

According to Plate Tectonics Theory, the outer crust is further subdivided into lithospheric or tectonic plates, that contain varying sections of oceanic and continental crust. These do move a few inches per year and its dynamic causes earthquakes, volcanic activity, submarine subduction zones (trenches) and mountain ranges. In chapter five we will discuss several *flaws* of this theory, which deserves either a revision of its foundations or to be discarded.

At the deepest crustal border, there exists a compositional change from crust material to mantle peridotite called the Mohorovičić discontinuity. The lithospheric plates carrying both oceanic and continental crust move on top of mantle's *asthenosphere.*

The Mantle

Upon once in a time, an attempt to retrieve a *sample* of material from the Earth's *mantle* by drilling a *hole* through the Earth's crust to the Mohorovičić discontinuity, or Moho was made. Project Mohole (1958-1966) represented, as one historian has described it, the Earth sciences' answer to the space program. If successful, this highly ambitious exploration of the intra-terrestrial frontier would provide invaluable *information* on the Earth's age, makeup, and internal processes.

The plan was to *drill* to the Moho through the seafloor, at those points where the Earth's crust is thinnest. Ocean drilling offered an advantage in that undersea samples, undistorted by atmospheric and surface actions, would provide better evidence of long-term geological activity than would samples drawn from land. Nevertheless, the US Congress; objecting to increasing costs of this scientific venture, discontinued the project at the end of 1966.

Since then very little practical and factual stuff, has come to the knowledge of the nature of the Earth's mantle. The current *model* says that it extends down 2890 km, making it the thickest layer of Earth. The mantle makes up about 84% of Earth's volume. Everything we know about it is made *indirectly* through extrapolations and assumptions, as no human study managed to go beyond the crust.

Our paltry knowledge about the mantle comes from *interpretations* of seismologic studies and glimpses we get of Earth's interior from mountain-building processes that occasionally, thrust rocks to the surface from depths as great as 95 kilometres (60 miles). Volcanic eruptions sometimes bring up rocks from depths of 200 kilometres (125 miles) originating in the upper mantle.

The "asthenosphere" was *postulated* as a mechanically weak layer beneath the lithosphere characterized by low seismic wave velocities (low velocity zone or LVZ) and high attenuation. It is generally considered to be the source region of mid-ocean ridge basalts (MORB). Because of the dramatic drop in seismic wave velocities and increase in attenuation of seismic energy, it would appear that partial melting must contribute to producing the LVZ.

These drops in seismic wave velocity can be explained by horizontal melt-rich layers embedded in the upper mantle, and it might account for the large viscosity contrast at the base of the lithosphere, that facilitates the alleged horizontal tectonic plate motions.

Despite its proven solidity, it is *assumed* that mantle minerals can undergo deformation and, over geologic time, large portions of the upper mantle known as tectonic plates do convect (i.e. flow as a response to variations in temperature and density). The problem here is the mantle's *rigidity* or stiffness.

Whether or not the Earth's mantle gets stiffer at great depths is an issue that has bitterly divided geophysicists. If it does, the lower mantle may be sealed off from the upper mantle, as the composition of some surface rocks implies; making *impossible* convection and mantle plumes. If not, the mantle may mix from top to bottom, as other indicators suggest. A neglected study settles the issue, showing to most geophysicists' satisfaction that the mantle does *stiffen* drastically with depth. (Kerr, 1996).

Notwithstanding the poor mantle's knowledge and some scientists *rejecting* this alleged mantle convection; the current accepted scientific view (dogma) is that there is a general convective circulation, with hot material upwelling towards the surface and cooler material going deeper. It is generally thought that this convection directs the circulation of the tectonic plates in the crust. The *unviability* of this concept will be discussed in chapter five.

Recently, analysis of a diamond volcanically coughed up from deep in the Earth and recovered in Brazilian river gravel has serendipitously revealed water-containing *ringwoodite*, a testimony to the presence of water inside Earth. Within this mineral, there is room between the atoms in this crystal lattice for a significant amount of water. The crystalline structure of ringwoodite could theoretically, accommodate up to 2% water by weight.

This leads scientists to affirm metaphorically, that there is a huge "ocean" in the lowermost part of the upper mantle transition zone (between 520 and 660 kilometres) where ringwoodite containing water, should be most stable. That portion of Earth's mantle alone could be holding 1.4×10^{21} kilograms of water, which is equivalent to the mass of all the water in all the world's oceans combined.

Some 2000 kilometres beneath our feet, there are enormous hundreds of kilometres' sized masses of mantle material behaving oddly, that have baffled scientists for the last four decades. Some scientists call them *blobs* and sit at the bottom of Earth's rocky mantle above the supposed molten outer core. By looking at the travel times of waves from many earthquakes, taken from thousands of instruments around the globe, scientists detected these huge zones, called *ultralow* velocity zones.

These slow-wave velocity zones are concentrated in two locations: One lies under the Pacific Ocean, and the other sits under Africa and part of the Atlantic Ocean. They appeared like massive mountains on the core-mantle boundary. There is little doubt that the blobs exist, yet scientists have no idea what they are. These blobs remain *enigmatic.*

Scientists can't even decide on what to call them. The nature of these blobs will be discussed in chapter five. The mantle is where deep focus earthquakes do occur. As mentioned earlier, the phenomena of some deep earthquakes are still not fully understood. Several scientific papers have been written offering tentative explanations.

On May 24, 2013, in the Sea of Okhotsk, Russia; a magnitude-8.3 earthquake occurred deep within the Earth's mantle about 378 miles (609 km) beneath the Earth's surface, at a crack where the Pacific Plate was pressing into the mantle, the layer of hot, plastic rock that sits below the crust.

This earthquake, described in the journal *Science*, is perplexing because seismologists don't understand how *massive* earthquakes can happen at such depths. As one researcher said: "It ruptured just like a breaking of the glass. And yet it's under this huge pressure, so that's something of a mystery. How does it happen?" This statement is an example of *tunnel vision* acquired by scientists through their specialization, which *limits* the ability to see things clearly, from a broader perspective.

This attitude often leads to a dead end. Most scientists agree that deep focus earthquakes shouldn't occur. This is blatant *ignorance*. They are taught to think all earthquakes, must be *tectonic* in origin. However, deep inside Earth the origin of mechanical vibrations (aka earthquake) does have *more* than one cause. In this case, it deals with what is known in Thermodynamics as *high-energy phase transformations*, something that chemical and mechanical engineers, are well aware of.

The Core

Scientists do *believe* that deep down inside the Earth, there's a huge hot ball of liquid and solid iron. This is the Earth's core. Per the standard but outdated hypothesis, when the Earth first formed, 4.6 billion years ago, it was a hot ball of molten rock and metal. And since it was mostly liquid, heavier elements like iron and nickel were able to “sink” down into the planet (neglecting its rotation) and accumulate at the core.

The core is believed to have two parts: a solid inner core, with a radius of 1,220 km (a little smaller than the size of the Moon) and then a liquid outer core that extends to a radius of 3,400 km. The core is thought to be 80% iron, as well as nickel and other dense elements like gold, platinum and uranium.

The inner core is solid and elastically anisotropic; that is, seismic waves have different speeds along different directions. The outer core is a hot liquid. Scientists believe that movements of metal, like currents in the oceans, “create” the magnetic field that surrounds the Earth and it protects us from the dangerous radiation of space.

However recent studies show that this conception of the core is *untenable* an it will be discussed shortly in this chapter. In fact, something is not quite right inside the Earth's core. Problematically, for researchers, the results of seismic measurements consistently show that these waves move through the Earth's solid inner core at much slower speeds than *predicted* by experiments and simulations.

Specifically, a type of seismic wave called a 'shear wave' (S) moves particularly slowly through the Earth's core relative to the speed *expected* for the material mainly *iron* from which the core is apparently made.

When seismic waves from earthquakes ripple through its solid centre, they hit a speed bump. The seismic vibrations should zip along about 30 percent *faster* than their actual speed, according to experiments and computer models re-creating the conditions inside the inner core.

Scientists have tried to explain this odd observation by *playing around* with the core's properties adding metal such as nickel, or suggesting that iron acts strangely deep inside the planet. In 2013, a new computer model of Earth's inner core explains the seismic wave slowdown via changes in iron's strength, just before the metal melts.

The findings were published in the journal *Science*. Scientists think the Earth's outer core is liquid, but the heart is solid iron and nickel, plus traces of elements such as sulphur and gold.

This new *model* indicates that inside the inner core, just before iron melts, the metal's strength *weakens* dramatically, according to researchers at University College London in the United Kingdom.

Weaker iron is less stiff, so one kind of seismic wave that passes through the core, called a shear wave, can travel more slowly. In the team's computer model of the inner core, when iron is at about 99 percent of its melting temperature, the seismic velocities match the speeds picked up by instruments that monitor earthquakes. *Why* this weakening of the iron happens is an open question.

"The proposed mineral models for the inner core have always shown a faster wave speed than that observed in seismic data," a co-author of the study, said in a statement. "This mismatch has given rise to several complex theories about the state and evolution of the Earth's core." This means that basically, the core is *not* made of an iron-nickel mixture as it is believed. In fact, another study shows that it *cannot* be possible. This issue will be discussed at the end of this chapter.

In 2013, a study published in *Nature Geoscience*, provides an experimental evidence that the Earth's inner core has rotated at a variety of different speeds and suggests that the Earth's solid inner core rotates at a *different rate* than the mantle.

Previous studies (J Zhang, 2005) arrived at the same conclusion. Associate Professor Hrvoje Tkalcic from the Australian National University College of Physical and Mathematical Sciences and his team used *earthquake doublets* to measure the rotation speed of Earth's inner core over the last 50 years. They discovered that not only did the inner core rotate at a different rate to the mantle but its rotation speed was variable. Sometimes is faster than Earth's mantle and sometimes it slows down.

Scientists have so far assumed the rotation rate of the inner core to be constant, because they lacked adequate mathematical methods for interpreting the data, says Associate Professor Tkalcic. A new method app lied to earthquake doublets pairs (repeating earthquakes that produce similar waveforms) of almost identical earthquakes that can occur a couple of weeks to 30 or 40 years apart, has provided the solution.

The best evidence for a *super-rotation* of the Earth's inner core is provided by comparing the signals of earthquakes occurring years apart but in almost exactly the same location. This study shows an Earth's inner body that is *independent* of the outer spherical shell or layer known as mantle.

Interestingly, English astronomer Edmund Halley, namesake of Halley's Comet, speculated that the inner shells of the Earth rotate with a different speed back in 1692. The true nature of this inner core will be discussed in chapter five.

Finally, during 2018 in a newly published study in *Science*, researchers from the Australian National University developed a technique to detect shear waves also called J waves, a type of wave that can only travel through solid objects in the inner core. Inner core shear waves are usually very difficult to be observed directly because they are so small and feeble.

The researchers ditched the first few hours of measurements after each quake, and started looking when all the signals had died down enough for minuscule J-waves to become apparent.

They found the inner core is indeed solid, but they also found that it's *softer* than previously thought, *departing* from the iron mechanical properties. That said, it doesn't seem to be completely rigid at all, it's actually a bit squishy.

Gravity inside Earth

We are pretty much familiarized with the force of gravity *on* the surface of Earth also, with its effects in the *outer* space in regards of rockets and satellites. Just like the gravitational force, the strength of the gravitational field around the Earth decreases as the square of the distance from the centre of the Earth. Twice as far means one-fourth of the field strength. But how far do we know the gravity behaviour beneath our feet? The force inside the Earth is proportional the distance from the centre.

Let's begin from the surface and then downwards. Technically on Earth's surface, the acceleration of gravity is associated with weigh (force) through the following equations:

$$\mathbf{F = ma}\text{, thus: }\mathbf{W = mg}$$

Were W and m being the weight and mass of the object and g the gravity acceleration on a particular spot on Earth. This value varies with the latitude being maximum in the equator and minimum on the poles. The average value of g is 9.80665 m/s.^{2} This variation is due to the *centrifugal force* of the Earth's rotation. In the Earth's crust, the mountains and geological structures below the surface can affect the average value of *g* calculated per the model of the planet's gravity field, causing *gravity anomalies*. Because the Earth rotates and its mass and density vary at different locations on the planet, gravity also varies.

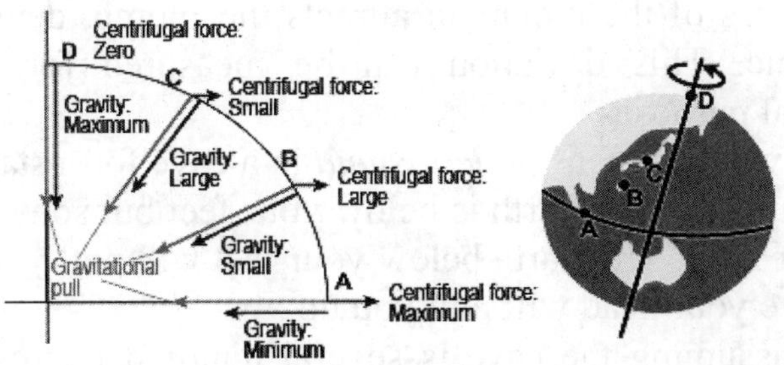

Fig. 38- Earth is not a perfect sphere. Due to rotation, it bulges in the equator and flattens in the poles. The total "weight" of a particle on Earth's surface is the result of the gravitation pull and centrifugal force. It decreases from the equator to the poles.

This is the reason why on the surface and close to it, the gravity of Earth *cannot* be considered as if the planet's mass was concentrated in its centre. Since gravity is associated with all mass, a mountain will exert an attractive force on us as does the planet…just significantly less.

These anomalies are caused due to different composition and density of the materials. Gravity anomalies tough being tiny, can be measured and are a powerful tool in geophysical prospection of oil and minerals, located beneath Earth's surface.

For example, the vertical deflection (VD); also known as deflection of the plumb line at a point on the Earth, is a measure of how far the gravity direction has been shifted by local anomalies such as nearby mountains.

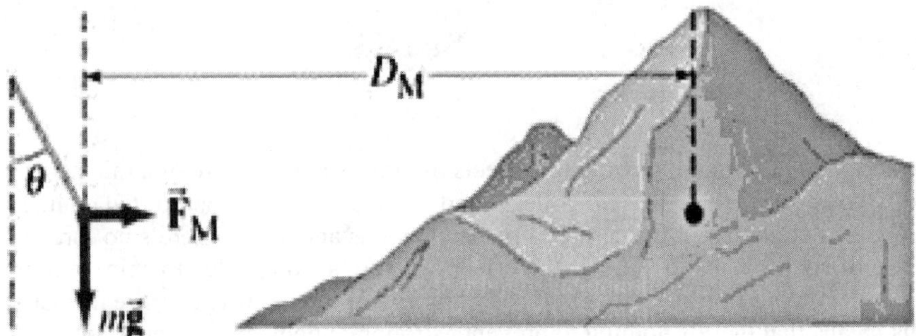

Fig. 39- The plumb is pulled away from its central position due to the gravitational attraction of a nearby mountain. It attracts the plumb with a force **F**, through a distance D. The angle θ can be measured, to calculate this force.

The mass of the mountain attracts the plumb, deviating from the vertical line. This deviation can be measured for surveying and geophysical purposes.

Now what happens *underground* in a cave for instance? If you are in a cave, most of the Earth is below your feet but some of the Earth is above your head. The Earth below your feet will pull you down and the Earth above your head will pull you up.

So, (assuming the cave is several hundred metres underground) there will be a tiny UP gravitational force which will counter-act the DOWN gravitational force.

If the cave would be in the centre of the Earth, there won't be any gravity at all. Equal quantities of the Earth's mass would be above and below you. The bodies *on* the Earth's surface bear the biggest gravity.

The gravity will become *smaller* if we move away from the Earth's surface to its centre.

- In a cavity at the center of Earth your weight would be zero, because you would be pulled equally by gravity in all directions.
- At the center of the earth. the gravitational field strength is zero.

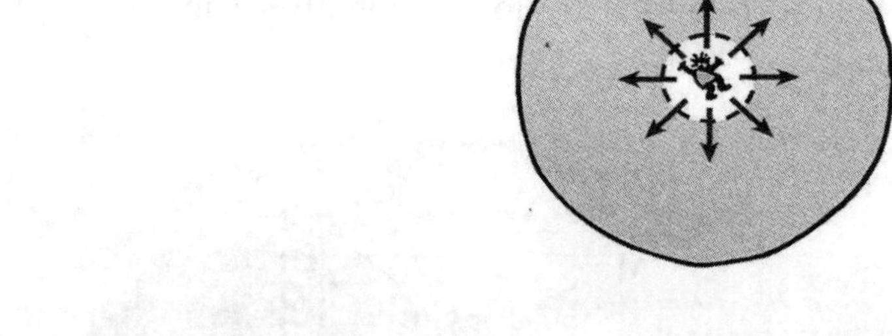

Fig. 40- Diagram showing the effects of gravity at the centre of a thick spherical shell. Any object would be "weightless" and no pressure would be exerted on it. There can be no net gravity at the centre of planets, moons and stars. There's no force at the centre of such bodies to counter the effect centrifugal forces due to spin. If a cavity was to develop inside an astronomic body, there would be no way to make it disappear. This means that there's nothing to prevent astronomic bodies from being hollow.

Mathematically g(effective) *inside* Earth is:

$$\mathbf{g_e = g\,(r)/R\ earth}$$

When r is 0 (in the centre of Earth), the effective gravity is *zero*. This is also known as the *Shell Theorem* of Newton. It says that no gravitational force is exerted by a hollow sphere on any object inside it. If we consider Earth as a thick shell, with a gaseous core, any object located in the Earth's centre will "float", exactly as it would do in space.

When an object starts to move on the radius from center to shell, both sides (perpendicular to radius) ratio are equal and opposite but it becomes stronger towards the *shell*. However, this theorem is *neglected* by mainstream science, which leads to errors in regards to pressure inside Earth and nonviable "convective currents" in the alleged outer liquid core.

Pressure inside Earth

This is a tricky question indeed and the term of pressure is often *misused.* Bear in mind that this concept originated on the *surface* of Earth, dealing mainly with fluids such as gas and water in the 1600's.

The concepts of stress, shear and strain which applies to *solids* and are of real importance here, haven't been developed yet. By far the bulk of our planet is hidden from view and direct measurements cannot take place.

It is believed that the interior of Earth, is under high pressures and temperatures. It is assumed that pressure inside the Earth as a function of depth.

The standard definition of pressure is *force* per unit area. It is usually more convenient to use pressure rather than force to describe the influences upon fluid behaviour.

The standard unit for pressure is the Pascal, which is a Newton per square meter. Another common unit used is the *atmosphere (atm)*.

Pressure P is defined as:

$$\mathbf{P} = \text{force/area} = \mathbf{F/A}$$

The higher the area the lesser pressure under the *same* force. For an object sitting on a surface, the force pressing on the surface is the weight of the object, but in different orientations it might have a different area in contact with the surface and therefore exert a different pressure.

That pressure depends on the area is sometimes overlooked. The usual reasoning is to think pressure as being punctual ignoring that, indeed, the same force causing it; could be widespread trough vast areas thus making pressure *lesser*.

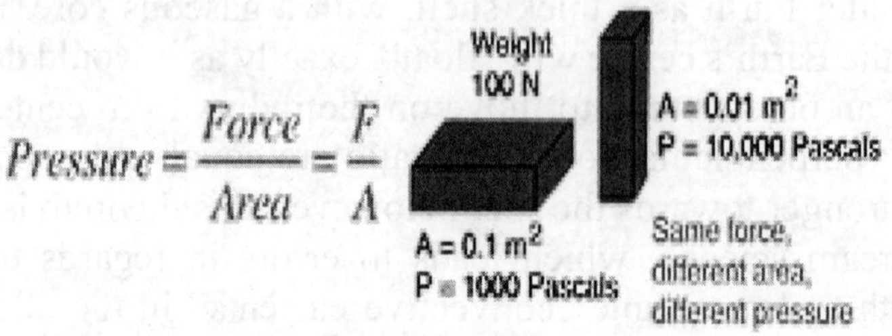

$$Pressure = \frac{Force}{Area} = \frac{F}{A}$$

Weight 100 N
A = 0.01 m²
P = 10,000 Pascals
A = 0.1 m²
P = 1000 Pascals
Same force, different area, different pressure

Fig. 41- The same force can exert different pressures. All depends of the area upon the force is being applied.

Calculating pressure in liquids

In the equation of pressure's definition, substituting F for mg and m is replaced for ρV, being V=Ah

$$\mathbf{P = h\rho g}$$

This equation represents the *hydrostatic* pressure due to the weight of any fluid of average density **ρ** at any depth **h** below its surface. Being density constant, pressure increases with depth.

For liquids, which are nearly incompressible, this equation holds to great depths. This reasoning is the one used by geoscientists to determine the pressure *inside* Earth.

In fact, the pressure inside Earth along its radius toward its centre, is calculated *assuming* "hydrostatic conditions" through the *Adamson-Williams* equation:

$$\mathbf{dp/dr = -g\,\rho = Gm\,\rho/r^2}$$

Where m is the mass of the sphere (in a fluid state) and r its radius. Pressure is then calculated, integrating this equation with g adjusted with depth.

According to this equation, the pressure at Earth's core is a staggering 3,400,000 times higher than the surface pressure or atmosphere.

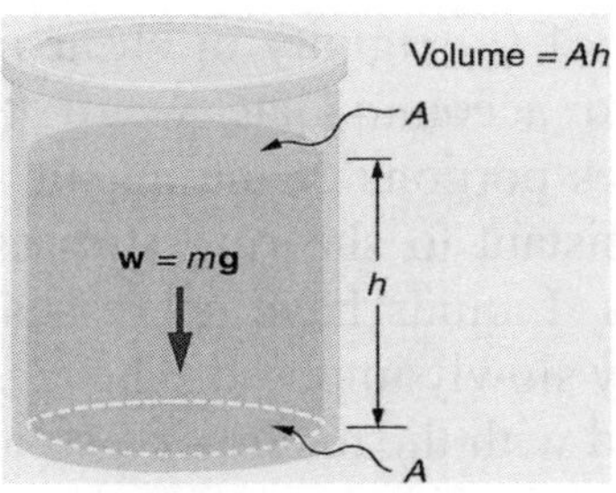

Fig. 42- Hydrostatic pressure. The bottom of this container supports the *entire* weight of the fluid in it. The vertical sides cannot exert an *upward* force on the fluid (since it cannot withstand a *shearing* force), and so the bottom must *support* it all. This does not happen in solids or rigid materials such as Earth' mantle and crust.

The usual explanation given by scientists is that in this case the force is primarily the *weight* of the overlying rock. So, in very simple terms, the deeper you go the more rock must be supported so the more force is required and the pressure goes up. This is technically known as lithostatic pressure. However, this explanation is too simplistic. Indeed, is an *oversimplification* of the real circumstances. This is because of the ignorance of what is known as the *mechanical state* of a rock body at a single moment in time and the concept of *surface forces* or the pushes and pulls exerted by surrounding material.

First, Earth's crust is very heterogeneous with a variable degree of *cohesiveness* of its materials. If the "rock" is a *loose* material like sediments such as gravel, sand or soil often mixed with water, then that's true, to some extent and the terms such as *bearing capacity* and *reservoir pressure* are used. But if the material is *rigid*, a continuous solid granite for example; this rock is "supported" along its depth, *sideways* by the same material. In other words, the vertical sides do exert an *upward* force, decreasing its "weight."

Second this reasoning applies the concept of *hydrostatic* pressure beginning on Earth's surface through its interior. However, different rules of "pressure" do apply, with gravity (and weight) diminishing with depth, following the Shell Theorem, making this fluidic reasoning *flawed.* The *only* exception is the local *molten* rock from volcanoes and underground chambers, several kilometres' depth; where the rules of hydrostatic pressure and thermodynamic phase equilibrium rightfully do apply, along high pressures and temperatures. But below these magma chambers, in the rigid mantle and going towards Earth's centre, the pressure is *unknown*.

The importance of (μ) rigidity or shear modulus in the mantle is often *downplayed* to accommodate weird "theories" such as plate tectonics that requires portions of the mantle to be "fluid". Rocks are *solid* and clearly resistant to shearing stresses. Rigidity is properly a measure of hardness. Liquids have no resistance to shearing stresses which are sufficiently slowly imposed. The *rigidity* of the mantle brings the mantle to respond with the nature of a solid to *seismic shear* waves (S) which pass through it. That is, the mantle does show *real* rigidity to seismic shear waves.

The term pressure is more adequate when dealing with fluids such as oil, gas or water. In the case of solids, the accurate terms are *strain* and *stress* and for the same purposes; especially inside Earth's mantle. This is because much of the materials which Earth's crust and mantle are made of, have the characteristic of *rigidity* or the resistance of an elastic body to a change of shape (deformation). This doesn't occur with fluids like gas or water.

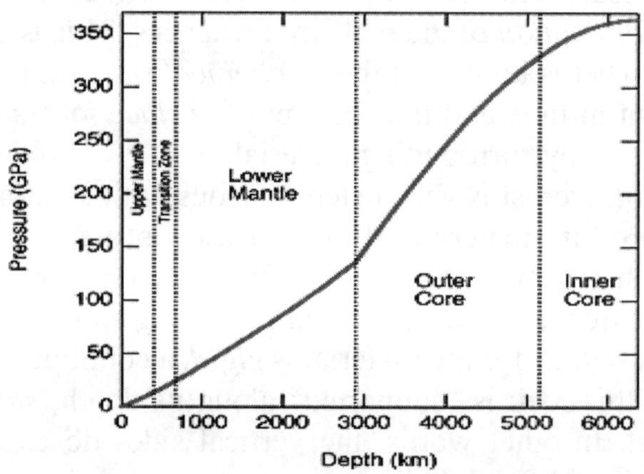

Fig. 43- Graphic showing the hypothetical pressure of Earth's interior, based upon the assumption of a hydrostatic pressure behaviour, along Earth's radius. It ignores the rigidity of Earth's mantle and Newton's Shell Theorem.

Strain is a dimensionless measure of *deformation*. It is a purely geometric concept that is meaningful only in the limit where solids are approximated as continuous materials. Stress is a measure of the *forces* that cause deformation. In the limit of small deformations, it is linearly proportional to strain for an elastic material. Stress and rigidity are interwoven with *strength*, which is a property of solids.

British scientist Sir Harold Jeffreys (1891–1989), one of the leading geophysicists of the early twentieth century, was fascinated (one might almost say obsessed) with the strength necessary to *support* the observed topographic relief on the Earth and Moon.

Through several books and numerous papers, he made *quantitative* estimates of the strength of the Earth's interior and compared the results of those estimates to the strength of common rocks. No hydrostatic approach is required anywhere, but it still is wrongly used in mainstream geophysics.

"The existence of any differences of heights on Earth's surface is decisive evidence that the internal stress is not hydrostatic. If the Earth was liquid any elevation would spread out horizontally until it disappeared. The only departure of the surface from a spherical form would be the ellipticity; the outer surface would become a level surface; the ocean would cover it to a uniform depth and that would be the end of us. The fact that we are here implies that the stress departs appreciably from being hydrostatic," Harold Jeffreys, Earthquakes and Mountains (1950).

Temperature inside Earth

What's temperature and heat? *Temperature* is operationally defined to be what we measure with a thermometer. Is the quantity of *thermal energy* or the energy of moving or vibrating molecules, inside a given system.

Traditionally, *heat* is this energy transferred between substances or systems due to a temperature *difference* between them.

As a form of energy, heat is conserved, i.e., it cannot be created or destroyed. It can, however, be transferred from one place to another. Heat can also be converted to and from other forms of energy. Low temperatures are regarded as "cold" whereas high temperatures are "hot".

The three most common temperature scales are Fahrenheit ° F, Celsius ° C, and Kelvin ° K. Temperature scales are created by identifying two reproducible temperatures. The freezing and boiling temperatures of water, at standard atmospheric pressure are commonly used. Is of paramount importance to understand *how* heat can flow inside Earth. Heat can travel in three different ways.

Conduction: occurs when two objects at different temperatures are in *contact* with each other. Heat flows from the warmer to the cooler object until they are both at the same temperature. Conduction is the movement of heat through a substance by the collision of molecules. At the place where the two-object touch, the faster-moving molecules of the warmer object, collide with the slower moving molecules of the cooler object. As they collide, the faster molecules give up some of their energy to the slower molecules. The slower molecules gain more thermal energy and collide with other molecules, in the cooler object.

This process continues until heat energy from the warmer object spreads throughout the cooler object. Some substances conduct heat more easily than others. Solids are better conductor than liquids and liquids are better conductor than gases.

Convection: In Nature regarding liquids and gases, convection is usually the most *efficient* way to transfer heat. Convection occurs when warmer areas of a liquid or gas rise to cooler areas in the liquid or gas. As this happens, cooler liquid or gas takes the place of the warmer areas which have risen higher. This cycle results in a continuous circulation pattern and heat is transferred to cooler areas. You see convection when you boil water in a pan. The bubbles of water that rise are the hotter parts of the water rising to the cooler area of water at the top of the pan.

Radiation: Both conduction and convection require *matter* to transfer heat. Radiation is a method of heat transfer that does not rely upon any contact between the heat source and the heated object. For example, we feel heat from the Sun even though we are not touching it. Heat can be transmitted though "empty" space by thermal radiation. Thermal radiation (often called infrared radiation) is a type electromagnetic radiation (or light). Radiation is a form of energy transport consisting of electromagnetic waves traveling at the speed of light.

A key concept to understand how temperatures inside Earth are estimated is *thermal conductivity.* Thermal conductivity of a material is defined as the *rate* of heat transfer, through a unit thickness of the material per unit area per temperature difference. Thermal conductivity is the ability of the material, to conduct the heat. Each material has its own capacity to conduct heat and it can be *experimentally* determined.

The thermal conductivity of a rock or mineral is the sum of conductivities caused by lattice vibrations (Kl) and by transfer of heat by radiation (Kr).

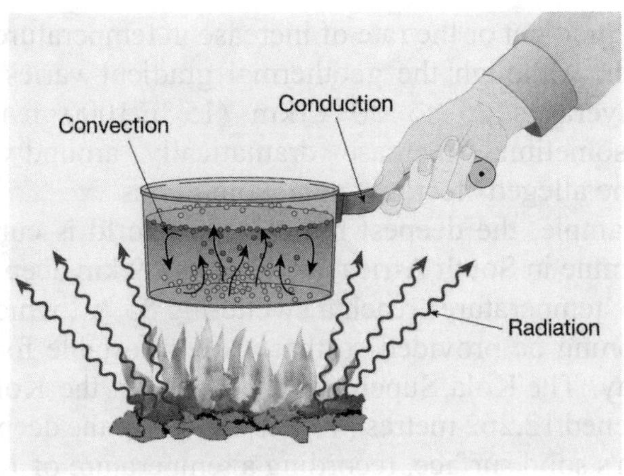

Fig. 44- Heat can propagate in three different ways. Conduction and radiation are known to happen inside Earth's crust and upper mantle. Convection is assumed (not proven) to occur mostly, in the upper mantle to explain the alleged horizontal movement of tectonic plates.

Below 750° K the thermal conductivity is due almost to lattice vibrations and generally it decreases with increasing temperature according to the formula:

$$\mathbf{Kl = (a + bT)^{-1}}$$

Where a and b are constants determinable by experiments. Above about 750° K, a significant amount of heat may be transferred trough rocks by radiation thereby increasing the thermal conductivity, above the value for lattice vibrations alone. An approximate expression for the radiative contribution to the thermal conductivity at temperature T is given by:

$$\mathbf{Kr = \frac{16n^2\, s\, T^3}{3e}}$$

Where n is the refractive index of the mineral, s is the Stefan-Boltzmann constant and e is the mineral's opacity. Thus, the temperature T can be calculated knowing the other parameters of the equation.

As geologists, have known for some time, if one digs down into the continental crust, temperatures will go up. This is known as the geothermal gradient or the rate of increase in temperature per unit depth in the Earth. Although the geothermal gradient varies from place to place, it averages 25 to 30 °C/km (15 °F/1000 ft.). Temperature gradients sometimes increase dramatically around volcanic areas, mainly in the alleged "tectonic plate boundaries".

For example, the deepest mine in the world is currently the Tau Tona gold mine in South Africa, measuring 3.9 km deep. At the bottom of the mine, temperatures reach a sweltering 55 °C, which requires that air conditioning be provided so that it's comfortable for the miners to work all day. The Kola Superdeep Borehole on the Kola peninsula of Russia, reached 12,262 metres (40,230 ft.) and is the deepest penetration of the Earth's solid surface, recording a temperature of 180 °C.

The only direct temperature readings from the Earth's interior (crust) comes from mines, oil & gas drillings and volcanic activity. Volcanic activity offers a unique window to somehow, assess directly the temperature at considerable depths of Earth's crust. Volcanic eruptions are caused by magma (a mixture of liquid rock, crystals, and dissolved gas) expelled onto the Earth's surface.

Under normal conditions, the geothermal gradient is not high enough to melt rocks, and with the exception of the outer core, most of the Earth is solid. Thus, magmas form only under special circumstances following thermodynamic laws and thus, volcanoes are only found on the Earth's surface in areas above where these special circumstances occur (volcanoes don't just occur anywhere).

Evidence comes from pieces of molten material or samples, brought up by erupting volcanoes. In the laboratory, scientists can determine the melting behaviour of specific minerals contained in rocks and find the temperature that matches the samples' melting point.

Laboratory measurement and limited field observation indicate these temperatures fluctuate between 650 to 1200 ° C. So far, the only direct evidence of the temperatures inside Earth reaches a few kilometres' depth. What about the rest?

The transmission of heat inside the Earth (K) involves three major processes: conduction (coefficient Kl), radiation (coefficient Kr) and subcrustal flow of material (magma).

While under the conditions in the Earth's crust, the heat transfer by radiation is negligible, it becomes increasingly important with increasing depth as T^3, the third power of absolute temperature, enters the equations.

Summarizing:

$$\mathbf{K_{earth} = Kl_{conduction} + Kr_{radiation}}$$

And here's when the *guesswork* comes in: conductivity Kl and thermal radiation Kr must be calculated theoretically for the Earth's interior. These coefficients are not based in facts whatsoever, not even remotely and there is no way to verify the validity of these.

The resulting values for both, depend not only on *assumptions* made as basis for the theory, but on assumed numerical values of physical "constants" in the Earth's interior. Consequently, at best the order of magnitude of the resulting value of K, can be given.

This is the main reason for the relatively great *uncertainty* in our knowledge, of the temperature in the deep portions of the mantle and consequently in the core. As one scientist described it aptly "the study of thermal processes within the Earth is one of the more speculative branches of geophysics" and it's true. But guesses are not good enough and these do lack scientific value. The idea that Earth's core formation was accompanied by conversion of "gravitational energy" to heat, is a *myth*. It doesn't follow the laws of Thermodynamics.

Scientists have applied indirect means to infer the temperature of the Earth's interior. But estimates thus far are inconclusive, and scientists *disagree* about how widely temperatures vary beneath the surface. Therefore, the alleged "temperature" of 6000 °C, which by the way, is the same as the surface of the Sun, attributed to Earth's core is pure nonsense. Indeed, it is pseudoscience.

The high temperatures attributed to Earth's centre are calculated to fit the *assumption* of a molten iron core, that's all. In fact, geoscientists do believe that there are three main sources of heat in the deep Earth:

- Heat from when the planet formed and accreted, which has not yet been lost;

- Frictional heating, caused by denser core material sinking to the centre of the planet; and

- Heat from the decay of radioactive elements.

Also, these alleged high temperatures inside Earth would allow the thermal energy necessary to sustain, the controversial “convection currents” in the mantle. However, it is physically *impossible* that iron or any metal molten core can develop as it will be explained in chapter three.

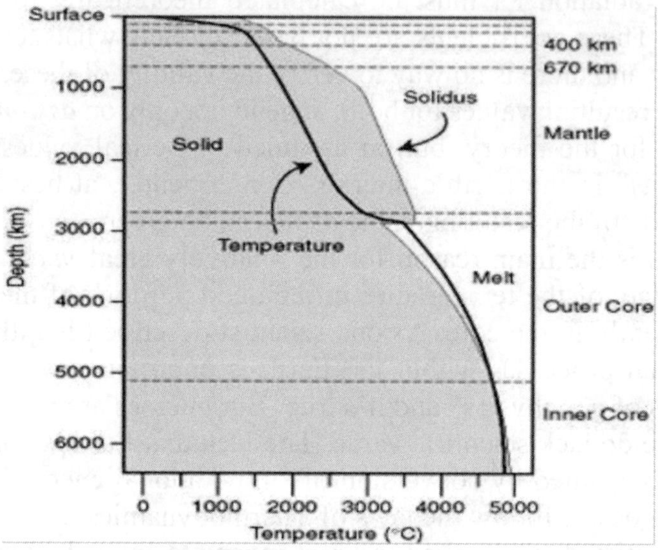

Fig. 45- Graphic showing the hypothetical temperature of Earth’s interior, based upon the assumption of a molten iron core; which is a leftover from a hypothetical Earth’s origin.

Simply stating that radioactive decay generates heat and concluding that buried radioactive material must be the Earth's heat source is *not* scientific. A hypothesis requires s*upporting data* (not assumptions), yet such data has not always been readily available.

A major issue is that experts in radioactivity often have a strong *bias* toward their field. Even today, many technical reports favoring radioactive heating rely on circular reasoning, ignoring the fundamental fact that the amount of radioactive material deep within the Earth remains *unknown.*

Geologists believe in the differentiation process of heavy elements “sinking” into the mantle and core, during Earth’s formation. The problem is heavy radioactive isotopes cannot be more concentrated in the mantle than the highest natural ore concentrations observed in the Earth's crust. Geological evidence shows that when these isotopes become too concentrated, they undergo rapid burn-off through *fission decay*. Therefore, no radioactive material is left to generate heat.

Succinctly put, geologists had considered Earth as a heat engine, slowly running down; without understanding basic Thermodynamics as usual. But in fact, there is NO evidence that the Earth's internal machinery is running down. Whatever the nature of the energy reservoir there is a continual drain on it for various thermal processes as well for mechanical deformations. Ultimately the supply of available thermal energy must be depleted.

This makes *impossible* any dissipation of heat from the mantle as the original source of the energy required to "drive" the alleged plate tectonics through convection, for long timescales because no heat would be left. There is no clear evidence of a *decrease* in the rate of various geological processes, hence the terrestrial reservoir of available driving energy (except heat already lost) must be large compared with the total drain throughout geologic history.

Finally, from a strictly scientific standpoint, any temperature within Earth, below the magma chambers in the upper mantle downwards, is *unknown.*

The Kola Deep Borehole

The Kola Deep Borehole was the result of a scientific drilling project of the former USSR, that started in May of 1970 under the code name of SG-3. After 5 years, when the depth of the SG-3 well exceeded 7 kilometres, the new Uralmash drilling rig was mounted one of the most modern and technologically advanced for those times.

Powerful, reliable, with an automatic tripping mechanism, it could withstand a string of pipes up to 15 km. The drill site is located on the Kola Peninsula, in the Pechengsky District of Murmansk Oblast, in the far northwest Russia and bordered by the Barents Sea.

Core samples of the Kola Superdeep Borehole can be seen at the repository in Zapolyarny, a town about 10 kilometres from the borehole's location. The project attempted to drill as deep as possible into the Earth's crust.

The reason, geologists chose Kola as the location for superdeep drilling is that the Fennoscandian Shield consists of very old rock, in some places the Precambrian crystalline igneous rock is exposed on the surface. Drilling deeper reaches even older rock and enables us to see even further back into the history of the Earth. The Kola borehole encountered 2.7 billion-year-old rocks at 12 kilometres' depth.

This superdeep drilling project, originated during the Cold War era, in which US and Soviet scientists and engineers, were trying to outdo each other in the subterranean realm just as they were trying to beat each other into orbit and to the Moon.

The US may have won the race to the Moon, but their 1958 initiated "Project Mohole" off the Pacific coast of Mexico lost funding and stopped drilling in 1966; but the Russians from 1970 to the early 1990s, kept on going. In that period, geologists and geophysicists had only scant indirect evidence and lots of educated guesses as to what was going on in the Earth's crust, and superdeep boreholes provided much needed information for a better understanding of the underlying geology by utilizing direct observation.

One of the benefits of digging so deep into the Earth was that the Russians were able to put geological-geophysical hypothesis to the tests. Even to this day, information gathered by this project is still being analysed and interpreted, mostly by Russian scientists. Kola ceased to be in the early 1990's because the temperatures down there were around an unsustainable 180°C (356°F), not the 100°C (212°F) expected.

The Kola Superdeep Borehole with 12,262 meters' depth, remains the deepest artificial point on Earth. This may all sound like a rather impressive feat of engineering and it is, but the borehole is barely a flesh wound on the Earth itself. With an equatorial radius of 6,378 kilometres (3,963 miles), this vertically ludicrous chasm is a mere 0.19 percent of the way down to the centre of the planet. However, the Kola Deep Borehole was able to deliver several surprises to the scientific community.

The first surprise they encountered was the lack of the so-called "basaltic layer" at about 7 km deep. The scientists discovered that the basic rock material deepest in the well was different quite from what most geologists had predicted. In fact, in 1926, the influential British geophysicist Sir Harold Jeffreys, has postulated that the Earth's crust consisted of three layers.

In Jeffreys' model, sedimentary rocks covered granite which lay atop basalt. Jeffreys believed basalt could be found everywhere on Earth, just above the mantle, if we dug deeply enough. Supporting "evidence" for this idea was a velocity increase in seismic waves at about 9000 metres depth. This, claimed Jeffreys, was where granite gave way to basalt. This was just a hypothesis at the time, based on seismic data.

Jeffreys didn't expect to be proved wrong in his lifetime but he was. Jeffreys' hypothesis was disproved when the Soviet drill bit found granite not basalt at the velocity anomaly. It made a kind of revolution in geological and geophysical science and completely changed the traditional notion of the Earth's crust interior, at least in Russia but hardly in the West.

Basically, the Soviet drilling team were expecting to find granites, and as they went deeper, basalts. But much to everybody's surprise, when they went deeper, they actually found… more granites. According to their data, furnished by a report prepared at the Institute of Physics of the Earth in Moscow, the first 7 kilometres are the granite strata of the upper crust, then the basalt layer begins.

Then the idea of a two-layer structure of the Earth's crust was generally accepted. But as it turned out later, both geologists and geophysicists were wrong. As it turns out, the seismic discontinuity was caused by the metamorphosis of the granites, not by basalts.

This proves that seismological analysis, regardless of its sophistication and mathematical furnishing, only shows a change in some physical properties of stone materials with depth, but never reveal the quality or nature of the materials involved.

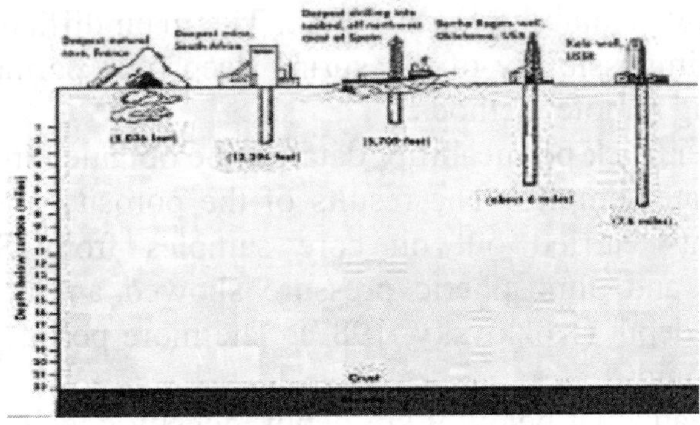

Fig. 46- The deepest artificial point on Earth the Kola Superdeep Borehole, compared with similar digs and in relation to Earth's crust and mantle.

In fact, the so-called seismic tomography which purportedly elucidates the Earth's interior, distinguishes only the wave-propagation velocity without giving any information about materials. Therefore, any seismological interpretation of what's the nature of Earth's interior is a conjecture, unless it is supported by independent sources.

Again, any statement that reads: "by measuring the speed of the waves at different depths, scientists can figure out which types of rocks the waves were passing through" is a *lie*. As if that wasn't enough, between 3 and 6 km deep, they also found water. Everyone figured that the granite would be as dry as a stone. By the knowledge, we had back then, water simply shouldn't have existed at those depths and yet, there it was.

It used to be natural to think, that with distance from the surface of the Earth, with increasing pressure, rocks become more monolithic (compact), with a small number of cracks and pores. Confining pressure tends to keep a solid mass of rock in the form of a single body and hinders the formation of fractures. SG-3 convinced scientists to the contrary.

Starting from 9 kilometres to the end of the borehole, the strata turned out to be very *porous* and *permeable*, literally crammed with cracks through which aqueous solutions circulated, this indicates that with depth at some point, pressure in the Earth's interior *decreases*, proving the hydrostatic assumption of pressure (and density) is *wrong*.

The studies of the Kola SG-3 revealed a presence of water fluids at a depth previously considered as unreachable for them Kozlovsky, 1987, Huenges et al., 1997. Rock permeability is a critical factor, as it controls fluid transport within the Earth's crust. The main difficulty arises from the present impossibility of measuring deep rock permeability in situ directly using remote methods.

However, rock permeability data can be obtained in the laboratory on appropriate samples. The results of the porosity and permeability measurements carried out on core samples from SG-3 at room temperature and atmospheric pressure showed an increase of their values with depth (Kozlovsky, 1987). The more porous a rock is, less dense it becomes.

This means that below 9 km depth, density does *not* increase and the conventional calculated velocities of the seismic waves S and P and their paths inside Earth are also *wrong*. Researchers also reported the extraction of mud, which was "boiling with hydrogen" The Earth's crust has gas.

Unexpectedly, helium, hydrogen, nitrogen, and even carbon dioxide (from microbes), were found all along the borehole.

Without a doubt, the biggest surprise was the discovery of life: microscopic plankton fossils in rocks over two billion years old, found four miles beneath the surface, contradicting the scientific ideas of the day. These "microfossils" represented about 24 ancient species and were encased in organic compounds which somehow survived the "extreme" pressures and temperatures, so far beneath the Earth's surface.

This raised numerous questions about the potential survival of life forms at impressive depths. Now, research has shown that life can exist even in oceanic crust, and even macroscopic life was found at over 1 km deep, but at the time, finding those fossils came as a shocker. With the Kola Deep Borehole findings, the lesson to be learned is simple: experience always beats theory. It has always been and always will be.

In Russia, scientists are well aware that without wells, geophysicists' sections of the Earth's crust are just that, models and n*ot* facts. In order for specific rock formations to appear on these schemes, *drilling data* is needed.

In 1984, the World Geological Congress was held in Moscow. Participants in the congress were waiting for an exhibition of the achievements of Russian geology, one of the stands was dedicated to the well SG-3. Experts around the world looked puzzled at the usual drilling head with erased carbide teeth.

A large delegation of geologists and journalists, went to the village of Zapolyarny in the Kola peninsula. Visitors were shown the drill in action, took out and disconnected the 33-meter sections of pipes. Piles of exactly the same drilling heads, like the one on the stand in Moscow, were rising around.

The delegation of by then the Academy of Sciences of the USSR, was hosted by a well-known geologist, academician Dr. Vladimir Belousov. During a press conference from the audience, he was asked a question:

- *What did the Kola well reveal the most important thing?*

- *Gentlemen! Most importantly, it showed that we know almost nothing about the continental crust,* the scientist honestly answered.

One wonders how many scientists in the West, would assume today the same position as Dr Belousov in regards of the knowledge of the continental crust. Let alone what lies beneath it and further down.

Little is known in the West, about the Kola Deep Borehole geological findings and it is unlikely that Western scientific textbooks, will be updated someday. There are only so far, two books published in

English (Springer-Verlag) dealing with this interesting subject, translated from Russian.

The myth of Earth having an iron core

It originated with the comparison of the mean Earth's crust density trough samples of different rocks which averaged 2.7 g/cm^3, against the Earth's mean density of 5.5 g/cm^3. This observation suggested that Earth, in regards to its bulk composition, is at least "partially differentiated".

This discrepancy should indicate that below Earth's crust, density must increase towards its centre. How much? This was found solving a simple first-degree equation, with one incognita:

$$\text{Earth's mean density} = (\text{crust density} + x)/2$$
$$5.5 = (2.7 + x)/2$$
$$\text{Where } x = 8.3 \text{ g/cm}^3$$

This density of 8.3g/cm^3 must be distributed between the mantle and the core or at least concentrated in the core.

In 1897, German physicist and geophysicist Emil Wiechert suggested and iron core for the Earth, in a molten state. Later it was assumed to be an iron-nickel core. Iron density is approximately 7.78 g/cm^3 which is close to the value of x and it is one of the main components along nickel, of meteorites known as chondrites.

It was believed at the time, that Earth originated with similar composition as those chondrites, by accretion of nebular cosmic dust. This lead the scientific community of those times (1900-1930), to believe that Sun and all stars consisted mainly of iron.

It is important to mention that the definition of density:

$$\boldsymbol{\rho = m/v}$$

Implies two variables: mass and volume of the object or material. In the case of Earth, its volume is considered a solid sphere, which is one possibility. If mass is constant, density will be dependant of the type of material and how it is occupying a specific volume. However, due to its size, Earth's mass must be found by indirect means using Newton's laws of motion and gravity.

Several inconsistencies with Earth's gravity field, which are shown in chapter five, make Earth's mass and density *unreliable*. Initially a given mass of material occupying an initial volume, can increase its density trough compaction and thus diminishing its original volume.

It is the same material with higher density accompanied by a decrease in volume of voids or intramolecular rearrangement. In the value of $\rho = 5.5$ g/cm^3, Earth's "solid" volume is well known and is obtained directly. As explained earlier, Earth's mass is obtained indirectly.

Given the above assumptions, coupled with further seismological interpretation of raw data and without any *independent* proof; the molten iron-nickel core came into being and it is considered as an *integral* part of the Earth' structure, despite being a product of a guesswork. Again, we must recall that geoscientists committed the deadly sin of *reification*, which means *mistaking*, with or without being aware of it, an abstract idea or concept with a physical object or thing.

Bear in mind that P and S waves can give clues of the physical state (phase) of the material, that causes changes inside Earth which allow to define boundaries (discontinuities). That's all. Nothing more nothing less. Thus, the current seismological *interpretation* of what material, caused the changes observed; is attributed to the *hypothetical* molten iron-nickel core proposed in 1897. This of course, is rooted in the outdated chaotic-accretion hypothesis of planetary formation.

There is NO *independent* evidence that the Earth's inner core is composed of a liquid/solidified iron-nickel metal alloy. There is also no independent evidence that the Earth is like an ordinary chondrite meteorite, even several scientists disagree with this idea.

This interpretation, leads to the *issue* of a sphere of 1.75×10^{11} Km^3 made of 1.41×10^{21} tons of iron-nickel that somehow, must have been "available", in the cosmic nebular dust when Earth and the rest of the Solar System formed.

Geoscientists didn't bother to consider whether this huge metallic sphere, would have been allowed to exist. It was left to astronomers, to deal with and that's the problem.

This hypothesis of a molten iron core is open to very *serious objections*. The iron core hypothesis would require that, about 33% of the mass of the Earth consisted of iron and other heavy elements. It is difficult to believe that the composition of the Earth is totally different from that of the material which the Solar System originated.

Composition of Earth cannot be understood in *isolation*. It must comply with the rules or the *Gestaltung* of the environment, that gave origin to Earth.

For instance, the reason why *silicon* is the most common element within the rocky planets, is because it was the most common element (next to hydrogen), in the original nebula that condensed to form the Solar System. This is simply how our Solar System was at the start, it had a greater ratio of silicon to all the other elements, except hydrogen.

Perhaps other planetary systems around other stars may have more aluminium or iron than silicon. Ours just happened to be a silicon-dominant one. Meteorites also have a significant portion of silicon. Chondrites are the oldest and most common of all meteorites found on Earth (85%).

The spectroscopic examination of stars and the evidence offered by the meteorites, seem to show that iron forms a *small* part of the bulk of the stars, and the presence of the several elements is the same in meteorites as in the crust of the Earth.

On the iron core hypothesis, there is complete *disagreement* so far as the iron content is concerned, as the following table shows:

Table of iron content (%)

	Crust	Sun	Meteorites (Chondrites)	Whole Earth
Oxygen	46.6	0.8	19.2	30.4
Hydrogen	0.2	73.6	0	0 .1
Silicon	27.7	0.07	14.5	15.1
Iron	4.8	0.16	8-18	28.8

Fig. 47- Iron content is excessive anomalous in relation to silicon, the most abundant refractory planet-building element, and in primitive and differentiated meteorites; during the early stages of the formation of the Solar System.

Even if we assume that the spectroscopic data for iron are not complete and that the percentage weight should be multiplied by a factor of 4 or 5, iron would still form a very much *smaller* portion of the Universe than the percentage for the Earth, demanded by the iron core hypothesis. In fact, metals constitute only about 0,15% of the Sun's mass.

Astronomers commonly express the Sun's composition by the percentage of total number of atoms. Ignoring the solar core, where it is thought hydrogen "fuses" to helium, the Sun's outer layers consist of more than 91 percent hydrogen and more than 8 percent helium.

The rest of the chemical elements constitute just 0.15 percent or so of the number of atoms in the Sun. There is only one iron atom for every 31,600 of hydrogen. The Sun is not hot enough, even at its centre, to make iron by the fusion of lighter elements.

Instead, exploding stars, called supernovae, make all the iron strewn in the Universe (0.12%) plus other heavier elements. In other words, the iron present in the Solar System's bodies, is a *leftover* from other stars.

Despite the many supernovae, billions of years older when our Solar System formed, the huge astronomical distances plus the interference of celestial bodies and other interstellar factors, would have made iron comparatively *scarce* in the cosmic dust as impurities and/or components of other grains through physical and chemical processes, in the interstellar medium that gave origin to our Solar System.

Once the Sun (98% of Solar System's mass) and other planets along minor bodies, got their share of iron, the probabilities to gather more than one thousand million of trillions of tons of iron, to make the Earth's core alone from cosmic dust, would have been improbable. Most likely, the *real* amount of iron in the whole Earth would be similar or below the percentage of iron found in Earth's crust (4.8%).

Standard scientific explanations or better said *speculations* say Earth (and the other planets in the Solar System) formed when smaller bits of matter in solar orbit came together gravitationally.

Because of those multiple "collisions", the kinetic energy of motion was converted to heat energy, resulting in a hot, molten mass that would become our Earth. While in this hot, fluid state, the denser, heavier material (iron and nickel) settled or sank to the centre and the less dense rocky material floated nearer the surface. This differentiation gave origin to the accepted Earth's iron/nickel core.

However, this hypothesis is *not consistent*. The differentiation process into a core, mantle and crust needed to support this hypothesis, would require that all heavy elements including the radioactive ones would be more abundant in the interior than in the crust. Also, it assumes a static non-rotating Earth, *ignoring* it spinning and centrifugal forces; which *organizes* and *shapes* the mass distribution of the planet, *outwards* not inwards.

Some scientists have shown however, that a distribution of radioactive elements throughout the Earth, greater in abundance than in the crust would prevent the Earth from cooling, because of the *heat* liberated in the process of radioactivity. Another issue is if the differentiation process was true, then why a *considerable* quantity of iron was left in the crust and it didn't sink? Whereas radioactive elements in the Earth's crust are sparse and managed in parts per million, the world's production of iron was more than two billion tons in the year 2017 alone. Due to iron's oxidation states Fe^{+2} and Fe^{+3} it is found in Nature in compounds with non-metallic elements such as oxygen, sulphur or carbon; not as a single element. An "alloy" of nickel and iron as allegedly exists, in the Earth's core is extremely unlike.

To make an alloy, the metal (iron) must be in its metallic state and typically molten, not compounded with another non-metallic element. In other words, iron must be *refined.* But this only happens in manmade industrial furnaces, not in Nature. While there are small quantities of some metals occurring as natural minerals that could have alloyed with other native metals, the quantity would be too small to be useful if ever found. Only gold, silver, copper and the platinum metals occur in Nature in larger amounts as elements.

Foregoing considerations lead one to believe that the iron core hypothesis is NOT plausible, and *other* hypotheses must be made to explain the high density of the Earth. Another reason orthodox science is still maintaining the iron core hypothesis, besides trying to fix the Earth's density, is because the iron core is used to "explain" Earth's magnetic field.

Earth's magnetic field, also known as the *geomagnetic field*, is the magnetic field that extends from the Earth's interior out into space. In space the Earth's magnetic field meets the solar wind, which is a stream of charged particles mainly a mixture of materials found in the solar plasma, composed primarily of ionized hydrogen (electrons and protons) that emanate from the Sun.

The magnitude of the Earth's magnetic field at the surface ranges from 25 to 65 microteslas (0.25 to 0.65 gauss). The Earth's magnetic field (magnetic dipole) is currently tilted at an angle of about 10 degrees with respect to Earth's rotational axis, it is as if there were a bar magnet placed at that angle at the centre of the Earth.

Our planet is largely protected from the solar wind, a stream of energetic charged particles coming from the Sun, by its magnetic field, which deflects most of these charged particles or ions. Some are trapped in the Van Allen radiation belt. If it weren't for the Earth's magnetic field we would be subject to bursts of radiation on the ground that would be, at the very least, unhealthy. The more serious, long term impact would be the erosion of the atmosphere. Therefore, life as we know today on Earth wouldn't exist. What causes the Earth's magnetic field?

Here's where the molten iron-nickel core hypothesis comes in. Since 1939, the geomagnetic field has been believed to originate by convective motions in the Earth's fluid, electrically-conducting core. The idea is that the convective fluid, interacts with the Coriolis forces produced by planetary rotation and acts like a dynamo which is a magnetic amplifier.

The idea seemed so logical and seemed to make such sense in explaining the origin of the observed geomagnetic field, that for decades no one seriously *questioned* whether long-term, sustained convection could in fact really occur in Earth's fluid core. This idea has been greatly *misunderstood* in geology and geophysics.

For decades, convection has been hypothesized to exist within the Earth's fluid core, and has become the stuff of textbooks, but there are serious *problems* with that concept, as discovered by maverick US scientist J. Marvin Herndon, who in this case, brought into focus the importance of discussing, debating, and challenging current thinking in science.

For stable convection to exist for extended periods of time in the Earth's fluid core, it is necessary that an *adverse temperature gradient* (the prevailing temperature gradient), must be maintained for extended periods of time, so that heat produced beneath the fluid core, will cause the fluid at the bottom of the core to be lighter; more buoyant, making it rise to the top of the core, bringing the heat with it.

To maintain that adverse temperature gradient for extended periods of time, requires the temperature at the top of the fluid core to be maintained at a lower temperature than at the bottom of the fluid core.

The top of the fluid core can only remain cooler than the bottom of the core, if the heat brought to top by convection and by conduction can be efficiently *removed*. Otherwise the top would remain hot as the bottom thus no convection happens. And therein lays one problem.

The Earth's fluid core is wrapped in an *insulating blanket*, a rock shell, the mantle, that is 2900 km thick, and which has a considerably lower heat capacity and lower thermal conductivity than the fluid core. Thus, heat brought to the top of the core *cannot* be efficiently removed by thermal conduction. In other words, in the fluid core, an adverse temperature gradient cannot be maintained for extended periods of time, and thus convection cannot be sustained for a long time.

But, it was discovered, there is another, even more serious, reason that convection is physically *impossible* in the Earth's fluid core. Because of the over-burden weight, the Earth's core is about 23% more dense at the *bottom* than at the top. The tiny, tiny amount of thermal expansion at the bottom *cannot* make the Earth's core top-heavy and *cannot* cause a thermally-expanded "parcel" from the bottom to float to the top of the core as required for convection. Thus, the Earth's core *cannot* engage in convection.

So, the implication is quite clear: Either the geomagnetic field is generated by a process *other* than the convection-driven dynamo-mechanism or there exists another fluid region within the deep-interior of Earth which can sustain convection for extended periods of time. The former of course, is the case.

Also, the convection-driven dynamo-mechanism *cannot* account for the planetary magnetic fields, so scientists try to explain Venus's weak field by its much slower rotation (243 Earth days for one rotation). But Mercury also rotates slowly (58.82 days), yet its field is about five times stronger than Venus's, while Mars rotates almost as fast as Earth (1.03 days), but its field is less than 1/10,000th of Earth's.

However, despite these *inconsistencies*, this setup of the solid core with an unstable liquid mantle on top is exactly the structure expected to occur in planets such as Earth and Jupiter. These planets' magnetic fields are similarly thought to be generated by convective dynamos, powered by the cooling and chemical separation of their interiors and the process can also be scaled up to account for the magnetic fields of fully convective objects like T Tauri stars, as well.

This process leads to the formation of a *Rayleigh-Taylor* unstable liquid mantle on top of a solid core. This convective region, as it apparently occurs in Solar System planets like the Earth and Jupiter, can produce a dynamo able to yield magnetic fields of strengths of up to 0.1 MG, thus providing a mechanism that could explain magnetism in single white dwarfs. However, an *alternative* hypothesis of the origin of Earth's magnetic field, backed by *experimental proof*, will be discussed in chapter four.

One more thing: when a team of scientists have measured the melting point of iron at high precision in a laboratory, and then drew from that result to calculate the temperature at the boundary of Earth's inner and outer core; this proves absolutely *nothing*.

Because the data doesn't come from a real physical system (core) through direct experience or *observables,* the first step of the scientific method. It comes from an ideal or speculative set of physical conditions (assumptions), that ought to be in the core. Thus these "conditions" are replicated in the laboratory yielding of course, results that "prove" the reality of Earth's core.

To prove any hypothesis, it must be anchored in real, *observable data* not fiction. If there were N different ideas or assumptions of what might be the core, there would be N "ideal" physical conditions assigned to this ideal core. Therefore, there would be N experiments carried out, simulating each of these ideal conditions yielding N different results, each one favouring or proving this ideal core.

All this fuss is because these scientists are trapped in the fallacy of *circular reasoning*, that is common in the world of academic science. They start with an assumption which they consider an incontrovertible fact (the iron molten core) which is treated as such in their subsequent analysis.

By following this trend of thought, they experiment, filter and interpret data arriving at the same conclusion which is the same assumption which with they have started. After more than century, the *idea* of an iron core still pervades the scientific community, despite it is unfeasible.

In fact, a recent study in 2018 shows that part of this iron core, the solid inner core, *shouldn't* exist. It is widely accepted that the Earth's inner core formed about a billion years ago, when a solid, super-hot iron nugget "spontaneously" began to crystallize inside a 4,200-mile-wide ball of liquid metal at the planet's centre.

The problem is that's not possible or, at least, has *never* been easily explained according to a new paper published in *Earth and Planetary Science Letters* from a team of scientists at Case Western Reserve University, comprised of post-doctoral student Ludovic Huguet; Earth, Environmental, and Planetary Sciences professors James Van Orman and Steven Hauck II; and Materials Science and Engineering Professor Matthew Willard.

They refer to this enigma as the "inner-core nucleation paradox." It goes like this: Scientists have known for more than 80 years that a crystallized inner core "exists". But the Case Western Reserve team asserts that this widely accepted idea *neglects* one critical point one that, once added, would suggest the inner core shouldn't exist. In fact, it is *impossible*. Here's why:

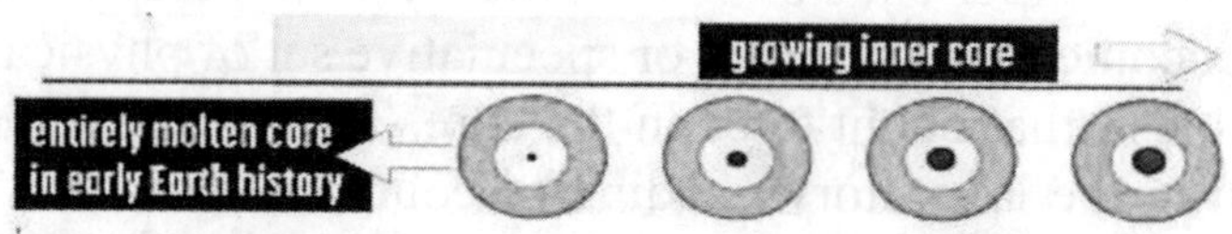

Fig. 48- Although nucleation theory has been discussed and tested for the last 100 years, it had not previously been applied to the formation of the Earth's inner core. By including the nucleation energy barrier (NEB), the Case Western University team calculated that the Earth's liquid metallic core would need 1,000°C of cooling below the standard freezing point before it could crystallize, which would take far too long for evolutionary core formation models to work. The time needed for the formation of the inner core blows out beyond the expected 1-billion-year age to at least 10 billion years, even longer than the alleged 4.5-billion-year age of the entire Earth.

While it is well known that a material must be at or below its freezing temperature to be solid, it turns out that making the first crystal from a liquid takes extra energy.

That extra energy, the *nucleation barrier* is the ingredient that models of Earth's deepest interior have *not been* included until now. To overcome the nucleation barrier and start to solidify, however, the liquid must be cooled well below its freezing point, what scientists call "supercooling."

Alternatively, *something* different must be added to the liquid metal of the core at the centre of the planet, that substantially *reduces* the amount of required supercooling. But the nucleation barrier for metal at the extraordinary pressures, at the centre of the Earth is enormous.

"Everyone, ourselves included, seemed to be missing this big problem that metals don't start crystallizing instantly unless something is there that *lowers* the energy barrier a lot," commented one of the scientists involved in the study. This "surprise" arose because, in this case, the hypothesis of the inner iron core was an *oversimplification*. Then how did the solid iron core form? It remains an open question.

What is clear is that a solid iron inner core, hardly fits within the current model of the inner Earth. Furthermore, the recent discovery of a solid but squishy inner core, coupled with the study showing slower speeds than predicted for the inner iron core mentioned early; demonstrate that this inner core is definitely, *not* made of iron at all.

It is *something different*. However, it is matter of time for scientists to come up with an *ad hoc* hypothesis to "solve" this issue and make things, even more confusing.

The final proof of the *unworkability* of the iron core, deals with Dynamo Theory. It came in 2016 when a group of scientists, from Linköping University (Sweden), Jožef Stefan Institute (Slovenia), Russia's National University of Science and Technology (MISiS) and France's École Polytechnique; led by the Russian Professor Dr. Igor Abrikosov, have helped to *disprove* the widely-held theory about the Earth's magnetic field.

It all started in 2012, when geophysicists in the UK published a supercomputer model which demonstrated that iron in the planet's core is more efficient at conducting heat than previously thought.

" Put simply, the paradox is that in this model, so much heat escaped from the core via conduction that there wasn't enough energy left over to fuel convection (when heat creates motion) in the liquid outer core. The implication: Earth's magnetic field shouldn't exist," Live Science explained.

In January 2015, US materials scientists Ronald Cohen and Peng Zhang of the Carnegie Institution (Washington) and Christian Houle from Rutgers University (New Jersey), published in one of the world's most prestigious scientific journal *Nature* an article, which argued that the theory of the formation of the planet's magnetic field, adopted after the UK study in 2012, is *untenable*.

Their calculations, justified by computer simulation, confirmed the classical theory, according to which the source of the magnetic fields of the planet is thermal convection, which exists in the core of the Earth.

The 2015 paper appeared to overcome this conundrum, because it apparently proved the traditional model.

"There was a big problem in how you generate a magnetic field, and now, because of our results, that problem has basically gone away," study co-author Ron Cohen told Live Science.

However, Cohen and co-authors Peng Zhang and K. Haule have now printed a *retraction* of their paper, based on experiments by an international team of scientists led by Russian Professor in Theoretical Physics, Dr Igor Abrikosov, which discovered an *error* in the calculations of US scientists.

It is noteworthy to mention that recall of an article and publication of a retraction in world's leading scientific journal is a highly unusual event, moreover, this happened for the *first* time in the sphere of physics research.

"L. Pourovskii, J. Mravlje, S. Simak and I. Abrikosov could not reproduce our findings, which led us to re-examine our computations. We found an error of a factor of two that is due to our neglect of spin degeneracy (two electrons per band), which would halve the electron–electron resistivity and increase the flow of electric current" they explained in their retraction.

As a result, the presence of the Earth's magnetic field remains as much a mystery as ever, representatives from Russia's National University of Science and Technology told RIA Novosti, in 2016:

"The researchers from the US agreed with Abrikosov's group, and in April 2016 they retracted their article from the Nature journal. The classical theory has been disproved again," MISiS explained.

"Now scientists must return to the calculations made in 2012, which show that heat convection cannot generate the Earth's magnetic field". MISiS finalized.

This news was widely reported in the media outlets of Russia, Iran and China. Not a single line was reported in the Western media and thus the public, was deliberately kept *ignorant*.

Bear in mind that the UK study of the impossibility of the alleged iron core to generate the Earth's magnetic field, led to the US team to "solve" this problem as it was disproving the prevailing scientific paradigm.

From this perspective, it might have been that the "error" made by the US team was not an error at all. As the reader, should recall from chapter one, that instead of reconsider their faulty hypothesis, scientists come up with weird, outlandish concepts or ideas to bridge the increasing gap between the observed data and the hypothesis.

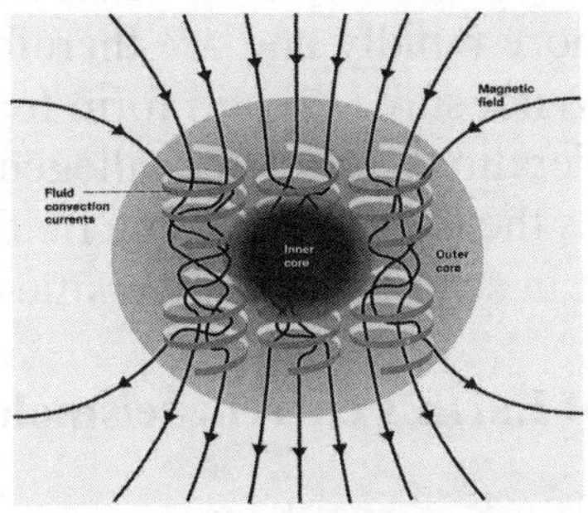

Fig. 49- A greater thermal conductivity for iron, challenges the notion that the planet's magnetic field originates solely from thermal convection of conductive iron in the outer core. At those higher levels of conductivity, the core would have cooled much faster than scientists previously thought, meaning either the planet started out much hotter than scientists can explain or some other force is stirring the core to sustain the dynamo. But what other force? There is none.

Even they can intentionally deceive their peers, through the *manipulating of data*, to fit their hypothesis and thus misrepresenting the way the Universe works; which could have been the case in this occasion. Not a single scientist in the West, bothered to *replicate* the findings of the US team. As mentioned in chapter one, the reproducibility of published experiments is the foundation of science. No reproducibility no science. It had to be Russian Professor Dr Abrikosov and his team, that when trying to replicate the US team findings, proved them to be *wrong*.

Currently, there exists *no* final definitive and generally accepted theoretical concept of terrestrial magnetism or any planet. The theory that existed until 2012 and has been considered as a classical theory maintaining that the Earth's core, is the source of the Earth's geomagnetic field, has NOT been supported by any calculations.

Although various other mechanisms for generating the geomagnetic field have been proposed, only the dynamo concept is seriously considered today despite the its aforementioned *unworkability*.

In fact, scientists' understanding of Dynamo Theory has been *complicated* by recent discoveries of magnetized rocks from the Moon and ancient meteorites, as well as an active dynamo field on Mercury places, that were thought to have perhaps cooled too quickly or be too small to generate a self-sustaining magnetic field.

It had been thought that smaller bodies couldn't have dynamos because they cool more rapidly and are therefore, more likely to have metallic cores that do not stay in liquid form for very long.

All these considerations make the alleged Dynamo Theory, that purportedly generates the Earth's magnetic field, a *bogus science* and as such; it is still taught in schools and universities to this day.

What is valid of Earth's current seismological interpretation?

All the information seismologists have about the inaccessible interior of the Earth is embodied in Earth models, which are not well constrained by observations and physical laws and thus open to *serious doubts*.

Given all the details mentioned in the previous paragraphs and stopping the imagination of seismologists running riot, what we do really have in regards of Earth's inner structure is an *interpreted cross section,* as it is indicated by changes in the calculated waves' velocities, under certain assumptions. Nothing more and nothing less. Based upon this information, strictly speaking Earth cannot be considered a solid or a 100% rigid body.

This cross section shows a 2900 km thick spherical *rigid shell* called mantle + crust, which is disrupted by two continent-sized unknown structures or blobs. Regarding Earth's inner structure, geophysicist Sir Harold Jeffries in his book *Earthquakes and Mountains* (1950) points out:

"It is found that we must regard the Earth as composed of a solid shell, reaching nearly half way to the centre, and resting on a liquid core capable of transmitting P but no S waves."

A solid *spherical shell* is an essential feature in the Hollow Earth Theory. Most detractors argue that Earth cannot be hollow because it would "collapse" which obviously it is not the case.

This is because this thick shell behaves like a *dome.* It can be viewed as multiple arches with multiple (gravitational) forces. This network of circular forces and lines of thrust create a net of *compression* forces over the surface of the dome or shell, which provide *resistance* to external stresses, thus avoiding collapsing.

At the centre of Earth there is an *independent* solid or semi-solid body 2400 km in diameter less rigid than mantle and spinning faster than it. This inner body, cannot be made of iron and the alleged pressure of 3,300,000 atmospheres acting upon it, would make *impossible* for this body, to spin faster than the mantle. Geoscientists couldn't figure this out but this inner solid body must be *unrestrained* with a degree of freedom in space, to spin with a different rate than the mantle.

Also, there is a friction at a solid-liquid interface (inner core / outer core). It is a 'damping' or 'dissipative' force, in part due to the viscosity of the liquid (internal friction), but also subject to other external factors such as the 'roughness' of the solid surface.

This would have dampened and stopped the spinning inner core, a long time ago. Between the mantle and this inner body, there is a 2000 km separation. However, as explained before, when we try to infer hidden properties of Earth's deep interior from surface measurements we are always faced with the problem of *non-uniqueness* of our conclusions.

Thus, this separation is filled not with liquid as it comes from the assumption of a molten iron core, but with gas as it will be explained in chapters four and five, along with the *real* cause of Earth's magnetic field.

On the other hand, in the current model, the path of P waves within Earth coming from the earthquake's origin to the seismological stations are presented, as if density increases with depth towards Earth's centre. But there is the fact that density, indeed *decreases* from a certain depth onwards as the Kola Deep Borehole shows.

In this case P waves, instead of waves curving in U to the surface, they would curve in the opposite direction. This causes P waves to travel around the core like if it were a *cavity*, right to the other side of the Earth.

At some point in the mantle, this decrease of the density causes the waves to split some to go down while others go up. So, there must be an area on the Earth's surface which receives less seismic waves than normal, hence the shadow zone. This would explain in part, why this region is not completely devoid of seismic waves.

During the 1990's new data bearing on the convection-cell *problem* have become available in the form of seismotomographic images of the

Earth's interior.

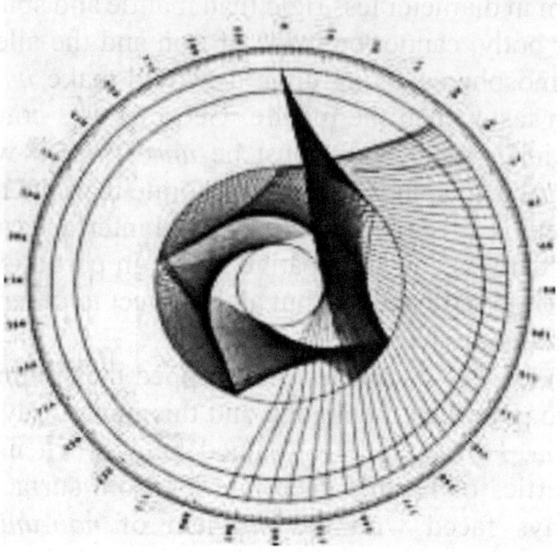

Fig. 50- Multiple reflections from P waves trapped inside the Earth's "liquid" core. The seismic phase P4Kab is a P wave bottoming in the upper outer core; ab indicates the retrograde branch of the PKP caustic. The seismic phase P4KPbc is a P wave bottoming in the lower outer core; bc indicates the prograde branch of the PKP caustic. The convoluted structure of these reflections, resembles a chamber within the Earth, surrounded by a thick rigid spherical shell (mantle).

These images show that (i) mantle diapirs as proposed by traditional plate-tectonic models *do not* exist; (ii) there is *no* discernible pattern of upper or lower mantle convection, and thus no longer an adequate mechanism to *move* plates; and (iii) the lithosphere above a depth of about 80 km is permeated by an *interconnected network* of low-velocity channels; which makes the "asthenosphere" concept a myth (Meyerhoff, 1994).

Ultimately, it must be highlighted that the *only* reliable work of seismologists, based in observables; is to locate the source, the nature and the size (magnitude) of seismic events or earthquakes.

However, certain seismologists study the relation between faults, stress and seismicity (seismo-tectonics), others interpret the mechanisms of rupture from seismic wave data (focal mechanisms), others integrate geoscientific information to define zones of seismicity (seismic zoning).

Some seismologists collaborate with engineers in an attempt to minimize the damage caused to structures (earthquake engineering) and finally others with the sophistication of seismic methods such as teleseismic delay-time methods, tomography and receiver-transfer functions; are able to determine, to some extent, local and regional crustal and upper-mantle structure with impressive resolution.

That's all they can do. Anything else that comes from seismologists, in regards to the Earth' inner structure can be only considered as a *guesswork* due to the uncertainty of their interpretation of the raw data. Developers of seismologic models stressed, that models *do not* represent real Earth (Kennett, 2019).

As an example: usually, positive seismic wave speed anomalies in the mantle were attributed (interpreted) to "cold slabs" or remnants of "subducted" tectonic plates. Seismic wave speed anomalies in the mantle have long been considered the "footprints" of these alleged tectonic plates. But a new study using full-waveform inversion (FWI) found that many of these anomalies do not spatially correlate with known "subduction" zones. Only 60% of the alleged subduction zones align with positive anomalies. This suggests that the wave speed anomalies have a *different* cause. (Schouten, 2024).

Chapter III

Revealing the true origin of the Earth and its interior, according to contemporary astronomical-astrophysical research that falsifies the old paradigm of planetary formation.

In every branch of knowledge, the progress is proportional to the amount of facts on which to build, and therefore to the facility of obtaining data.

James Clerk Maxwell (1831-1879)

Basic preliminaries

In geology, minerals, rocks and fossils can be studied and analysed in laboratory to gather some knowledge about these. But their origin and relationships with the *environment,* must be studied in the field. Any student must see this. Following this principle, we will switch to the domain of Astronomy, where Earth is viewed as a planet orbiting a star (the Sun) and follows strict physical laws and observable astronomical processes.

However, many geologists do believe that determining our planet's structure and composition (trough models), is critical for understanding the origin and evolution of Earth. But they often ignore that Earth's origin is an *astronomical process* grounded in observable facts not hypothetical occurrences.

Our Earth as a planet is bound to the same astronomical rules and organisational patterns, that gives the *raison d'être* to other planets and celestial bodies in general.

Through History, astronomy has been a hotbed for scientific debates and still is today, though with a less dramatic tone of killings, tortures, trials and imprisonment. It is important to have a briefing of what astronomy deals with, to know several key concepts and sanitize the astronomical panorama of several misconceptions.

The educative system along TV and popular science outlets, filled the reader of fancy information about the origins of the Solar System along Earth and other topics. We are going to work hard, to forget all that and get a new and more *accurate* picture, derived from the state-of-the-art of recent scientific research and new emerging paradigms.

Astronomy is the oldest science, with the first observations of the heavens conducted by our early human ancestors. Historical records of astronomical measurements date back as far as Mesopotamia nearly 5000 years ago, with later observations made by the ancient Chinese, Babylonians, and Greeks.

However, the invention of the *telescope* in the 17th century, was required before astronomy could develop into a modern science. Late developments such as *spectroscopy* in the 19th century and radio astronomy coupled with *space crafts* and space telescopes in the 20th century, improved and shaped the astronomy we know today.

Humans seek to explain their world with models; one of the earliest is that the affairs of humans and the world, are controlled by the positions of the stars and planets. Although astrology is now regarded as a pseudoscience, it was the original motivation for the mapping of the stars and the assignment of constellations.

Astronomy is more than simply a mapping of stars and planets, into outlines of gods and magical creatures. It is the scientific study of the contents of entire Universe. The study of celestial objects (such as stars, planets, comets and galaxies) and phenomena that originate outside the Earth's atmosphere (such as the cosmic background radiation). It is concerned with the evolution, physics, chemistry and motion of celestial objects, as well as the formation and development of the Universe.

Historically, astronomy has included disciplines as diverse as astrometry, celestial navigation, *observational astronomy* and the making of calendars. Professional astronomy is nowadays often considered to be identical with astrophysics. Since the beginning of the 20th century (1916), the field of professional astronomy *split* into observational and theoretical branches.

Observational astronomy is focused on acquiring and analysing data, (observables) mainly using basic principles of physics. Theoretical astronomy is oriented towards the development of computer or analytical (mathematical) models to describe astronomical objects and phenomena. When the subject is the Universe, Cosmology is an often-used term for this sort of theoretical activity.

The two fields should *complement* each other, with theoretical astronomy seeking to explain the observational results, and observations being used to *confirm* theoretical results (often these are falsified).

Trouble arises when observations do not confirm, instead *contradict* theoretical results as in the case of the Big Bang, discussed in chapter one. These conflicting observations are either *dismissed* or swept under the carpet, by the orthodoxy or mainstream scientific community.

Also, it turns out that the history of astronomy is littered with ideas that once seemed incontrovertibly right and yet later proved to be bizarrely wrong. Not least among these are the ancient ideas that the Earth is flat and at the centre of the Universe. But there is no shortage of bizarre ideas in modern era. A very common flaw of astronomers is to *believe* that they know the truth even when data is scarce or contradictory.

The subject of importance in this chapter is the Solar System where our Earth belongs to. Therefore, *understanding* how it really originated can throw light on how Earth could be formed and thus elucidate its internal structure, which is the main subject of this book. Before going into details, let's first examine the concepts of nebula, plasma and star; which are linked to Earth's formation.

Nebula: named after the Latin word for "cloud", nebulae are not only massive clouds of dust, hydrogen and helium gas and plasma; they are also often "stellar nurseries" i.e. the place where stars are born. For centuries, distant galaxies were often mistaken for these massive clouds. Such descriptions barely scratch the surface of what nebulae are and what their significance is. Between their formation process, their role in stellar and planetary formation and their diversity, nebulae have provided humanity with endless intrigue and discovery.

For some time now, scientists and astronomers have been aware that outer space is *not* really a total vacuum or emptiness. Conventionally, it is made up of "gas" (which is a misnomer) and dust particles known collectively as the Interstellar Medium (ISM). Approximately 99% of the ISM is composed of "gas", while about 75% of its mass takes the form of hydrogen and the remaining 25% as helium.

The interstellar "gas" consists partly of neutral atoms and molecules, as well as charged particles (aka plasma), such as ions and electrons. Bear in mind that this so-called gas is extremely dilute, with an average density of about 1 atom per cubic centimetre. In contrast, Earth's atmosphere has a density of approximately 30 quintillion molecules per cubic centimetre (3.0×10^{19} per cm^3), at sea level.

Plasma: it is often *underrated* in modern astronomy. It is well known that the Universe is plasma. Only a very *insignificant* part of the matter in the Universe is in the *solid state* and a negligible part is in liquid state Plasma is commonly defined as the fourth state of matter. Space is *filled* with plasma (wrongly considered as gas). In fact, plasma is the *most* common type of matter in the Universe (99%). It is found in a wide range of places from fire, neon lights, and lightning on Earth to galactic and intergalactic space. The only reason that we are not more accustomed to plasma is that mankind lives in a thin biosphere largely made up of solids, liquids, and gases to which our senses are tuned.

For example, we don't experience fire as a plasma; we see a bright flame and feel heat. Only *scientific experiments* can show us that plasma is actually present in the flame. How? The flame is a plasma but it is net neutral. If someone places electrodes in the flame and pass a current through the flame, then a magnetic field will affect it. In fact, "Plasma is a collection of charged particles that responds collectively to electromagnetic forces" (from the first paragraph in Physics of the Plasma Universe, Anthony Peratt, Springer-Verlag, 1992).

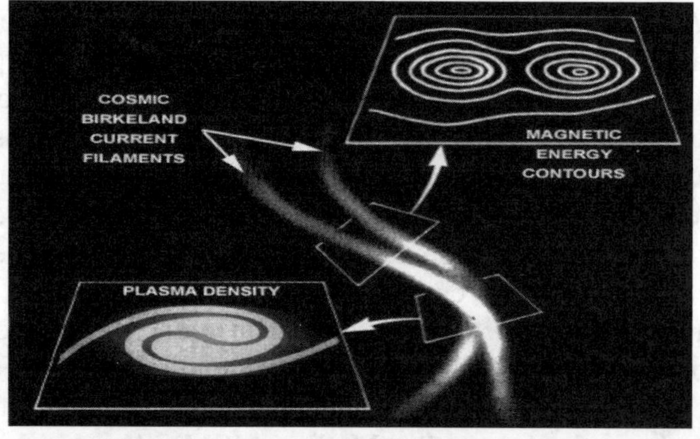

Fig. 51- Plasma Astrophysics and Plasma Cosmology are usually underrated by conventional or gravitational Astronomy, which often leads to a dead end (unsolved problems). *Dissipative structures* in hot interplanetary plasma are formed due to a resonance interaction of long-wave longitudinal perturbations with nonlinear Alfven waves propagating along the magnetic field. This causes dissipation of gravitational binding energy, decreasing its role in forming cosmic structures.

A plasma region may also contain a proportion of neutral atoms and molecules, as well as both charged and neutral impurities such as dust, grains and larger bodies from small rocky bodies to large planets and of course, stars. The defining characteristic is the presence of the free charges, that is, the ions and electrons and any charged dust particles. Their strong response to electromagnetic fields, cause behaviour of the plasma which is *very different* to the behaviour of an ionized gas. Also, all particles charged and neutral respond to a gravity field, in proportion to its local intensity. As most of the Universe consists of plasma, locations where gravitational force dominates that of *electromagnetism* are relatively *sparse*.

Notable US chemist and physicist Irving Langmuir (1881-1957), an uncommonly careful *experimentalist*; was the one who coined the term "plasma" in the 1920s. He studied the electrical properties of ionized gases. His work laid the foundation for understanding how plasma behaves under different conditions, which became *crucial* in explaining the behaviour of interstellar and intergalactic matter. Langmuir's research showed that plasma can sustain *long-range* electromagnetic interactions, making it the perfect candidate for explaining *large-scale* cosmic phenomena.

One more thing and this is of key importance: in astrophysics, plasma is sometimes referred to merely as an “ionized gas” which is ambiguous. While technically correct, this terminology is incomplete and outdated. It is used to *disguise* the fact that plasma seldom behaves like a gas at all.

In space, plasma does not simply diffuse, but *organizes itself* into complex forms or structures and will not respond significantly to gravity, unless local electromagnetic forces are much weaker than local gravity. Plasma is *not* matter in a gas state; it is matter in a *plasma state.* Again, a significant behavioural characteristic is plasma’s ability to form *structures*, large-scale cells and filaments.

In fact, that is why plasma is so named, due to its almost *life-like* behaviour and similarities to cell-containing blood plasma. However, astronomers *fail* to understand that ordinary and ionized gas; *cannot* behave in a such way, ever.

Star: the standard definition of star is a celestial sphere of “gases” (indeed is plasma) hydrogen and helium, held together by its own gravity. All stars are not identical. Stars range in age from those nearly as old as billions of years to those being born every day. As they age, they pass through many known changes in size and colour.

Small stars are called dwarfs, and large stars are called giants. The *colour* of a star is dependent directly on its *temperature*. Hotter stars are white and blue whereas cooler stars are orange and red. Yellow stars, like our Sun, have an average temperature. Stars are the most widely recognized astronomical objects and represent the most fundamental building blocks of *galaxies*.

Most stars are observed to be members of *binary* star systems, and the properties of those binaries are the result of the conditions in which they formed. The age, distribution and composition of the stars in a galaxy; trace the history, dynamics and evolution of that galaxy. Moreover, stars are responsible for the manufacture and distribution of heavy *elements* such as carbon, nitrogen and oxygen (nucleosynthesis) and their characteristics are intimately tied to the characteristics of the planetary systems, that may coalesce about them.

Stars are *believed* to be giant nuclear reactors. Nuclear fusion is an atomic reaction that apparently, fuels stars. Mainstream science believes that in the core of a star, gravity produces high density and high temperature (which is untrue, as it will be discussed later).

The density of gas in the core of our sun is 160 g/cm^3, much higher than the densest metal, and the temperature is 15,000,000 K (27 million degrees C). At this temperature, the hydrogen and helium gases become a plasma. That is, the electrons separate from the nuclei to give a mix of positively charged ions and electrons.

In fusion, many nuclei (the centres of atoms) combine together to make a larger one (which is a different element). Fusion inside stars transforms hydrogen into helium, heat, and radiation. Heavier elements are created in different types of stars as they die or explode. This process is called stellar nucleosynthesis. All of the atoms in the universe began as hydrogen. It is estimated that there are 100 billion galaxies in the Universe, each possibly containing 100 billion stars and thus, countless *planets*.

New answers to old questions

Once the concepts of nebula, plasma and star are understood, we can check out some issues that currently is facing astronomy. In fact, despite being a modern science more than four centuries old and with all the scientific and technological advances accomplished all this long, still there are "mysteries" or *unsolved problems* (gaps in knowledge) pertaining astronomy and astrophysics.

There is an acknowledge of poor understanding, regarding certain astrophysics' processes and controversial issues arise from time to time.

Of course, this is known in the academic milieu but for the general public, the media and popular science outlets informs very little and thus the overall impression is that, astronomical science is "making progress". However, the problem lies in the *paradigm* or mindset, that conventional astronomy and cosmology, are currently using when dealing with the inevitable issues in their fields.

The current paradigm means that the driving force, that runs and shapes the Universe and its components; is *assumed* as being gravity alone (Gravity Model) coupled with the materialistic-mechanistic approach, where material particles, colliding and bouncing each other, driven by randomness will "eventually" form complex structures; from planets to galaxies. Only gravity could gather clouds of gas and dust into a star or a planet. Only gravity could produce galaxies and massive clusters of galaxies.

Other variables such as *electricity* and *magnetism* for instance, are considered as playing a very little role. Also, the interstellar medium, considered before as empty vacuum, now is regarded as composed mainly by rarefied "gas" and cosmic dust. Charged particles and *plasma* have minor or no importance and overall, the Universe is considered electrically "neutral".

However, a neglected scientific discipline known as *Plasma Cosmology* has proved that indeed, in the Universe, charged particles fill all of space as electrically conductive plasma. Plasma behaves differently than gas. The Sun is plasma. Stars are plasma. Galaxies are plasma. The filaments of magnetized and radiating matter between stars and between galaxies are plasma.

The isolated islands, we once imagined in space do not exist. A web of electromagnetism *connects* planets, moons, stars and galaxies. These freely *moving* charged particles are much more strongly affected by electric and magnetic fields, than by gravity.

Electric fields across cellular boundaries in space (called double layers), can accelerate charged particles to *great* speeds and *powerfully* alter the behaviour of galaxies stars and even planets.

Traditional models claim that the force of gravity is solely responsible for the formation of stars, star clusters and galaxies. In fact, in a paper published in the journal *Astrophysics and Space Science* (1978), by Swedish physicists Hannes Alfvén and Per Carlqvist; clearly demonstrated the *impossibility* of matter condensing by weak gravitational forces alone. Instead, they show that an electro-gravitational model that involves electromagnetic condensation and compression of matter, followed by gravitational attraction is necessary.

For more than 30 years, *plasma physicists* have had an electrical model of galaxies. It works with real-world physics not fancy theories or untestable hypothesis. This model is capable to successfully *account for* the observed shapes and dynamics of galaxies *without* recourse to invisible dark matter and central black holes.

Unfortunately, conventional scientists, due to their *tunnel vision* have misread most of these “phenomena”, because of not having knowledge of jet phenomena and their *true* cause. Plasma physics explains simply the powerful electric jets, seen issuing along the spin axis from the cores of active galaxies.

Far back in 1963, Swedish electrical engineer and plasma physics pioneer Hannes Alfvén (1908-1995) *predicted* that galaxy distribution in the Universe would prove to be filamentary. Much later, it was discovered that 99% of all matter in the Universe, was in the form of plasma. Then in 1991, the first galaxy distribution survey revealed that galaxies were indeed distributed in a filamentary structure, *confirming* Alfvén’s prediction.

Of course, astronomers and cosmologists were caught by surprise. No-one, not even the brightest of them, could have expected that result as it would *not* conform to the expected distribution of galaxies, as predicted by the Gravity Model (Big Bang Theory).

On the other hand, a branch of astrophysics, known as *Astrophysical Magnetism* tells us that, contrary to what the popular science books (stressing gravity) say, magnetic fields are omnipresent in space and play a *major* role in some of the most important astrophysical phenomena.

Unlike the familiar magnetic fields on Earth, magnetic fields in space are not the passive result of electric currents, but are *dynamically active*, self-generating the currents needed to sustain them through inductive processes, in the highly-conducting plasma that fills space. Such fields are *detected* via synchrotron emission, polarisation, or Faraday rotation in the radio frequency and appear between 0.2 and 10 GHz.

These high-energy fields, power exploding stars and erupt as intensely energetic jets from galactic cores, often *misinterpreted* as black holes. Without magnetic fields, stars would form in a very different way, there would be no stellar winds, no cosmic rays, no accretion disks or jets in active binary stars and active galactic nuclei, no pulsars and perhaps no neutron stars at all; as magnetic fields seem to be essential to the Type II supernova explosions, that form them.

Nearly all the complex phenomena on the surface of the Sun are driven by magnetic fields. Modern telescopes can now view these events at high resolution and across the full electromagnetic spectrum.

Unexpected, elaborate structures have sprung to life in radio, infrared, ultraviolet, and X-ray wavelengths. These surprises of the space age, call for a re-examination of long-held theories about how the Universe works. As one notable magnetist astrophysicist stated:

"Magnetism is to astrophysics what sex is to psychology. At first everyone pretended that it didn't exist. Then suddenly it explained everything. Perhaps now we see it in its proper perspective as one of several fundamental drivers of behaviour".

Bear in mind that gravity is, what in physics is known as a *weak force*. Electromagnetic force is 1036 times stronger than gravity. A huge difference. However, conventional astronomers and space scientists would say that, on the very large scale; as in astronomical systems, the gravitational pull is the "dominant force" determining the motions of moons, planets, stars and galaxies. They deny that there is anything electrical in outer space at all, the Universe is electrically "neutral".

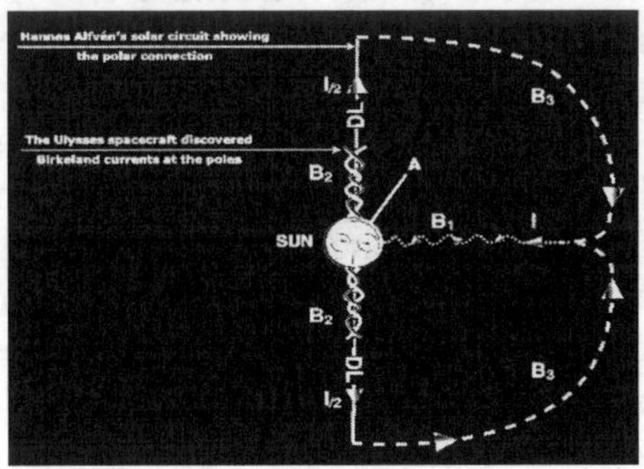

Fig. 52- Plasma physicists Kristian Birkeland and Hannes Alfvén managed to show that there is a sophisticated electric and magnetic structure in the cosmos. For example, Alfvéns Heliospheric Circuit model shows the Sun as having a field much like that of a magnet, as does Earth. (In fact, his model has been confirmed lately by observations of spacecrafts such as Ulysses). The Sun acts as a unipolar inductor and B is the Birkeland current.

Besides ignorance, this is due to the *misuse* of the Law of Conservation of Charge and where and how, it is supposed to work. This law *only* applies for isolated systems.

If outrightly applied, this law would mean an equal balance between positive and negative charges in the Universe, therefore there is no excess of any that would generate electric and magnetic fields in space. This *contradicts* the experience of the wide range of phenomena, that plasma cosmology and astrophysical magnetism, deal with on a daily basis.

Conventional astronomers don't have a clear distinction or knowledge where and how, positive and negative charges are involved and what is then, the total charge to be conserved. Therefore, a neutral charge for the Universe is *wrongly* assumed to favour the Gravity Model.

Consequently, conventional astrophysicists and cosmologists *never* contemplate an electrically-magnetically driven model, because they *assume* strict electrical neutrality throughout the Universe. Things got so far that even this electrical astronomical approach, is considered "pseudoscience" by many, despite the overwhelming research, a Nobel laureate (Alfvén) and compelling evidence proving *otherwise.*

An example of how scientific research disproves this misconception of an "neutral" Universe, is the work of US astrophysicists J.H. Taylor and J.M. Cordes in 1993, who have studied the distribution of electrons in interstellar space in our galaxy. Because electrons repel each other, we might expect this distribution to be widely dispersed spherically but this is not the case.

Careful *observation* shows, electrons keep closely to the disk and are concentrated in the central bulge of the galaxy and in the spiral arms, where most of the stars are concentrated. The diagrams of Taylor and Cordes are quite clear on this point. This observation presents a strong evidence in favour of *electrically charged* celestial bodies in the Universe, because the explanation is that negative electrons are attracted to positively charged stars.

The Sun has a positive charge. In the corona of the Sun, oxygen ions O^{+5}, are accelerated radially away from the Sun faster than protons. Conventional astronomy *fails* to explain this "mystery". Since oxygen ions are 16 times heavier than protons, this unexpected phenomenon cannot be explained by simple kinetic diffusion that can overcome the gravity and the escape velocity of the Sun. The explanation is that the oxygen ions, with the higher positive charge are being accelerated away, by a positive *electric field* from the Sun.

On the other hand, a shocker for the "neutralists" advocates of the Universe came in 2011, when a cosmic jet 2 billion light years away was carrying the highest *electric current* ever seen: 10^{18} amps, equivalent to a trillion bolts of lightning. Canadian physicist Philipp Kronberg of the University of Toronto in Canada and colleagues, measured the alignment of radio waves around a galaxy called 3C303, which has a giant jet of matter shooting from its core. They saw a sudden change in the waves' alignment coinciding with the jet. "This is an unambiguous signature of a current," said Kronberg.

This team *believes*, as it is usual in conventional astronomy, that magnetic fields from a colossal black hole at the galaxy's core are generating the current, which is powerful enough to light up the jet and drive it through interstellar gases out to a distance of about 150,000 light years.

However, a lack of knowledge often means a lack of vision. Moving charged particles constitute an electric current, and that current is wrapped in a magnetic field. When more charged particles accelerate in the same direction, the field gets stronger. That is a familiar idea to electrical engineers, but when astronomers find moving charges in space; they are baffled and refer to them as "winds" or "shock waves." In fact, the "solar wind" is indeed a Birkeland current.

Something else not considered when researchers attempt to explain structure in the Universe, is that for charged particles to move, they must move in a *circuit*. Energetic events cannot be explained by local conditions alone. The effects of an entire circuit must be considered. There is a connectivity with an electrically active network of "transmission lines" composed of Birkeland current filaments. Space probes *confirmed* this situation in our Solar System.

Filaments expand and explode, throwing off plasma that can accelerate to near light-speed. Jets from opposite poles of a galaxy, end in energetic clouds emitting X-ray frequencies. These phenomena are based in *plasma science* supported by extensive research and *not* gas kinetics, gravity or particle physics. Astrophysicists see magnetic fields but not the underlying electricity, so they are at a loss to explain them.

They don't understand the *Gestaltung* that drives these powerful jets, emitted by galaxies. This is understandable because as any scientist, they have developed a tunnel vision in their specialization, which makes them unable to see the whole picture of what *really* is going on. There is more: "One of the most profound mysteries of the Universe, consists in the fact that everything in it rotates", wrote the English astronomer Sir Arthur Eddington. Indeed, regardless of whether it spins clockwise or counter clockwise, *everything* in the Universe within a wide range of degree; moves and *spins*:

From small asteroids to entire galaxies. There is some recent evidence that the Universe itself may be *rotating* too. Gravity, electromagnetism, momentum and inertia; ensure that bodies big and small act upon each other, causing everything to move and *spin*.

However, conventional astronomy has given little attention to the *rotational physics* of celestial bodies and its key importance in explaining, some "mysterious" phenomena as we shall see in chapter four. Also, another neglected factor is the key concept of *vortices*, which has just recently been considered, to explain planetary formation; in a far more effective way than the standard and deficient hypothesis of accretion and collisions. A *vortex* is a region in which a turbulent circular movement of fluid particles, flows in a *spiral* toward the vortex centre around an instantaneous axis.

Fig. 53- Fluid vorticity, defined as the curl of the fluid velocity field ***u***, is essential to understanding early Universe structure formation. For a charged fluid, the definition of vorticity is generally extended by adding the magnetic field to the fluid vorticity, which can be called the generalized vorticity.

The speed of fluid flow is fastest at the centre of the vortex, and decreases with distance from the vortex centre. Tornados and whirlpools are examples of vortices. Differences in pressure and/or density are the main causes of vortices. The whole physical system of our Universe, is essentially governed by *flow dynamics* or the physics of turbulence.

Turbulent systems are characterised by the eddies, currents and vortices of “flow”, which *distributes* and *organizes* material. Vortices are self-sustaining *Energie Gestalten* (energetic structures), that occur everywhere naturally expressing minimum energy and stable configurations.

As we can see there are three main players in the cosmic scene that gives origin, sustaining and ultimately an end to each one of the components of the Universe. This can be called The Cosmic Trinity: *gravity, electromagnetism* and *rotational physics/vortices*. There are two competing forces which, depending of the circumstances, might overcome or balance that of gravity.

Nevertheless, brainwashed by the role of gravity in the cosmos, astronomers *failed* to recognize the pervasive role of charged particles and electric currents in space, also the influence of spin/vortices. Therefore, their model of how the Universe works, based only in gravity is *incomplete*, thus despite it may explain several well-known facts, some of its predictions had failed and new facts or phenomena; show that the current Gravity Model is *flawed.*

However, to bridge the gap between theory and observed facts, mainstream scientists introduce weird concepts and hypothesis *ad hoc*, to “improve” the flawed model, instead of abandoning it and make a new and reliable model based in all the *observed facts*, as the scientific method clearly indicates.

Gravity alone, the *weakest* universal force is *not* enough, to account for all we see in the Universe. It had been realized the importance of applying electromagnetism and plasma physics to the *problem*, faced by conventional astronomy/astrophysics, in figuring out the radiogalaxy and galaxy formation; derived from the fact that the Universe is largely a *plasma Universe*.

An unnecessary and misleading hypothesis

It is noteworthy to mention that the law of gravity, which relies exclusively on the masses of celestial bodies and the distances between them, works very well for explaining planetary and satellite motions, *within* our Solar System. All the ingredients that enabled to elaborate the law of gravity, came from our Solar System or better said: from the *particular distribution* and values of masses (planets, Sun) in the Solar System. But when astronomers tried to apply this law to galaxies and clusters of galaxies, it turns out that nearly 95% of the mass necessary to account for the observed motion is missing.

For instance, a study of radial velocity plots (radius from the centre versus stars' speed of rotation) for stars in the Milky Way galaxy; revealed that the speeds flatten out rather than trail down.

This implies that velocity continues to increase with radius, contrary to what Newton's Law of Gravity predicts for and which is observed in the Solar System. This along other anomalies suggested that much more mass than could be seen, was required to keep the Gravity Model valid. The notorious concept of "dark matter" was born.

According to mainstream science, we only perceive barely a 5% of the Universe's mass. The rest is invisible or undetectable by any current instruments, is optically inert, does not cause scattering or shadows, has no electrical or magnetic properties, and does not cause friction. Its distribution is *arbitrarily* assigned wherever it is required, by observed deviations from theoretical expectations.

This is a foul play of convenience instead of science. In other words, is pseudoscience. It sounds absurd that in order to explain the behaviour of the matter we can detect, the Gravity Model needs to imagine twenty-four times more matter than we can see, in special locations, and of a special *invisible-intangible* type.

One must wonder why anyone, would adopt such a strange and *unnatural* concept and take on its formidable problems. An unprejudiced and sincere scientist, would rather claim the honest *Ignoramus Amendment*, the "we do not know (but may, in the future)".

It seems much more reasonable and scientific, to reconsider if this hypothesis is wrong or investigate whether *other* factors, the known physics of electromagnetic forces and electric currents, can bring about the observed effects; instead of having to invent what doesn't exist (*ad hoc* hypothesis). This is also a violation of one of the principles used in the scientific method as stated by Newton: "We are to admit no more causes of natural things, than such as are both true and sufficient to explain their appearances". Plain and simple.

In fact, as the study of Taylor and Cordes showed, the stars have a *positive charge* and this includes the stars in the central bulge; it follows that the negative electrons, are attracted to some extent towards the centre of the galaxy, i.e. there is an electric field which decreases from the centre outwards. There is an attractive force between this negative electric field and the stars circulating in the galaxy. The explanation is put forward here, that it is this attractive (electric) force which is causing the characteristic star velocities in a galaxy and this velocity effect is NOT caused by the gravitational force of the 'missing mass'. We need to consider some aspects of the "missing mass" theory in contradistinction to the new hypothesis. In most galaxies, but *not* all galaxies, the orbital velocities of the stars and the inter stellar gas are approximately *constant* over a substantial range of radii of orbits.

They are said to have a 'flat curve'. This velocity pattern is contrary to Kepler's Third Law, which when velocities are translated as periods. It says: "the square of the period of any planet is proportional (not constant) to the cube of the semi-major axis of its orbit". Therefore, this *anomaly* is explained by an "extra" gravitational force provided by a "missing mass" or "dark matter", that keeps constant these orbital velocities.

However, the missing mass theory *fails* on its own terms, because if the orbital velocities were characterized by gravity, the rotation curve should be *smooth* in the way that planetary orbital velocities make a smooth (elliptic) curve and not the *scatter* as the graphics of the Taylor and Cordes study shows. These stars are *not* on a smooth curve; they would be on a smooth curve if their velocities were determined purely by gravitational considerations.

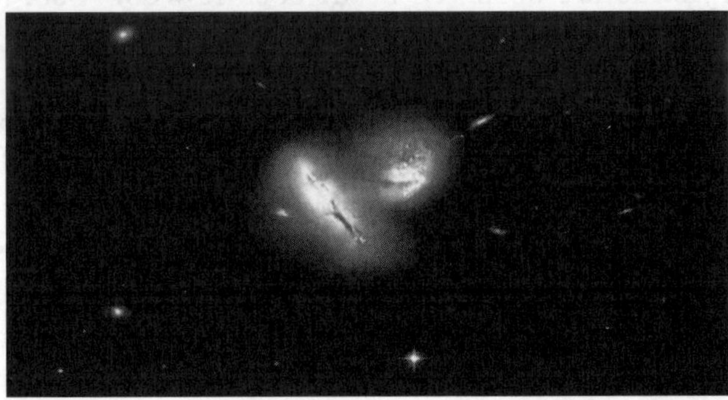

Fig. 54- The current cosmological model only works by postulating the existence of dark matter, a substance that has never been detected, but that is supposed to constitute approximately 95% of all the universe. But a simple test suggests that dark matter does not in fact exist. If it did, we would expect lighter galaxies orbiting heavier ones to be slowed down by dark matter particles, but we detect no such slow-down.

Indeed, there is considerable scatter or dispersion. This contrasts with the smooth curve of the orbits of planets, whose velocities are

Keplerian, i.e. determined by the laws of gravity. The scatter of star velocities, comes about because the stars are electrically *positively charged* in varying degrees and *electrical forces* are added to the gravitational force in *controlling* the orbits. Whereas in gravity, the mass of a celestial body is *constant* thus orbital velocities, follow Kepler's laws and are decreasing smoothly at the inverse of the distance squared. Electrical currents in space are *variable* and follow erratic patterns thus electrically charged star velocities will appear dispersed in the graphs, counteracting the gravity pull.

On top of that, in 2019 as the second galaxy astronomers have found *without* any dark matter, the new finding named NGC 1052-DF4 (DF4 for short) *confirms* that the first discovery, NGC 1052-DF2 (DF2 for short), was not a *mistake*. Upon its discovery, DF2 was indeed a huge *surprise*, and it came as tough news in the current ideas about galaxy formation and dynamics, because "dark matter" is a *vital* part of the conventional and flawed understanding of galaxies.

In the Lambda Cold Dark Matter model (dogma), galaxies form primarily through the "gravitational collapse" of dark matter halos, followed by the accumulation of baryonic matter (ordinary atomic matter). However, this model *struggles* to fully explain the observed distribution of galaxies, their rotation curves, and the intricate filamentary structure observed in the cosmic web. Also, it is *useless* to "explain" the formation of spherical galaxies.

The *key* feature of these two galaxies is that they are *spherical* and the *ad hoc* hypothesis of dark matter, originated from *spiral* galaxies. Indeed, the problem lies in not understanding or ignoring the *Gestaltung* of a galaxy. Astronomers assumed *all* galaxies would behave gravitationally, in the *same* way, regardless of the *shape*.

In other words, the astronomers did fall in the logical fallacy of *oversimplification*, discussed in chapter one. Spherical galaxies act as *point masses* and have spherical (uniform) gravitational fields. Pretty much like the Sun, Earth and planets, were these are considered point masses for orbital calculations and Keplerian laws do apply. On the other hand, spiral galaxies *don't* act as point masses.

These have *complex* shapes creating very complex *nonspherical* gravitational fields. If one considers, the nonspherical shape of the spiral galaxy, the velocity of the edge stars *matches* calculations using Newtonian gravity with the *existing mass* of the stars (Cameron Regipsol, 2016). No dark matter is needed, when the job is done properly. Bear in mind that this procedure has absolutely *nothing* to do with the Modified Newtonian Dynamics (MOND), which suggests that Newton's law of gravity, becomes "irregular" when the gravitational pull is very weak, as is the case in the outer regions of the galaxy.

Either way, whether due to electrical currents or misapplication of Newton's gravity law or both; dark matter is *unnecessary* and very *misleading*. Indeed, to believe in dark matter is like to believe in flying pigs.

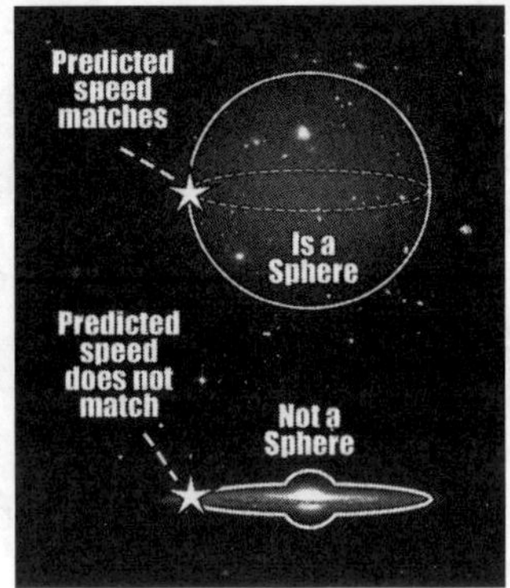

Fig. 55- The invention of dark matter owes its existence to the failure by mediocre astronomers, to recognize that most galaxies don't act as point masses, but are complex shapes creating complex, non-spherical gravitational fields. The space industry recognizes the complicated gravitational fields in our own Solar System, when sending probes into planets, moons asteroids and comets; to guarantee successful missions.

Another *myth* ingrained in conventional astronomy/astrophysics is the concept of "dark energy," introduced in 1998. Dark energy is a huge problem, allegedly it accounts for about 70% of all the energy the Universe, and scientists have absolutely no idea what it is. The issue was that galaxies, while keeping their form as individual structures, had been *interpreted* to be moving away from one another at an accelerating rate, so the Universe is not "expanding" at constant speed.

Scientists jumped to this conclusion on measurements of the distances to supernova explosions in distant galaxies, which *appear* to be farther away than they should be if the Universe's expansion were not accelerating. How could a force be accounted for that would cause this; what could make this happen if there already was a binding force (gravity) to hold things together? It was declared (without proof) that a new 'anti-gravity' force was required. This force, when dominant, would act to push bodies away from one another and therefore, accelerating expanding space could be "explained". The solution for this was to postulate an *invisible*, *undetectable* and *untestable* dark energy.

Dark energy in the same fashion as dark matter, is inferred on the basis of circumstantial evidence, but "science says" this mysterious unknown stuff had to be real. Why? because, mediocre astronomers and astrophysicists swear that without dark matter/energy, the behavior of stars, planets, and galaxies would be "inexplicable." The problem is that *inference* is NOT proof. Conventional astronomers/astrophysicists cannot just think something up, out of the blue, to accommodate a prevailing theory. That's not science. That's not the way it works. That's simply *unfounded* speculation.

Let's suppose that little gnomes come in the night and do strange things in my garden. Next day, the upturned chair in the garden, is *inferred* "proof" of the existence of little gnomes. I first postulate that the little people exist. Then I "predict" that strange things will be observed. When I see the upturned chair, I offer this "observational consequence" as proof I was right. The gnomes really do exist and I claim I have proved it.

Can you see the *fallacy* of this reasoning? It could have been a gust of wind or the neighbours' dog that upturned the chair. This is how these "scientists" think they are understanding the way Universe works and The Nobel Committee is stupid and incompetent enough, to award them the prize. A hypothesis is falsifiable *only* if there is a conceivable *test* to which it can be put, such that a negative result would *disprove* the hypothesis. Neither dark matter nor dark energy can comply with a conceivable test.

Conventional astronomy/astrophysics tells us not to believe in things that we can *observe* and *measure*, as Plasma Astrophysics and Plasma Cosmology does on a daily basis and can properly explain several features of the Universe, which mainstream science is unable to do so (the so-called unsolved problems).

We are forced to believe in the existence of invisible and untestable entities, for the simple reason that the inferred "hypothesis" of conventional astronomy/astrophysics, require these *fictions*. They ignore as we saw earlier, that there is *more* than just gravity, intervening in the ordering of the Universe. Also, these astronomers don't consider that their working premises are *flawed.* In fact, dark energy, and its cousin dark matter, are not showing up in any empirical tests.

Moreover, a recent study: *Supernovae evidence for foundational change to cosmological models* (Seifert, *et al*) published in *Monthly Notices of the Royal Astronomical Society* (2025), shows that dark energy doesn't exist. This study uses an improved analysis of supernovae light curves to show that the Universe is "expanding" in a more varied, lumpier way. The new evidence supports the "timescape" model of the so-called cosmic expansion, which doesn't have a need for dark energy because the differences in stretching light aren't the result of an "accelerating" Universe but instead a consequence of how we *calibrate* time and distance. Dark energy is a *misidentification* of variations in the kinetic energy of expansion, which is not uniform in a Universe as *lumpy* as the one we actually live in. However, lots of money is being *wasted* in futile space programs such as ESA's Euclid mission launched in 2023, purportedly designed for mapping the extragalactic sky for years and provide unprecedented data, to give "new insight" on the nature of dark energy and dark matter.

Several years before the Euclid endeavour, another mission with the same purpose was launched: eROSITA (extended ROentgen Survey with an Imaging Telescope Array). Its driving science was the detection of large samples of galaxy clusters up to redshifts $z > 1$ in order to study the large-scale structure of the Universe and test cosmological models including dark energy.

Apparently, eROSITA's venture wasn't good enough and the Euclid mission had to be launched. Probably it would not meet the expectative and another mission will have to be scheduled and so on. This shows us, how chasing *unicorns* can be a profitable *business*, promoting careers and supporting livelihoods.

The Solar System

Once we have dealt with some astronomical preambles, we can proceed with the origin of the Solar System and therefore, the Earth. We now know the essentials to understand the Solar System. It comprises the Sun, planets, moons and an assortment of debris. The planets and debris orbit the Sun. Larger chunks of debris are known as asteroids or minor planets. Debris that contains volatile material become comets, if their orbit brings them closer to the Sun.

The Sun is the richest source of electromagnetic energy (mostly in the form of heat and light) in the Solar System. The Sun's nearest known stellar neighbour is a red dwarf star called Proxima Centauri, at 4.3 light years away. The whole Solar System, together with the local stars visible on a clear night, orbits the centre of our home galaxy, a spiral disk of 200 billion stars we call the Milky Way.

Planets are divided in terrestrial and Jovian. The terrestrial planets are the four innermost planets in the Solar System: Mercury, Venus, Earth and Mars. They are called terrestrial because they have a compact, rocky surface like the Earth's.

The planets Venus, Earth, and Mars have significant atmospheres while Mercury has almost none. Jupiter, Saturn, Uranus, and Neptune are known as the Jovian (Jupiter-like) planets, because they are all gigantic compared with Earth and they have a gaseous nature like Jupiter's. The Jovian planets are also referred to as the gas giants, although it is believed that some or all of them, might have small solid cores.

The planets, most of the satellites of the planets and the asteroids revolve around the Sun in the same direction, in nearly circular orbits When looking down from above the Sun's north pole, the planets orbit in a counter-clockwise direction. The planets orbit the Sun in or near the same plane, called the *ecliptic*.

The Sun contains 99.85% of all the matter in the Solar System. The planets, which originated out of the same disk of material that formed the Sun, contain only 0.135% of the mass of the solar system. Jupiter contains more than twice the matter of all the other planets combined.

It is noteworthy to mention two interesting things that happened in our Solar System, in the 1990's:

First: *Comet Shoemaker-Levy 9 (SL9),* which was discovered by a US couple: geologist Eugene Shoemaker and astronomer Carolyn Shoemaker along Canadian astronomer David Levy in 1993. Shortly after its discovery, it was determined to be in a highly elliptical path near Jupiter and on a collision course. SL 9 passed by Jupiter within the Roche limit, then it was broken into at least 21 separate fragments which were dispersed several million kilometres along its orbit.

The size and mass of the original body and the individual fragments, were estimated ranging from 2 to 10 km in diameter for the original body and from 1 to 3 km for the largest fragments. It was the first collision of two big-sized Solar System bodies ever to be recorded, in modern times.

Between 16 July 1994 and 22 July 1994, the fragments impacted the upper atmosphere of Jupiter. It produced different visible or detectable effects such as giant fireballs, plumes that rose to an altitude of 3300 kilometres, debris fall-out which created gigantic dark stains reaching the enormity of four times the size of Earth, all this, as well as the effects in the infrared, ultraviolet, x-rays, and other observations, less spectacular but by no means less important. This was the first time that scientists, had an opportunity to witness a *cataclysmic collision* of two extra-terrestrial bodies within our Solar System. The impacts were observed by virtually every large ground-based observatory, thousands of small and amateur telescopes, and several spacecrafts, including Hubble Space Telescope and the probe Galileo.

This collision between SL9 and Jupiter, *demolished* the scientific belief that our Solar System is a quiet and orderly place, where planets and other smaller objects or leftovers; have been orbiting smoothly for eons, since the beginning and nothing could alter significantly this order. However, besides the common meteorite collisions, sudden and considerable changes: *catastrophic cosmic events* might occur that considerably affect or even modify the Solar System components.

Many ancient cosmological "myths" referred to a battle in the sky, in which the planet god slays a sky monster usually a dragon or a serpent, could be based on similar facts observed when SL9 collided with Jupiter in 1994.

In general, comets have inspired dread, fear and awe in many different cultures and societies around the world and throughout time. They have been branded, superstitiously, with such titles as "the Harbinger of Doom" and "the Menace of the Universe" for a good reason.

Second: The discovery of *extrasolar planets*. Until the mid of the 1990's, there was a hypothesis of the origin of the Solar System looking good enough, that astronomers thought "it simply had to be true." It was the core accretion hypothesis (the updated Laplace Nebular Hypothesis).

It looked nice because it fit well with Darwinism's slow, gradual accumulation of infinitesimal changes. It was nice too, because it "explained" several features of our Solar System's arrangement: rocky planets near the Sun, icy bodies farther out.

Core accretion hypothesis predicted that any system of 'exoplanets' around another star, would look pretty much the same. It even made predictions of how the exoplanets, would behave. The only thing left was an opportunity to *test* these predictions. Then, astronomers started finding those exoplanets and they looked *nothing* like those in our Solar

System. Gas giants the size of Jupiter whipped around their stars in tiny orbits, where core accretion hypothesis said gas giants were impossible.

Other exoplanets traced out wildly elliptical orbits. Some looped around their stars' poles. Planetary systems, it seemed, could take any shape that did not violate the laws of physics. Some of the latest batch of extrasolar planets have orbits tilted so extremely that they're moving *retrograde*, that is, going around in directions opposite the way their stars are spinning.

It's fair to say that these discoveries have literally turned the long-held view of *how planets form* on its head. The findings have triggered controversy and confusion, as conventional astronomers *struggle* to work out what the old theory was missing.

They are trying ideas, but are still far from sure how the pieces fit together. The field in its current state is *unable* to give a reliable account of how our own Solar System fits into the grand scheme of things, let alone predict what else might exist. As the following statement says:

"We are finding planets that are two or three times the size of Jupiter, and Jupiter is the size of a small star so we are finding planets as big as stars... We have found that planets can get bigger by some process and we don't understand that process...What that tells you is that our concept of how planetary systems, based on how our own Solar System is put together, is probably not applicable for many of these other solar systems" – Source: Kepler Mission

Flaws in the conventional hypothesis of the origin of the Solar System

The origin of the Solar System is an issue if we follow the flawed Gravity Model. Despite the vast amount of new information gathered about the Solar System in recent years, the basic "explanation" of how it began has remained the *same*. However, it is not impossible to figure it out and get to known how our planet formed; if we get the full picture.

Earth our home, is the third planet, orbiting at an average 150,000,000 km from the Sun and so far, the only place in the known Universe, confirmed to host life. Earth is part of other celestial objects (such as stars, planets, comets and galaxies).

To grasp an idea of Earth's origin and thus its internal structure from an astronomic perspective, the starting point will be: "the present is the key to the past" or *actualism*; a slogan used in geology which is applicable to other sciences as well.

Physical laws are practically constant in time and space and *current observable* processes, should be consulted before resorting to unseen or *hypothetical* processes. It necessarily follows that all past processes, acted at essentially their current rates thus past astronomical events such as planetary formation, can be explained by phenomena and forces *observable today*.

Therefore, it would be totally absurd to postulate something that is not observable nor testable in the present and then, pretend that in a distant past it has happened and thus, it left some connection or "traces" with the observable facts we are dealing with in the present. The idea is to *observe* how a star and a planetary system, *actually* forms and develops, to get clues of how Earth originated as it will be discussed through this chapter.

The current official explanation of how Earth along the Solar System formed and got its internal structure, follows a modified version of the "Nebular Hypothesis" originally proposed by notable French mathematician Pierre Simon Laplace in 1796, which is mechanistic at its core. It is better known as The Solar Nebular Disk Model (SNDM) which is the widely-accepted (though is wrong), *paradigm* for Solar System formation.

According to this hypothesis, the Sun and all the planets of our Solar System, began as a giant cloud of molecular "gas" and dust (nebula). Then about 4.5 billion years ago, something happened that caused the cloud to "collapse". This could have been the result of a passing star or shock waves from a supernova explosion. Whatever it was, the result ended up in a "gravitational collapse" at the centre of the cloud. In other words, the origin of the Solar System happened by chance.

From this alleged collapse, pockets of dust and "gas" began to collect into denser regions. As the denser regions pulled in more and more matter, conservation of angular momentum caused it to begin rotating, while increasing pressure caused it to heat up. Most of the material ended up in a ball at the centre, while the rest of the matter flattened out into disk that circled around it.

While the ball at the centre formed the Sun, the rest of the material would form into the protoplanetary disc. It is *believed* that collisions were common in the young Solar System and most of planetary formation can be "explained" by the dramatic cosmic impacts, early in the history of the Solar System. However, this scenario *doesn't* follow the scientific method, it cannot be tested therefore it is *pseudoscience*.

The story goes like this: planets formed by "accretion" from this disc, in which dust and gas gravitated together and gradually "coalesced" or collapsed to form ever larger solid bodies with cores. Core accretion means that a core forms first and then accumulates lighter elements. The Solar System's planets grew as small grains colliding chaotically, coalescing into bigger ones; colliding yet more until formed planetesimals. The planetesimals then collided until they "formed" planets as varied as the Earth and Jupiter.

Due to their higher boiling points, only metals and silicates could exist in solid form closer to the Sun and these would eventually form the terrestrial planets of Mercury, Venus, Earth and Mars. Because metallic elements, only comprised a very small fraction of the solar nebula, the terrestrial planets could not grow very large. In contrast, the giant planets (Jupiter, Saturn, Uranus and Neptune); formed beyond the point between the orbits of Mars and Jupiter where material is cool enough for volatile icy compounds to remain solid (i.e. the Frost Line).

The ices that formed these planets, were more plentiful than the metals and silicates that formed the terrestrial inner planets, allowing them to grow massive enough to capture large atmospheres of hydrogen and helium. Leftover debris that never became planets congregated in regions such as the Asteroid Belt, Kuiper Belt, and Oort Cloud.

Fig. 56- If the nebular hypothesis were true, astronomers should observe clouds of debris elsewhere in the galaxy collapsing, as the solar nebula did. No astronomer has ever observed the process of cloud collapse or clumps in clouds that have been observed. None of the clumps in the clouds observed are gravitationally bound (collapsing). Because the clumps are so far from being gravitationally bound. The clumps are in fact, dissolving or expanding.

The Solar Nebular Disk Model (SNDM), includes four stages of formation: First, an accretion stage; second a planetesimal formation stage; third a planetary core (planetary embryos) stage; and fourth a planetary migration stage. The planetary migration stage is necessary because, according to this model and as weird as it might sound; once planetary cores have formed they are in the wrong places to resemble a planetary system so planetary cores, must be made to "migrate" to their proper location, near the Sun.

But this raises another question: *what* stops the planet from ploughing into its parent star? Protoplanets are not sitting stationary in the gas disks as they bulk up. Due to gravitational interactions with the disks, the protoplanets swirl rapidly inwards toward their central stars in what scientists call "Type 1" migration.

Models predict that this death spiral can take as little as 100,000 years. However, the *huge* amounts of energy, that would require a protoplanet to "migrate" are often neglected.

From the surface this hypothesis may look somehow convincing, but the *least* understood process is how planets are "formed" which is precisely the point of this chapter. Overall, under scrutiny this hypothesis has several *inconsistencies* and has been severely criticized in the past.

Bear in mind that the gravitational collapse alone, *cannot* explain the origin of the spin in the Solar System in the first place. Spin originates conservation of angular momentum, not the other way around as the SNDM lead us to believe. Gravity only works radially to the centre, not tangentially to it.

Theoretically, in a perfectly randomized system (gas cloud) there would be torques in all directions as the collapse continued, and overall torque on a perfectly centred mass, would even out to zero. Therefore, no torque and no spin. A star produced in this kind of collapse, would have no spin and therefore *no planets*.

The "collapsing" of the nebula or dust cloud is another *problem.* To begin with, gravity does *not* work on individual atoms, molecules and particles of dust very well. Never was and never will. In the vacuum of space, gravity has very little *control*.

It is much more likely that the pressure of the gas would cause the material to *disperse* into space (Kinetic Theory of Gases). This is why Mars has a very little atmosphere and the Moon has none. The planet and the Moon do not have sufficient gravity to hold the atmosphere.

So how does one get molecules like helium, argon, iron and gold; to coalesce together to form larger particles until those particles are large enough to be visible? How do these visible dusty particles continue to "attract" (instead of dispersing) one another and more importantly; stick together without some glue? Nevertheless, the SNDM still is the official version currently accepted by academics and scientist in general.

The condition for a cloud of interstellar gas to "collapse" and form stars, was first formulated by English physicist and astronomer James Jeans in 1902, while working at Trinity College, Cambridge; though it was partly based in the work of US astronomer J. H. Lane in the late 19th century, who used the ideal gas law to estimate the internal temperature of the Sun.

Jeans treatment considers *only* gravitation and some basic thermodynamics but *ignores* other important effects discussed before such as rotation, turbulence, internal macroscopic velocity gradients, plasma and magnetic fields.

Notice also that Jean's approach is an *oversimplification* of the real situation, not only because he has ignored the other factors mentioned before, but also because he has neglected any external pressure on the cloud such as the external pressure of an encompassing giant molecular cloud (GMC), on an embedded dense core or proto-sun. Another problem whit the Lane and Jeans approach, is with the argument that as a gaseous mass "collapses" on itself, its temperature will rise; this is known as Lane's Law. However, this notion is directly *violating* the First Law of Thermodynamics.

This often-neglected law, clearly states that stars *cannot* be self-gravitating masses, as the system *cannot* do work on itself and increase its own temperature. But as Lane and other theoretical physicists claim, this is exactly what is supposed to happen which is a clear and huge *blunder*.

Technically, this increase of system's temperature is known as negative work (-W) and it *only* can be done on this gaseous mass, by its surroundings. An *outsid*e energy, must enter the system to perform a negative work. But in this case, according to astrophysics, the system essentially undergoes "self-compression", the temperature increases with decreasing radius.

Furthermore, the fundamental equation of a star collapse that was introduced by Eddington, Jeans and Chandrasekhar which is based on this blunder, also constitutes a *violation* of thermodynamic principles.

Therefore, these astrophysical equations are flawed and scientifically invalid. These have *misled* astrophysics to this day.

The so-called "gravitational collapse" that allegedly do occur in nebulae and stars is a *myth*. It has no scientific basis whatsoever as many scientists do believe. Gravitational collapse is not a *Gestaltung* working feature of Nature.

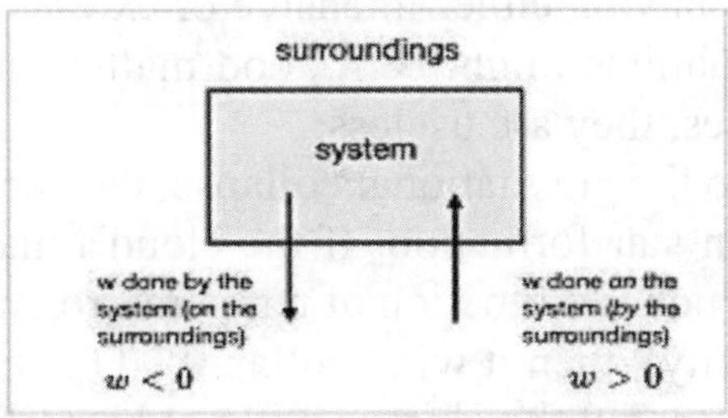

Fig. 57- The First Law of Thermodynamics deals with the conservation of energy, stating that any changes in the total internal energy of the system such a nebula or star with mass *m*, must be due to exchanges of either heat or work with the surroundings. For a "collapse" to occur, the system must gain energy. This energy is transferred to the surroundings in the form of work which modifies the internal energy of the nebula or star and *cannot* be done by the system itself (self-gravitation).

Bear in mind that if application of Thermodynamics shows us, that a particular physical process or chemical reaction is *impossible*, there is no point in attempting to make it proceed. However, astronomers and theoretical physicists, often ignore this rule. And guess what? gravitational collapse *is* the (flawed) basis of singularities, "neutron stars" and the overhyped black holes.

On the other hand, let's examine what in Thermodynamics extensive and intensive properties are known for: *Extensive* properties, such as mass and volume, *depend* on the amount of matter being measured. *Intensive* properties, such as density and colour, *do not* depend on the amount of the substance present. Any equation involving thermodynamic properties of a *system*, must have extensive (or intensive) properties on *both* sides. For example:

The ideal gas equation:

$$\mathbf{PV = nRT}$$

For a specific gas (s), P can be expressed in terms of intensive properties:

$$\mathbf{P = \rho_0 R_s T}$$

This equation is physically sound because it has *intensive* properties on *both* sides (P and T). It is valid for many gases in many situations. That's why this equation and what it represents, has a wide range of *industrial applications*. Whereas several equations used in astrophysics don't comply in having either intensive or extensive properties on *both* sides. These equations might be a good mathematical exercise but for practical purposes, they are useless.

In regards to the gravitational collapse, the concept of Jeans mass is currently used in star formation. If the cloud's mass is larger than this critical mass (which is a function of temperature, mass of the molecules, and overall density), then it will "collapse". Otherwise, it will continue to swirl and clump, but the clumps will not be permanent, and they will dissolve in the cloud. This process is given by the equation:

$$M_J = \left(\frac{5kT}{Gm}\right)^{3/2}\left(\frac{3}{4\pi\rho}\right)^{1/2}$$

$$\text{if } M_{\text{cloud}} > M_J \quad \rightarrow \text{collapse!}$$

Where M_j is the cloud's mass, T is temperature, ρ density, m the mass of the particles comprising the gas. G, k and π are constants. We clearly see that in one side of the equation, M_j is an extensive property whereas the other side T and ρ are intensive properties. Extensive properties *cannot* depend on intensive properties and vice versa, it is a *violation* of the basic rules of Thermodynamics.

Therefore, this equation is *invalid* and does not represent any physical phenomenon of Nature. The Jeans length or the length scale of a collapsing cloud of self-gravitating gas, is even worse because in the equation the variable R (length) which is neither extensive or intensive, depends on the intensive properties: temperature and density.

To any equation, per the Second Law of Thermodynamics, the temperature must always be *intensive*, but in the fundamental equation of star collapse developed by Eddington, Jeans and Chandrasekhar; the temperature is governed by *mass* which is extensive property and by a radius which is neither extensive nor intensive property, the other terms of the equation all represent constants. Therefore, it is obvious that this fundamental equation is *invalid* from a physics point of view and should have been rejected decades ago. It is a clear mess with the wrapping of science and this shows us, the level of incompetency, scientists can have while doing their "research".

The *temperature* of the system must always be *intensive* because to state otherwise, invalidates the use of all physics and *laboratory* evidence that *must* be applied in astrophysics. Moreover, it is infinitely important to recognize that the potential energy (U) within a thermodynamic system, can play *no* role in determining its temperature.

This also *invalidates* the alleged heating of the planetary interior, mainly through gravitational compression during their origin, purported by the "accretion hypothesis" and widely used in planetary sciences. Also, the origin of the high temperatures inside stars by the same process. In fact, this is at the heart of the problem and the reason why "gravitational thermodynamics" should have no place in astrophysics at all. But mainstream science, with its characteristic obsessive compulsion for gravity, clings to it.

On the other hand, the current accepted "accretion" hypothesis of planetary formation, is scientifically no longer tenable. Scientists have been tossing about the term "accretion" without stopping to think what it means and what it entails. *Research* shows that the claim, that the idea of planets accreting from collisions of dust and small rocks, has *not* been demonstrated by *application* of the law of physics trough *laboratory experiments*.

"It turns out to be surprisingly difficult for planetesimals to accrete mass during even the most gentle collisions", *Scientific American*, May 2000. "But little is known about how microscopic dust particles, can grow 14 orders of magnitude bigger to become a giant planet within the relatively short lifetime of the disk" the journal *Nature*, November 2005.

And regarding the gas giants, "Talk about a major embarrassment for planetary scientists. There, blazing away in the late evening sky, are Jupiter and Saturn the gas giants that account for 93% of the Solar System's planetary mass and no one has a satisfying explanation of how they were made", *Science*, November 2002.

For decades, experiments have *failed* to show that mere *collision* of particles, can make them stick and grow into larger bodies under conditions *believed* to exist in the early Solar System. Have scientists therefore, considered "accretion hypothesis" to be falsified? (as they should). The answer is NO.

Instead, the concept of "gravitational instabilities" (an *ad hoc* hypothesis), was introduced to explain how colliding particles might be forced to adhere despite their natural tendency *not to*; which is absolutely ridiculous and *unscientific.* This demonstrates to the reader that in occasions, scientists do behave as *stupidity* was a virtue. In fact, things went so bad that despite decades of attempts, no computational realization of standard formation theories, has reproduced the mass and orbital distribution of both the terrestrial planets and the asteroids.

Writing in the *Monthly Notices of the Royal Astronomical Society*, Izidoro et al (2015), shows that this is *impossible*. However, mainstream scientists ignore these issues and keep misleading the public. The formation of a *solid*, massive terrestrial planet either by accretion or gravitational instabilities, has NO scientific basis whatsoever.

In fact, ordinary evidence suggests that if neighbouring, sun-orbiting rock particles hit at low speeds, they would simply bounce apart without sticking; if they hit at high speeds, they would shatter each other instead of combining.

A team of US scientists J.F. Kerridge and J. Vedder in 1972, designed an experiment with silicate particles hitting each other at speeds of 1.5 to 9.5 km/s (typical of collisions in today's asteroid belt), to test whether any sticking or impact welding occurred. They found *none*; the particles shattered.

To avoid shattering during collisions, Kerridge and Vedder proposed much lower hypothetical approach velocities. The velocity became an adjustable parameter, which might hypothetically allow accretion to occur. Other researchers ran computer simulations at the lower velocities and concluded that with these *hypothetical* conditions, accretion was possible.

But this type of 'confirmation' is an example of *ad hoc* assumption formulation in which the *lower* velocities required by accretion theory were *assumed* to justify the theory. This is reasoning in a circle. But still there is another issue. Beyond the 1-meter particle size, problems develop which not even theoretical models have solved for meter sizes, coupling to nebula turbulence, makes destructive processes more likely.

Global aggregation models show that in a turbulent nebula, small particles are swept up too fast to be consistent with *observations* of protoplanetary disks. Even computer modelling designed to demonstrate accretion, shows that particles 1 meter and larger are more likely to be *destroyed* than grow.

Some scientists do believe that an extended phase may therefore exist in the nebula, during which the small particle component is kept "alive" through collisions driven by turbulence, which frustrates growth to planetesimals until conditions are more favourable for one or more reasons. This 'extended phase' has been detected neither empirically nor in theoretical modelling. Neither support the belief that "accretion" could occur. This shows that there is absolutely *no way* for "accretion" to occur and form a *solid* planet like Earth, because it is not a *Gestaltung* working feature of Nature. The so-called "planetary scientists" should get this straight, once and for all.

There is a special feature in the Solar System that conventional science, finds hard to explain. Once the planets have formed per the SNDM, for some reason we find that there is *no* dust or debris between planets except for the asteroid belt found between Mars and Jupiter.

So, although there should be massive evidence for the formation of planetesimals and other debris between the planets, left over from the nebular condensation, we must pretend that the larger planets swept up this debris when they were forming; *except* for the asteroid belt even though we know this is not possible.

In regards of debris it is noteworthy to mention the case of Saturn's rings, which puzzled scientist until the middle of the 19th century.

The famous Scottish physicist James Clerk Maxwell, made a thoroughly study that would exhibit *stability* of the rings, surrounding the planet Saturn and the importance of some aspects of *rotational physics*. In his essay published in 1859, while working as Professor of Natural Philosophy at Marischal College, Aberdeen; Maxwell showed that the rings could not be solid, but rather a *swarm* of particles.

A solid ring would become unstable and break up. He carried out a careful mathematical treatment and concluded that the rings could not be solid or liquid, since the mechanical forces acting upon rings of such immense size, would *break* them up.

He suggested that, instead the rings were composed of a vast number of *individual* solid particles rotating in separate concentric orbits at different speeds, instead of "coalescing" into single planets.

According to Maxwell's calculations, for example, the gaseous ring that Laplace supposed, had given rise to Jupiter would break up into about fifty separate bodies, distributed along the present orbit of Jupiter and showing *no* tendency to coalesce or *accrete*. The mechanistic hypothesis of particles conglomerating into a *solid* body, would be effectively counteracted by the *dynamical rotational factors* such as dissipative collisions and shear, involved in the *revolution* of the particles around the central massive body. The same applies for asteroids, always orbiting as a *swarm* but never "coalescing" or "accreting" into a planet, asteroid or whatever. Perhaps the *most* important issue to be resolved in the SNDM is that of the distribution of *angular momentum*. The material that formed each of the bodies in our Solar System, had some rotational motion or angular momentum since its inception. That rotational motion continued even as the material was "condensing" to form planets and moons.

The angular momentum problem is a problem in astrophysics identified by British-Australian astronomer and astrophysicist, Leon Mestel in 1965. It was found that the angular momentum of a protoplanetary disk is misappropriated when compared to models during stellar birth. The problem for SNDM is that, it predicts that most of the mass and angular momentum should be in the Sun. In other words, the Sun should spin much more rapidly than it does (one revolution every 24.47 terrestrial days).

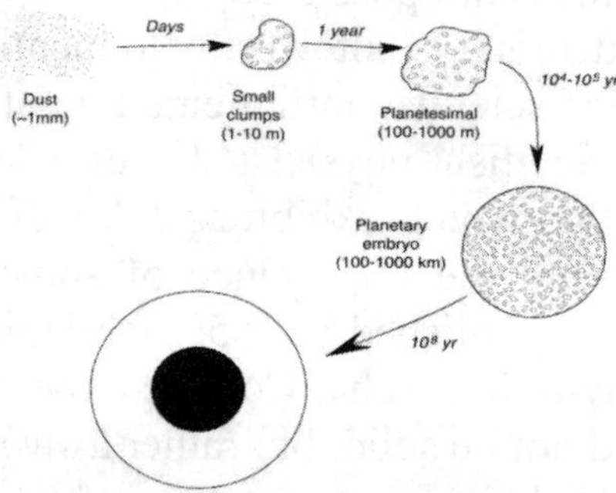

Fig. 58- Accretion hypothesis apparently is key in "explaining" planetary formation and it is a common argument against the Hollow Earth Theory. It starts with accretion of cm sized particles to follow with physical collisions to form kilometre-sized planetesimals. Then it is followed by gravitational accretion on 10-100km scale to form a molten protoplanet, from the heat of accretion. However, research has shown that the rocks would not stick, but most likely simply zoom past each other or collide and recoil like snooker balls.

Whilst the Sun makes up almost 99.9% of the Solar System's mass it only accounts for around 1% of its total angular momentum. Jupiter itself has 60 percent of the planetary angular motion. A mechanism is therefore required to transport angular momentum away from the central proto-sun and redistribute it in the outer planetary disk. So far, the SNDM failed to provide such reliable mechanism. There are of course many other problems we encounter in our Solar System, that cannot be explained by the Solar Nebular Disk Model.

For one, why did moons form and remain at the distances they hold from their planet, without falling into the planet when all the debris surrounding the Sun, got swept up by the gravity of the planets? Why do planets have still allegedly, molten cores since gravity alone, cannot account for this kind of heat and keep it during billions of years?

The SNDM has proved pretty much *ineffective* in understanding the origins of the Solar System, let alone the formation of planets, including Earth which is the main point of interest in this chapter.

One thing is obvious, the original theory of planet formation, that they formed by the slow accumulation ('accretion') of dust particles orbiting a new star, is clearly *wrong*. Any honest scientist could have told us that, because it was already well known that the 'accretion' hypothesis didn't have a workable mechanism.

But why bother with hypothetical astronomical issues about the past, when we can actually *observe* the formation of stars in real time? This allows us to have clue of how a planetary system originates. Present-day processes provide a sufficient explanation for past astronomical phenomena, although the rate of activity of these processes may have varied.

Many newly formed stars are surrounded by what are called *protoplanetary disks*, swirling masses of warm dust and gas that can constitute the core of a developing stellar system.

Exoplanets form directly from larger structures in the primordial disks of gas and dust orbiting young stars. The formation of exoplanets occurs, at the same time as the formation of stars.

New information (data) comes from observing developing star systems

More accurately a protoplanetary disk is a mixture of hydrogen and helium gas (99% by mass) and dust (1%), orbiting a newly formed star, from which planets are formed. Disks are common by-products of star formation, and range in mass from 0.001 to 0.3 Solar masses (10^{27}–10^{29} kg) and in size from several tens to almost 1,000 Astronomical

Units (10^{12}–10^{14} m). Protoplanetary disks have temperatures in the range 10 K < T < 10,000 K.

Inside the disk, matter slowly moves inwards, and dust particles "grow" to centimetre-sized pebbles: the first steps toward the allegedly formation of kilometre-sized planetesimals according to the SNDM. Protoplanetary disks typically disperse after 2–3 million years, through which they "coalescence" of their matter into planets and photoevaporation by the stellar radiation. During the lifetime of a protoplanetary disk, both solid and gas phase chemistry is active, shaping the initial composition of planets, asteroids, comets and Kuiper belt objects. Many of the chemical properties and architecture of the current day Solar System, were set during the nebular-protoplanetary disk phase.

Protoplanetary disks evolve through a variety of processes, including viscous transport, photoevaporation by the central star, grain "growth" and dust settling, and dynamical interaction with sub-stellar and planetary-mass companions.

The formation of planets occurs at the *same* time as the formation of stars. Proof of the existence of such disks didn't come until 1994, when the Hubble telescope examined young stars in the Orion Nebula. Protoplanetary disks may potentially become celestial bodies such as planets and asteroids.

The properties of protoplanetary disks are closely tied to the star-forming environment in which they are created. Understanding this environment can therefore tell us a great deal about which disk features are most likely to occur whenever a young star is born. Observations of protoplanetary disks are challenging due to their small size, low masses, and cool temperatures. However, several basic *facts* have been clearly established.

Thanks to recent developments in *infrared astronomy*, astronomers can now *directly* study protoplanetary disks around. These modern observational techniques are still not powerful enough to observe planetary formation, only faint glimpses of it. To figure out this planetary formation, astronomers rely on the same worn out and old hypothesis. Concepts such as dilute gas, gravitational collapse or contraction and core accretion are still used, despite being proved *useless*.

Conventional hypotheses of planet formation, are usually worked out within two accepted paradigms: core accretion and gravitational instability. *Core accretion* is the "bottom-up" approach: Large objects form from smaller ones, eventually building up to exoplanets. *Gravitational instability* is the "top-down" method.

It may operate in cold, massive disks in which random gas over-densities, start growing under their own self-gravity, which neither pressure nor rotational support can initially resist. If this "collapse" of dense regions can continue deeply into the nonlinear regime, and their density come to far exceed the nebula's, such clumps become self-gravitating objects and, over time, contract, cool, and look like planets.

However, several astronomers are beginning to take in to account, *more* physical variables and *new* concepts from *other* branches of science; that can help elucidate stars and planets formation. Over the last decade or so, progress in observational techniques have allowed astronomers to find answers to at least some of the open questions, regarding stars and planets formation.

With observations in the optical/near infrared with high-contrast imaging instruments, new high-resolution images have begun to show protoplanetary disks of gas and dust in unprecedented detail. The Herschel Space Telescope (active 2009-2013) and the Atacama Large Millimetre/submillimetre Array (ALMA), an international astronomy facility, localized in Chile (completed in 2013), have allowed researchers to trace interstellar dust and gas as closely as never before.

ALMA's early observations of young protoplanetary disks, some only about one million years old, reveal surprisingly well-defined *structures*, including prominent rings and gaps, which appear to be the hallmarks of planets. Astronomers were initially cautious to ascribe these features to the actions of planets since other natural processes, could be at play.

Depending on the star's distance from Earth, ALMA was able to distinguish features as small as a few Astronomical Units (An Astronomical Unit is the average distance of the Earth to the Sun about 150 million kilometres, which is a useful scale for measuring distances on the scale of star systems). Using these observations, the researchers could image an entire population of nearby protoplanetary disks and study their AU-scale features.

The researchers found that many *substructures* such as concentric gaps and narrow rings, are common to nearly all the disks, while large-scale spiral patterns and arc-like features are also present in some of the cases. Also, the disks and gaps are present at a wide range of distances from their host stars, from a few AU to more than 100 AU, which is more than three times the distance of Neptune from our Sun.

These features, which could be the imprint of large planets, may explain how rocky Earth-like planets can form and grow. For decades, astronomers have puzzled over a major hurdle in planet-formation theory: Once dusty bodies "grow" to a certain size about one centimetre in diameter, the dynamics of a smooth protoplanetary disk would induce them to fall in on their host star, never acquiring the mass necessary to form planets like Mars, Venus, and Earth.

The dense rings of dust we now see with ALMA, would produce a safe haven for rocky worlds to fully mature. Their higher densities and the concentration of dust particles would create perturbations in the disk, forming zones where planetesimals would have more time to "grow" into fully fledged planets. This has allowed for detailed reconstructions of the structure of the opaque clouds in which new stars can form, as well as the identification of the earliest known examples for regions of slightly higher density or "blobs" that have only just begun their formation.

Fig. 59- The protoplanetary disc around the young star TW Hydrae is the closest known example to Earth, at a distance of only about 170 light-years. As such it is an ideal target for astronomers to study discs. This system closely resembles what astronomers think the Solar System looked like during its formation more than four billion years ago. Methanol, a derivative of methane, is one of the largest complex organic molecules detected in discs to date.

Several protoplanetary discs observed by ALMA, show dust concentrations consistent with particle trapping in giant vortices. The formation and survival of vortices are of major importance for planet formation, because vortices act as particle traps and are therefore preferred locations of planetesimal formation. These vortices are self-sustaining *Gliederung Energies* that occur everywhere naturally, assisting the emergence of higher quality and more complex systems like a planet, with coherence and efficiency.

To understand the underlying physical mechanisms of the protoplanetary disk, currently astrophysicists run computer simulations, considering key factors such as *gravity, hydrodynamics effects including turbulence and magnetic fields*. The protoplanetary disk surrounding a young star is a complex, three-dimensional situation, and detailed simulations are only possible using high-powered supercomputers.

Of course, they only give reliable *physical* results when they stay in *contact* with the ever more detailed *observational data.* Indeed, new concepts coming from *Fluid Mechanics* such as vortices, turbulence, drag and Coriolis force; are shedding new light in *planets,* stars and galaxies formation.

A paper published in *The Astrophysical Journal* (2017) by US researchers Darryl Seligman and Greg Laughlin, points out that *vortices* and *turbulence* play a critical role in creating ideal conditions for the *birth of planets*, which is a challenge to the prevailing theory of planet formation, gravitational instability and accretion.

The first theories of planet formation, lacked detailed treatments of vortices and other complex features that *modern approaches* include. Using three-dimensional simulations of the dust and gas that orbits young stars, the study shows that turbulence is a significant *obstacle* to "gravitational instability", which scientists have used since the 1970's to explain the early stage of planet formation.

Gravitational instability proposes that dust will settle into the middle of the protoplanetary disk, around a newly-formed star. It is thought that the dust will gradually become denser and thinner until it reaches a critical point and "collapses" into kilometre-size clumps, which later collide to form planets.

But new research by San Francisco State University professor Joseph Barranco, shows that *turbulent forces* keep the dust and gas *swirling* and prevent it from forming a dense and thin enough layer, for gravitational instability to occur.

Therefore, the formation of a *solid* planet is unlikely to occur. While previous studies have used two-dimensional models to simulate the orbiting dust and gas around young stars, these *failed* to take account of a crucial force that causes turbulence: the *Coriolis Effect*.

The first to use three-dimensional models Barranco, investigated the Coriolis Effect, the same mechanism that produces cyclones and tornadoes on Earth and vertical shear. Vertical shear occurs because the faster-moving dust, settles into the middle of the orbiting plane, with the slower-moving gas above and below it.

The velocity difference between the dust and gas causes waves to form, like when wind blows over the surface of water. "What happens to the dust and gas after a period of turbulence is still an open question" Barranco said. "But it could be that in this vortex of a hurricane-like storm, dust can get trapped, seeding the beginnings of planet formation."

Observational data coming from protoplanetary discs, would be needed to test this conclusion. The trouble is that observations of the innards of protoplanetary disks are hard to come by; the gas and dust in a disk block most of the light that would otherwise show the planet-forming action as it happens.

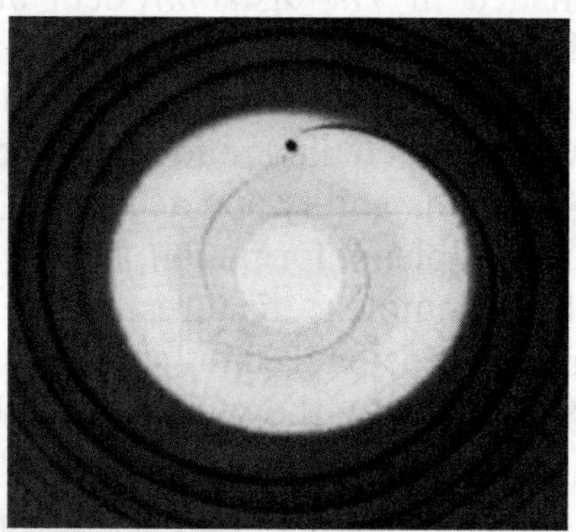

Fig. 60- A model fed with <u>real</u> parameters of Fluid Mechanics, shows a disk with an embedded planet as vortices spinning revolving around a star. They can also arise as the direct consequence of a planet's motion as it carves out a turbulent wake. Circular vortex flows generate interesting self-organizing phenomena of particle motions, that are able concentrate a large amount of dust and grains on the outer part of the vortex; to form a planet. No accretion and collisions are needed.

But millimetre and sub-millimetre radio waves emanating from cold dust in the disk's outer regions can escape relatively unscathed, offering astronomers a way to look inside. Arguably the best facility in the world to perform such observations is the Atacama Large Millimetre

Array (ALMA) observatory, a network of radio telescopes in Chile's notable Atacama Desert, mentioned before.

In 2017 in a study reported in *The Astrophysical Journal Letters,* Stefan Kraus, an astronomer at the University of Exeter in England, led the team that used ALMA to peer deep into the protoplanetary disk of V1247 Orionis, a young star around 1,044 light-years away. ALMA's radio eyes revealed curious *vortices spinning* among density waves that *revolved* around the star, a little like the spiral arms around the Milky Way.

The team's subsequent simulations suggest that these features, arose from an *unseen planet* and that they could give rise to more worlds. In this scenario, the birth of just one planet in a disk could spawn new features, such as *vortices*, which would catalyse further planet formation. Bear in mind that vortices are *hollow* not solid structures.

Vortices can form spontaneously in any turbulent fluid, but in protoplanetary disks, they can also arise as the direct consequence of a planet's motion as it carves out a turbulent wake. The spiral arms in the ALMA images of V1247 Orionis are probably a planet's wake, spinning out vortices like small eddies in the wake of a speeding motorboat.

In the old paradigm of core accretion, to make planetesimals, microscopic sized dust must form and somehow collide and stick together within the protoplanetary disk, long enough to build these solid kilometre-scale objects. However, to do this, dust must overcome a *major obstacle* called "the radial drift problem."

This problem arises when dust grains experience a drag from gas in the disk. This drag should cause the particles to lose momentum and rapidly spiral inward like water down a drain, eventually falling into the star and leaving nothing in the disk from which to make planets.

Because planets obviously do exist, something must be missing from this picture. *Vortices*, this latest work suggests, would be that missing piece, because it has the capabilities of *Gestaltung,* allowing the formation a protoplanet which the hypothesis of "accretion" by random collisions, lacks. These vortices are coherent, elongated and *long-lasting* structures, characteristic of high Reynolds number in turbulent flows.

In another study, published in the journal *Physical Review Letters* (2013), a team of Fluid Mechanics experts led by computational physicist Philip Marcus at the University of California, Berkeley; shows how *variations* in "gas" (plasma) density lead to instability, which then generates the whirlpool-like vortices needed for stars to form. This explains how *spin* originated in a protoplanet disk, which the SNDM was *never* able to account for.

Astronomers wrongly accept that in the first steps of a new star's birth, dense clouds of gas "collapse" into clumps that, with the aid of angular momentum (which they fail to explain its origin), spin into one or more Frisbee-like disks where a protostar starts to form. But for the protostar to grow bigger, the spinning disk needs to lose some of its *angular momentum* so that the gas can slow down and spiral inward onto the protostar.

What has been hazy is exactly, *how* the cloud disk sheds its angular momentum so mass can feed into the protostar. This shows that the Jeans instability, which apparently causes the "collapse" of interstellar gas clouds and subsequent star formation, simply doesn't work. The leading theory in astronomy, relies on *magnetic fields* as the destabilizing force that slows down the disks. One problem in the theory, has been that gas needs to be ionized, or charged with a free electron, in order to interact with a magnetic field.

However, there are regions in a protoplanetary disk, that are too cold for ionization to occur, thus the disk is very stable and it has been unclear how disk matter destabilizes and "collapses" onto the star. The researchers said current models, also *fail* to account for *changes* in a protoplanetary disk's gas (plasma) density based upon its height.

When these changes were *accounted for*, a new picture emerged because these changes in density, creates the opening for violent *instability*. When they accounted for density change in their computer models, 3-D vortices emerged in the protoplanetary disk, and those vortices spawned more vortices, leading to the eventual disruption of the protoplanetary disk's angular momentum.

These vortices *destabilize* the orbiting gas, which allows it to fall onto the protostar and complete its formation. The researchers note that *changes* in the vertical density of a liquid or gas do occur throughout Nature, from the oceans where water near the bottom is colder, saltier and denser than water near the surface to our atmosphere, where air is thinner at higher altitudes.

These density changes often create instabilities that result in turbulence and vortices such as whirlpools, hurricanes and tornadoes. Jupiter's variable-density atmosphere hosts numerous vortices, including its famous Great Red Spot. Indeed, vortices are *Gestaltung* working features of Nature at play.

Given all these details from the recent research, the origin of our Solar System is no longer an unsolvable problem or a dead end as many scientists believe. The same applies for planets. All this fuss that started about star and planet formation, was because astronomers together with astrophysicists, have paid too much attention to purely mechanistic and random interactions with gas and dust driven by gravity. They haven't pay attention to natural *self-organizing patterns* that are observed in plasma flows, electromagnetic fields and vortices at *cosmic scale*.

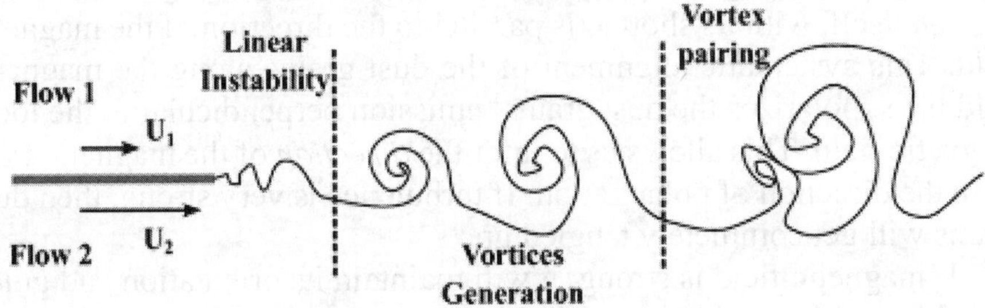

Fig. 61- Observable shear induced instability and turbulence in fluid plasma is the real mechanism, behind the formation of protoplanetary discs and planets; not hypothetical gravitational instability or "collapse" let alone accretion. Turbulence and vortices are crucial for protoplanetary disk dynamics. Relevant turbulent processes include the transport of angular momentum, mass, and heat, which were critically important to the formation of the Solar System.

On the other hand, one of the *major problems* in conventional star-formation theory, has been to understand why star formation is apparently so inefficient. Only about 1% of gas in molecular clouds is converted to stars. Now there is consensus that *turbulence*, is crucial for preventing too much gas from collapsing into stars.

Also, the dusty molecular gas in stellar nurseries is permeated with *magnetic fields*, which impede the inward pull of gravity and slow the rate of star formation. Tell-tale magnetic fields which accompany all *electric currents*.

The simplest (oversimplified) models of star formation, predict that gravity should be much more effective in forming stars as it "compresses" the gas in molecular clouds, but *observation* shows that it is *not* the case. Turbulence makes bigger discs, *relevant* for planet formation.

This explains why some stars have planets whereas other stars don't. Observations of polarised light from star forming molecular clouds in our own Galaxy (such as the Orion nebula), show that magnetic fields are present in these clouds and are having a strong influence on the gas flow in them. In this theoretical picture, the observed magnetic fields are thought to be crucial in *controlling* the turbulent motion of gas in the molecular clouds, from which the stars form and in slowing the spin of regions of gas which are apparently "collapsing" under their own gravity.

How can we learn about the *magnetic fields* of distant objects? One way is by measuring dust polarization. An elongated dust grain will tend to *align* itself, with its short axis parallel to the direction of the magnetic field. This systematic alignment of the dust grains along the magnetic field lines, polarizes the dust grains' emission perpendicular to the local magnetic field. This allows us to infer the *direction* of the magnetic field from the direction of polarization. If turbulence is very strong, then dust grains will get completely tangled up.

If magnetic field is strong, it will maintain its orientation and guide turbulence along with the dust grains. Space *magnetic fields* can explain one of the long-standing challenges in stellar astronomy: why stars rotate so slowly. As the disk contracts, its rotation should speed up to conserve angular momentum.

The region with the fastest rotation should be at the centre in our case, the Sun. But we don't see this. The Sun rotates much more slowly than it should, and the Solar System's angular momentum is concentrated in the Jovian planets. How's that? Gravity Model was *unable* to figure it out. Given their stars' large masses, as they collapsed to form, they should spin up to the point of flying apart, preventing them from ever reaching the point that they could ignite fusion.

To explain this rotational slow down or braking, astronomers have considered the possibility of an interaction between the forming star's *magnetic field*, and forming accretion disc. This interaction would slow the star allowing for further collapse, to take place.

Until very recently, astronomers were unable to directly observe circumstellar discs around newly formed stars. Since dust discs will be warmed by the forming star, systems with these discs will have extra emission in the infrared portion of the spectra. According to the magnetic braking theory, young stars with discs should rotate more slowly than those without.

This prediction was confirmed in 1993, by a team of US astronomers led by Suzan Edwards at the University of Massachusetts,

Amherst. Numerous other studies confirmed these general findings. Indeed, the molecular clouds are observed to be permeated by *magnetic fields,* which can in principle, strongly affect the evolution of angular momentum during the "core collapse", through magnetic braking.

Although the details of it aren't fully fleshed out, it is clear that these magnetic braking effects, play a significant effect on slowing the angular speed of stars whilst keeping the original angular speed of the forming accretion disc. The "mystery" of what process could have accomplished this transfer of angular momentum, during the formation of a Solar System; including our own, is practically solved.

However, one must proceed with caution. Nearly 50% of all known stars are *binary,* these are stars which orbit one another. This means our Sun might have a companion star, if so this fact needs to be accounted for, in the angular momentum issue of our Solar System.

Vortices, dusty plasmas and planetary formation

Back to turbulence. What makes this interesting for physicists is that turbulence in one system, is exactly analogous to the turbulence in another. The reader should recall the concepts of Golden Ratio (Phi), *Gestalten Energie* and spiral patterns, discussed in chapter one. So, the airflow around an insect's wings is similar to the flow of honey around an orange. And the turbulence in an inertial confinement fusion machine, is the same as the turbulence inside stars.

In fact, the dynamics of heavy inertial particles in turbulent flow is a universal problem that appears in astrophysics, oceanography, engineering and atmospheric sciences. In recent years, aerodynamicists have discovered an entirely new phenomenon of turbulence, called *preferential concentration*, that plays an important role in everything from aerosol production to crystallisation reactors.

Any turbulent flow is made up of eddies of various sizes that rotate at different rates. When the flow contains particles, they are inevitably *influenced* by these vortices. It's easy to imagine that the influence of these eddies, average out to create random motion.

But in recent years, aerodynamicists have found something different. The particles become influenced by inertial forces in the rotating vortices. This causes the particles to be pushed out of the eddies, into the regions of low vorticity between them. The mechanisms which drive preferential concentration, are centrifuging of particles away from vortex cores and accumulation of particles in convergence zones. Inertial particles spin out from the centre of eddies; if the particle and fluid time constants are commensurate, so that the eddy persists on this spinout time scale, then the particles will concentrate in regions, where straining dominates vorticity.

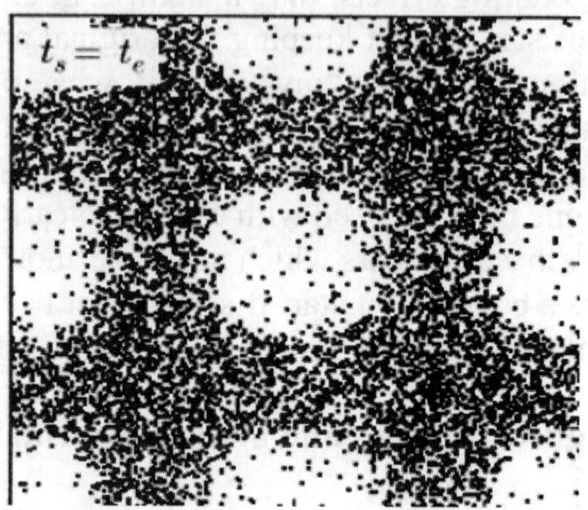

Fig. 62- A numerical simulation shows that, when the gas stopping/drag timescale is comparable to the turbulent eddy turnover time, *preferential concentration* of inertial particles does occur in the periphery of the vortices. The centre remains practically *hollow* (white circles). Preferential concentration occurs in a wide range of flows including plane and axisymmetric free shear flows, wall-bounded flows, homogeneous turbulent flows, and complex shear flows.

Experimental and numerical studies demonstrate that preferential concentration occurs in a *wide* range of flows. The effect of collisions is found to be *negligible* for the most part, although in some cases they have an interesting antidiffusive effect. This turns out to be hugely important. It means that particles of a certain mass become *concentrated* or coalesce in the regions between the vortices and *not inside* as was previously thought.

Applying this finding to the turbulence flows in planetary formation and the fact that the material is practically dust with little differences in sizes and plasma. It results that dust particles and the prevalent eddies in the protoplanetary vortex, with the same fluid time constants, entails the particles to become influenced by inertial forces in the rotating vortices.

This causes the particles to be pushed out of the eddies, into the regions of low vorticities; concentrating in the periphery of these vortices. Therefore, it becomes clear that is impossible the formation of any solid or massed protoplanet since its mass will be distributed and consolidated in the periphery of the vortex, acquiring a *shell structure*; expressing minimum energy and stable configuration.

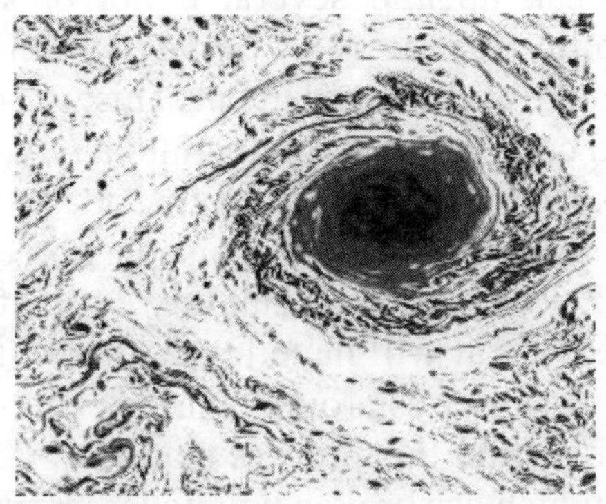

Fig 63– Experiments show that a key attribute of turbulence, is the vigorous local spinning of the fluid flow. Centrifugal forces eject heavy particles from coherent vorticial structures, leaving a void inside the vortex. These structures reveal a preferential concentration mechanism, where inertial particles cluster around the void. In protoplanetary disks, such vortices have a higher density of pressure and matter (gas and dust) than the rest of the disk; creating the perfect conditions for the birth of a new planet.

Only diffuse material and plasma, would be left inside the vortex. In this scenario, there is no room for "accretion" or "gravitational instability" to occur as it is commonly believed. Protoplanets would be roughly, *thick spheroidal shells* since its inception. This includes the one which gave origin to Earth.

When comes to planet formation, from the *electromagnetic* and *plasma physics* perspective, the concept of dust is often oversimplified. A considerable amount of this "dust" clouds, out of which stars condense, are ionized particles. This ionization is produced mostly by hydromagnetic transfer of kinetic or gravitational energy into electric current which ionize, *not* by ultraviolet starlight as is generally accepted. Most cosmic low-density plasma, do have filamentary structure produced by electrical currents.

There are often stable current layers, in which the magnetic field may change 180 degrees within a few radii of the circular motion of plasma particles in the presence of this magnetic field. In this case stars are usually formed along filaments.

Perhaps the most efficient (and Nature is nothing if not efficient) method to "concentrate" matter over cosmic distances is that of the electromagnetic "pinch effect" caused by parallel electric current filaments in plasma. In fact, a major filament, such as those involved in star-making, can break up into several minor or sub-filaments by a filamentation instability due to any *variation* in the electric current flow, or a temperature change along the filament. This will cause the magnetic field to pinch in and compress the plasma into a *ball*.

Laboratory experiments do confirm these vortex electrical filaments, causing the pinch effect. This phenomenon is called the Bennett Pinch or Z-Pinch. This instability will often leave a large filament in the centre but it will have lesser filaments around it, rather like a conduit with lesser wires around a large central wire that goes to make up a cable. It is from this process that *planets* can develop when the major filament, which has broken up into lesser filaments, undergoes a Bennett pinch. The pinch begins at the outside and works its way in. As each lesser filament pinches, a *planetary ball* forms, with the large central filament becoming a ball and lighting up last of all.

So, on this model, the Sun formed last in the Solar System. The spin axes of stars formed in this manner are *aligned* with the filaments. Such alignments have been *discovered* in groups of stars (Lerner, 1991).

Finally, within this ball or vortex of plasma, *preferential concentration* tends to force the heavier elements and mostly ionized dust, *out* of regions of high vorticity, forming a shell. Inside are left the lighter ones such as hydrogen, helium and traces of plasma. Electrostatic and van der Waals forces will start to cohesion this material in the outer shell.

If this *protoplanet* has enough mass to attain self-gravity and the turbulence decreases, it could attract material and increase the shell's thickness to a certain extent until a *balance* of rotational and gravitational forces, is reached coupled with the least electrical interaction; putting the planet on a Keplerian orbit.

The nearly spherical shape of the planet is a matter of efficiency. In Nature, a sphere is the largest volume to surface area ratio also, in a sphere, every point on the surface is equidistant to the centre of gravity. Thus, a sphere is the most *stable* shape for a celestial body because matter cannot be "placed" anywhere with less gravitational potential energy (U_g). Due to its *spinning*, an outward centrifugal force flattens the planet in the poles and bulges it at the equator. Bear in mind that since its inception, the material of this spherical shell due its inertia, spin and consequent centrifugal forces; cannot "collapse" due to the mutual gravitational pull of the shell's "walls" and become a solid body by gravitational accretion, as many believe.

The apparently unsolvable problem in conventional astronomy of planet formation, in regards of how high-speed submicron sized dust grains, from the interstellar medium would grow 14 orders of magnitude in size and overcome the meter-size barrier to form planets, *before* the lifespan of the protoplanet disk ends; is no more a problem.

This is because planets originate from *vortices* which trap dust, organize it and shape protoplanets along plasma physics and gravity. Also, a planet's size depends mainly on the original size of the *vortex* from which it was formed, during the lifespan of the protoplanetary disk.

This scenario of planetary formation is well grounded in *observed facts* gathered from the latest research and it considers more variables than the old and failed hypothesis/models, put so far to explain Earth's origin. The chaos-accretion collisions hypothesis is a dissipative or *destructive* rather than a formative process and it is *not* an observable fact. The vortex accretes or *concentrates* interstellar gases in its *periphery* like a giant whirlpool and it is vortex action that is causing the accumulation of matter or dusty plasma (outwards) *not* gravity (inwards), as it is wrongly assumed.

The final blow to the mythological accretion hypothesis, was given by Russian astrophysicist Vadim Nicolaevich Tsytovich in 2001, when he published his paper: *Evolution of voids in dusty plasmas*. It shows that formation of dust voids is a general phenomenon in rotating dusty plasmas, the *raw material* in planet's formation. Whereas both dust structures and dust voids are formed in some plasmas, this *cannot* happen in a purely homogeneous (continuous) plasma.

There can be no internal architecture if the dust structures are not *separated* from one another by voids. A *structure* must be separated from the rest of the mass by a space, for otherwise it is no longer a distinct structure. This is another example of *Gestaltung* at play,

Therefore, there are NO planetary "solid/liquid cores". Only *hollow* or diffuse ones. Also, s*pin* is the real causative force in forming any agglomeration and *organization* of cosmic matter, mainly dusty plasma in clusters and depleted regions, observed in Nature. It is the *seed* for the formation of all galaxies, stars and planets. Also, this is the answer of Eddington's question of why everything in the Universe, rotates.

Protostar chemistry and planets

Recent research shows new planets form more commonly in star systems with relatively high concentrations of elements heavier than hydrogen and helium. It is believed such heavier elements are necessary to form the *dust grains* and planetesimals that build planetary "cores".

Moreover, evidence suggests that the disks of dust that surround young stars, don't survive as long when the stars have lower concentrations of heavy elements, or lower "metallicities" in astronomers' jargon. Probably, heat radiation coming from the star, is the reason for this shorter lifespan as it causes clouds of dust to *evaporate*. Our cosmic history has several defining epochs, one of which is the point at which star systems began to form planets. Heavy elements such as carbon, silicon and oxygen; first needed to be created from huge star explosions called supernovas and the stellar cores of the first generations of stars, before the first planets could form.

A *minimum* quantity of heavy elements must be present in in circumstellar disks, before planets can form. Because these heavy elements, must be produced by the first stars in the Universe, the first planets could only form around *later* generations of stars.

A planet as massive and dense as the Earth, could only form once stars and supernovae had enriched the gas with an abundance of *heavy elements* that is at least 10 percent that in the Sun. This would imply that many generations of stars, had to form and evolve before habitable planets could form. One important consideration for planetary formation, is the dispersal rate of the circumstellar disk of gas and dust around a host star. Two of the more prominent mechanisms for dispersing a planetary disk are giant planet formation and photoevaporation by the host star.

Photoevaporation appears to be the more dominant process. Observations show that low-metallicity disks have shorter lifetimes, which is bolstered by data showing higher-metallicity disks are better "shielded" from evaporation by a host star's radiation. Disks with higher metallicity tend to form a greater number of high-mass giant planets. Is important to notice that for a star system to form planets, the time required for dust grains to settle *cannot* exceed the disk's lifetime.

Since the settling time for dust grains depends on the density and temperature of the disk, which are related to the distance from the host star, the critical metallicity is also a function of *distance* from the host star. When closer to the host star, light elements are "evaporated" leaving the heavy ones forming terrestrial proto-planets. Further away from the host star, remnant lighter elements can form Jovian gaseous planets.

The formation of planetesimals can *only* take place, once a *minimum* metallicity is reached in a protostellar disk. Also, certain amount of *water* vapour may have been trapped *inside* during the formation of planetesimals, the rest would condensate in the outer disk, where most of water ice reservoir is stored. Thus, more water would be added on terrestrial planets by impact of icy bodies forming in the outer disk. Some of the earliest planets may have formed at a distance of 0.03 astronomical units from their parent star. One astronomical unit, or AU, is the distance from Earth to the Sun, or about 93 million miles (150 million kilometres). For comparison, our Solar System's innermost planet, Mercury, orbits at just under 0.4 AU.

Given the high temperatures likely at 0.03 AU (estimated at roughly 2,370 degrees Fahrenheit, or 1,300 degrees Celsius), the first planets probably were too hot to host life as we know it. Also, is probably that the first Earth-like planets, may have formed in the habitable zones of stars slightly more massive than the Sun.

Since more massive stars burn out faster, it is possible that any life that evolved on these planets may have already perished with the death of its host star, which may have lived only 4 billion years compared to the 10-billion-year lifetime expected for the Sun. Overall, this would explain the relatively close distance of terrestrial planets to the host star, which kind of stars may have planets harbouring life and why some stars, don't have any planets at all.

Chapter IV

Polar enigmas and clues to a new understanding of the inner Earth. What is the *real* cause of geomagnetism and the magnetic field of celestial bodies?

S*peculations? I have none. I am resting on certainties.*
Michael Faraday (1791-1867)

Basic preliminaries

In this chapter, we will explore some interesting *anomalies* about the North and South poles of our planet and other topics. The aim is to lay down a fundamental basis to get a new version of Earth's internal structure. Our study will concentrate in the polar regions of the Arctic and Antarctic.

Before going into this matter, the reader will get a clear picture of the physical/geographical factors; that govern these regions for a better understanding of these anomalies. Also, in this chapter enters the human factor in regards of the poles' exploration and how this information, has been handled to the public. In fact, during the last two centuries, the polar regions attracted the attention of many adventurers, explorers, scientists and even politicians; especially in the region or continent of Antarctica.

We must define exactly, what is a pole and what means a polar region. In our case, we are dealing with what technically is known as the *geographical* North and South poles. Earth's geographic poles are fixed by the *axis* of Earth's rotation. This axis, when extended into space defines the celestial poles, in reference to stars. On maps, the north and south geographic poles are located at the congruence of lines of longitude. In theory, if somebody steps on the North or South pole, all meridians or lines of longitude will coincide beneath a person's feet. In the northernmost point on the globe, all directions are south from the North Pole. It defines geodetic latitude 90° North.

The *geomagnetic poles* (dipole poles) are the *intersections* of the Earth's surface and the axis of a bar magnet, hypothetically placed at the centre the Earth by which we *approximate* the geomagnetic field. There is such a pole in each hemisphere, and the poles are called as "the geomagnetic north pole" and "the geomagnetic south pole", respectively. On the other hand, the *magnetic poles* are the points at which magnetic needles become *vertical*. There are the *magnetic north pole* and the *magnetic south pole*.

Geographic poles and magnetic poles are *not* located in the same place, in fact they are hundreds of miles apart. As are all points on Earth, the northern magnetic pole is *south* of the northern geographic pole (located on the polar ice cap). The north magnetic pole is not static, it *wanders* about 34 miles (55 kilometres) a year. Not long ago, it was located near Bathurst Island in northern Canada approximately 1000 mi-

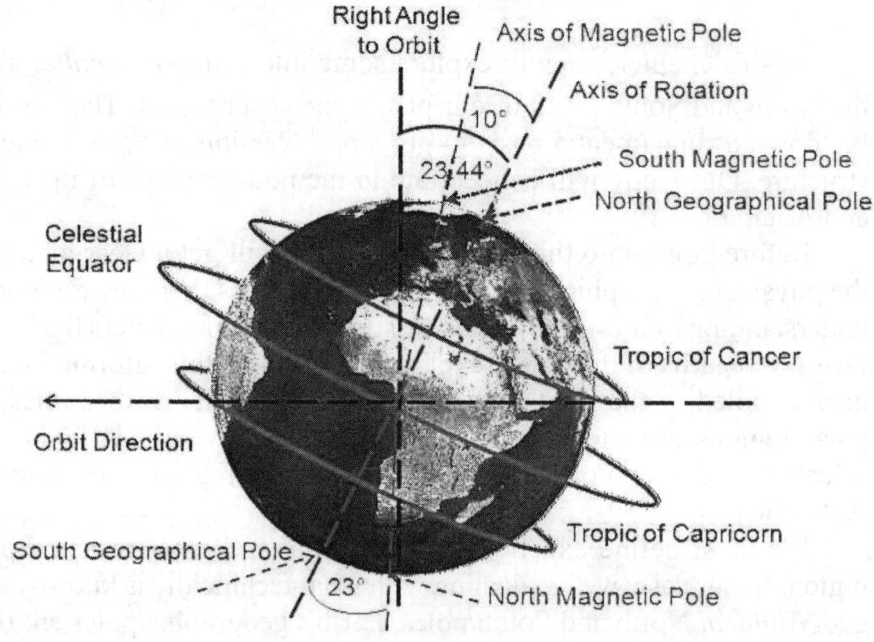

Fig. 64- Earth is like a giant magnet with a North and South Pole. However, the magnetic North and South Pole are not aligned with the Geographic North and South Pole. The Geographic North Pole is defined by the latitude 90° N and is the axis of the Earth's rotation. The Magnetic North Pole is where the Earth's magnetic field points vertically downward. The separation between both is about 10-11 degrees.

les (1,600 km) from the geographic North Pole.

It crossed the international date line in 2017, and is leaving the Canadian Arctic on its way to Siberia. The constant shift is a problem for compasses in smartphones and some consumer electronics. Airplanes and boats also rely on magnetic north, usually as backup navigation. Therefore, navigation charts and other aids, must be frequently *updated*. Navigators using magnetic compass readings must make corrections, both for the distance between the geographic poles and the magnetic poles and for the shifting of the magnetic poles.

However, the magnetic poles may undergo displacements of 40-60 km from their average or predicted position, due to magnetic storms or other disturbances of the ionosphere and/or Earth's magnetic field. The southern magnetic pole is displaced hundreds of miles away from the southern geographic pole on the Antarctic continent and it moves as its northern counterpart. This is known as *polar wandering*.

Although fixed by the axis of rotation, the celestial poles undergo slight wobble-like displacements, in a circular pattern that shift the poles approximately six meters per year. This is known as *precession* of the Earth’s axis and one of its effects is that the positions of the south and north celestial poles appear to move in circles against the backdrop of stars, completing one cycle every 25,772 years.

Located on shifting polar ice, the North geographic pole is technically defined as that point 90° N latitude, 0° longitude. The South geographic pole is technically defined as that point 90° S latitude, 0° longitude. Early explorers used sextants and took celestial readings to determine the geographic poles.

Modern explorers rely on GPS coordinates to accurately determine the location of the geographic poles. Earth's magnetic field shifts over time, reaching the point of completely *reversing* its polarity. There is evidence in magnetic mineral orientation that, during the past 10–15 million years, reversals have occurred as frequently as every quarter million years. Although Earth's magnetic field is subject to constant change (periods of strengthening and weakening) and the last magnetic reversal occurred approximately 750,000 years ago, geophysicists assert that the next reversal, will not come within the next few thousand years.

The present alignment means that at the northern magnetic pole, a dip compass (a compass with a vertical swinging needle) points straight down. At the southern magnetic pole, the dip compass needle would point straight up or away from the southern magnetic pole. This is because the needle aligns with the magnetic lines coming out (up), in the south magnetic pole and getting in (down) at the north magnetic pole. In regards to the Arctic region, four poles can be located there:

Besides the well know geographic and magnetic poles, there are the less known North Pole of Cold and the North Pole of Inaccessibility.

The Pole of Cold is the place which is colder than the North Pole. On the meteorological maps, it can be found under the name of the Pole of Cold. It is situated in the Yakut settlement of Oymyakon of the East Siberia. It is inhabited by 521 people and because it is situated far from the ocean and in high latitudes the climate here is severely continental with harsh, long winters and short, hot summers.

During winter season the temperature here can drop as low as -50/-60°C. The absolute minimum here was registered in February of 1933 (-67,7 °C). This is the coldest place on Earth, where people live in such unwelcome conditions on a constant basis.

The last but not the least is the North Pole of Inaccessibility, which is the most remote point of the Arctic. It is situated as far from the mainland as possible: it is situated in the 170th meridian of the eastern longitude, in almost 600 km from the geographical North Pole. Because this place is situated that far, it is considered to be mostly inaccessible. Still in the year 1941 Soviet polar explorers have managed to organize the first air expedition to this part of the world. This expedition was called "USSR-N-169″. The notable importance of the North Pole of Inaccessibility will be discussed later in this chapter.

One fundamental and determinant variable in the polar regions is the *climate,* with very cold temperatures. Flora, fauna and humans are heavily influenced by it. The most inhospitable place on Earth's surface is the Antarctic continent. It is important to understand the climate factor, when we explore some anomalies in the polar regions, that are *inconsistent* with what is generally known of the climate attributed to polar regions.

The *temperature* of any place in our planet is dictated by solar radiation, time of the day and the curvature of Earth or more specifically, the *latitude.* The key factor here is the solar or *Sun angle*. Sun angle is the angle between the horizon and the Sun; the angle from the ground to the Sun. In higher altitudes or mountainous regions, along the variables mentioned before, *air density* determines the temperature, but this is not the point to discuss, in this chapter.

The Sun is low in the sky in the morning and evening, and thus the Sun angle is low in the morning and evening, which means *low* radiation is coming. The Sun is high in the sky at about noon, and thus the Sun angle is greatest at about noon. The time when the Sun is highest in the sky is called solar noon. Is here when the solar radiation is *maximum,* at any given place.

Because the Earth is a *sphere*, sunlight does not spread over the Earth's surface evenly. Since the distance between the Earth and Earth's surface, being minimum and reaching the freezing point in the polar regions, the Sun is so large (more than 148,800,000 kilometres), the light plus thermal energy that reaches the Earth always comes in as *parallel* rays from the same direction and the Earth is curving away.

Sunshine and Temperature on Earth: Zenith Angle

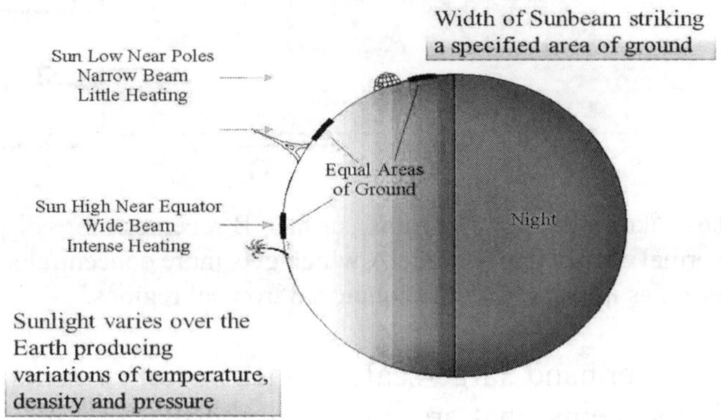

Fig. 65- The same amount of sunlight causes different temperatures on Earth's surface due to its curvature. Temperature will depend on the zenith angle between the sunbeam and Earth's surface, being maximum at the equator and minimum, reaching the freezing point in the polar regions.

Sunlight reaching the equator (0° latitude) is consistently more concentrated and this starts to decrease when the latitude is moving towards the poles (90° latitude), reaching its minimum.

The more concentrated the light, the more it hits a given area, then higher the light intensity and the more *thermal energy* that receives. That's why equatorial and tropical areas are always warm and polar areas are so *cold.* At the poles, the Sun can never be directly overhead (a sun angle of 90 degrees), only a few degrees above horizon. As one gets closer to the equator, the solar angle gets larger and larger. Also, in the poles, solar light must go through more atmosphere, thus *losing* more thermal energy, than in the equator.

The above explanation is the reason Earth is divided in three climatic regions: tropical, temperate and *polar*. Each region has its specific mean temperature, specific flora and fauna are *constrained* in each of these regions, except for some migrating species. Human activities are very restricted in polar regions. Only a small number of compounds, inhabited with a few, are found there.

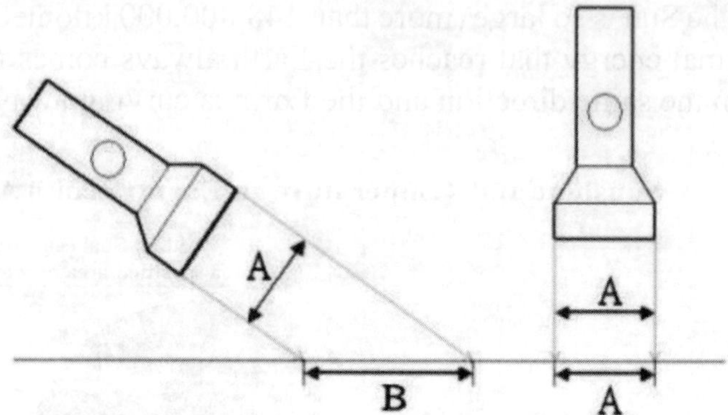

Fig. 66- Like in the polar regions, surface B receives more dispersed light and have less thermal energy than surface A which gets more concentrated light thus more thermal energy, as it happens in the equatorial/tropical regions.

On the other hand, large-scale surface *ocean currents* are driven by global wind systems, that are fueled by thermal energy from the Sun. These currents *transfer* heat from the tropics to the north and south hemispheres, influencing local and global climate. The warm Gulf Stream originating in the tropical Caribbean, for instance, carries about 150 times more water than the Amazon River.

The current moves along the US East Coast across the Atlantic Ocean towards Europe. The heat from the Gulf Stream keeps much of Northern Europe, significantly *warmer* than other places equally as far north. But, upon reaching cold regions, such as the Arctic and Antarctic Polar Circles, ocean water *loses* heat to the atmosphere and becomes cold and dense. When ocean water freezes, forming sea ice, salt is left behind causing surrounding seawater to become saltier and denser.

However, Atlantic water is present, beneath the Arctic Ocean but covered by low salinity water, deriving from runoff from sea ice melting, which prevents the remaining *heat* from the Atlantic layer, to *reach* the sea surface and the ice. This is key to understand because this layer, acting as *insulator*, makes impossible the melting of the arctic ice cap due to the so-called "global warming". This layer *prevents* any heat from oceanic currents, to reach the North pole.

Finally, the time of year can affect Sun angle. The Earth's axis is not perpendicular to the plane of the Earth's orbit around the Sun. Because of this tilt and the fact that the orientation of the tilt is always the same, sometimes the North Pole is angled toward the Sun and sometimes the South Pole is angled toward the Sun.

When the North Pole is angled toward the Sun, it is summer in the Northern Hemisphere and winter in the Southern Hemisphere. When the South Pole is angled toward the Sun, it is summer in the Southern Hemisphere and winter in the Northern Hemisphere. Due to the locations of the poles, in winter the night lasts for roughly *six months* and in summer day lasts the same.

Therefore, in the past, expeditions and explorations always were scheduled and organized, to accomplish the planned targets during the polar summer. Those men were keen in trying to avoid the risk to face and endure, a very cold and long night.

In regards to human activity, polar regions are very *distinct* to any other. These are in contrast with all other terrestrial regions on Earth (not including the subsea). The reader should think about remote places such as the mount Everest base camp, the Sahara Desert or the Colorado Great Canyon, for instance.

Plenty of people have been there and with enough finances and skills, anyone is free to go. There is accessibility and reliable information available. Excepting the polar regions, the places where one can find some difficulties to mobilize on Earth, for instance is in certain country borders or close to military bases, but that's normal.

The truth is that only a few people could and still can reach the North and South poles and its surrounding areas. Those people are *not* what we can call, ordinary citizens. Almost all of them, mostly scientists and military personnel, are and have been there on behalf of governmental institutions and the like, who gives them, the means and financial support.

Currently, whatever information obtained regarding the poles, is *controlled* and channeled by those governing bodies. In the case of Antarctica, international treaties have been signed to ensure human activities, such as fishing or tourism are practically *banned.*

The official narrative is, as well as conducting globally important science, governmental institutions help protect the Artic and Antarctica's pristine environment. Some fishing vessels are allowed to venture in the peripheral waters free of ice, of the polar zones.

Few tourists might attempt to visit only the periphery of the South pole, because this kind of visits are restricted, expensive and difficult.

A tourist operator, Quark Expeditions, can get tourists only into the north of the Antarctic peninsula nearly 2000 miles away from the South Pole.

There are around 75 research stations or bases belonging to different countries, that have been established in Antarctica. For some reason, only the *global powers* got their research stations in or close to the South Pole. Other countries had to content to have their stations, located on the coast or the peripheric islands off the continent, thousands of miles away from the South Pole.

The US have the biggest of all research stations: the McMurdo base, 850 miles from the South Pole. Also, the Amundsen-Scott South Pole Station, belongs to the US since November 1956.

Despite being theoretically prohibited on paper (Antarctic Treaty, 1959) actually, the military is heavily present in Antarctica. It is said that their personnel are primarily dedicated to such aspects as search and rescue, transportation, infrastructure services, and so on.

As far as military weapons, most of the military are limited to small arms, in keeping with the limitations of the Treaty, although more substantial ordnance is sometimes noticed. The military aircraft are mostly for transportation of cargo and personnel as well as search and rescue. It could be that there are good reasons that so much research is done in Antarctica. It has the cleanest air in the world, and atmospheric monitoring is done here to provide neutral data sets.

Antarctica is the land area where polar night lasts longest; when combined with extremely dry, thin, clear air, a unique geographic situation with the celestial south pole in zenith, low temperatures, and minimal background radiation. It is obvious why Antarctica has been such a success as an observatory for astronomy, astrophysics and related disciplines.

The regions that extend from 66.5° north latitude (Arctic polar circle) and south latitude (Antarctic polar circle), in direction to the North and South Pole respectively, are the *polar zones*. Within each polar zone are two distinct sub-regions, the ice cap and the tundra.

The tundra is a level or rolling treeless plain that consists of black mucky soil with a permanently frozen subsoil, and has a dominant vegetation of mosses, lichens, herbs, and dwarf shrubs. In other words, it is a *barren* and very cold land.

Bear in mind that climate is frigid and life is scarce on these regions, with living conditions being severe and harsh on them all year long. This is because they are the farthest from the Equator and thus, receive *much less* sun radiation and warmth than any other place on our planet, as explained before.

Polar regions are pretty much permanently covered by *ice* so they have no soil, mostly in Antarctica. Survival is tough in polar regions.

Fig. 67- By definition of polar regions, the Arctic habitat is a freezing cold area at the top of the Earth, above the Arctic Circle. It's made up of the Arctic Ocean. As a result of the freezing cold temperatures, many areas in the Arctic are completely covered by ice in the form of sea ice, glacial ice, or snow all year round. Almost all areas in the Arctic are covered by ice in some form or another for long periods of time. However, anomalous regional warm temperatures and currents, have been found nearby the North pole.

If humans couldn't survive in the far north, neither can most land animals. This is key to understand. The animals we typically associate with the North Pole, like polar bears and arctic foxes, almost never fully leave the safety of solid ground, and if they do, they tend not to travel far. Despite all that, polar regions contain ecosystems.

Numerous species of fish, crustacean and mollusc inhabit the waters beneath the ice which means that there is plenty of food for carnivorous birds and mammals to eat. Seals, killer whales, sea lions, walruses and narwhals; can all be commonly spotted feeding on the fish in the Arctic and Antarctic circles.

However, as humans are concerned, the northern tribes of the Arctic, the Inuit, have settled in the far north of Canada, Greenland, and Russia but since there is no land, no agriculture or resources in the far north; nobody can truly colonise the Arctic, as far as the official story goes.

Where facts do not fit theory

In the late 19th and early 20th centuries, the "Heroic Age" of polar expeditions, people took great risks to increase the world's knowledge of these remote parts of the globe. Although not all explorers were exclusively motivated by scientific discovery, the research they conducted was not just a pleasant side-line but often in addition to a good measure of luck, a prerequisite for an expedition's success.

Several reputable and experienced explorers from different countries, ventured to reach the North and South poles and thus, make history. Most notable of them were Norwegians Dr Fridtjof Nansen and Roald Amundsen. Those were the old times of private initiative and courage, where personal prestige along national pride were at stake and all of this going on with very little involvement from the government and corporations.

While some of them failed, and even perished; others did succeed in such risky endeavour. They were public figures. However, perhaps their greatest achievement was the valuable *information* they gathered during their hazardous journeys through those inhospitable places, mainly in the North Pole. This information was collected from the lucky ones who managed to get back home and delivered their statements to the *press* and *interviews*. In other words, this information has been historically confirmed and documented.

Nansen's book *Farthest North* (1897) has far more reliable account of what *really* is going in the North Pole area, than mainstream science would have led us to believe. Also, several other books were written about the arctic endeavours during that bygone era. Two of these books were *The Phantom of the Poles*, written in 1906 by William Reed and *A Journey to the Earth's Interior*, by Marshall B. Gardner, published in 1920. These books are detailed accounts of a diverse range of *anomalies* that explorers have found during their trip to reach the North Pole.

These anomalies have *no* explanations from the official scientific standpoint, which washes them away as either mere curiosities or fairy tales. These anomalies are very little known by the public, mostly because current textbooks and the mainstream media don't mention these.

According to the standard scientific knowledge, the Arctic Ocean is located completely within the Arctic circle and is an isolated Mediterranean basin; with only limited communication with the world's oceans, principally the Atlantic Ocean via the Fram Strait and the Barents Sea, and the Pacific Ocean via the Bering Strait. The Arctic Ocean is bordered by the US state of Alaska, Canada, Russia, Iceland, Norway, and Greenland.

The ubiquitous feature of the Arctic Ocean is the sea ice, that covers the entire Arctic basin during the winter months and only retreats off the shallow water shelf areas in the summer months, creating a permanent cap over most of the central Arctic basin. Icebergs and ice packs will be encountered in the Arctic during *any* season.

Temperature ranks from 0° C in summer to -40° C in winter. The deepest sounding obtained in Arctic waters is 18,050 feet (5,502 metres), but the average depth is only 3,240 feet (987 metres). It is the smallest of the world's oceans, centering approximately on the North Pole.

The Arctic Ocean includes its marginal seas the Chukchi, East Siberian, Laptev, Kara, Barents, White, Greenland, and Beaufort. To some oceanographers, also the Bering and Norwegian seas are the least-known basins and bodies of water in the world ocean as a result of their remoteness, hostile weather and perennial or seasonal ice cover.

However, one of the most *puzzling* facts of Arctic exploration is that while the area is oceanic, covered with water, which is frozen over or partially open; many explorers remarked paradoxically, that *open water* exists in greater measure at the points *nearest* to the Pole, while further south there is more ice. In fact, some explorers found North pole area very *hot* going at times, and were forced to shed their Arctic clothing. There is even one record of an encounter with naked Eskimos.

Polar explorers not only mention fauna (animals) but flora (vegetation) unlike tundra in the *extreme north*. Also, many animals, like the musk-ox that, contrary to expectations, migrates north in the wintertime which it would do only if it reached a *warmer* land there. Repeatedly, Arctic explorers have observed bears heading north into an area where here cannot be food for them. Foxes also are found north of the 80° parallel, heading north, obviously well fed. Without exception, Arctic explorers agree that, strangely, the further north one goes, after a certain latitude, the *warmer* it gets. This obviously shouldn't happen near the North Pole as explained before.

The *fact* is that a north wind brings warmer weather. Coniferous trees were found drifting ashore, coming from the far north. Butterflies and bees were found in the far north, and even mosquitoes, but they are not found hundreds of miles to the south and not until Canadian and Alaskan climate areas conducive to such insect life, are reached.

Similar things happen in Antarctica. Unknown varieties of flowers were also found in the extreme north. Birds resembling snipe, but unlike any known species of bird, were seen to come from the north and to return there.

Hare are plentiful in a far northern area where no vegetation grows but where vegetable matter, is found in drifting debris from the more northern open waters. Eskimo tribes have left unmistakable traces of their migration by their temporary camps, always advancing northward. Southern Eskimos speak of tribes that live in the far north. They hold the belief that their ancestors, came from a land of paradise in the extreme north. As the Arctic explorers sailed further north, the north winds became *warmer* and warmer. The weather was mild and pleasant.

Sometimes they faced a *foehn* phenomenon which has nothing to do with the down slope wind, that occurs in the downwind side of a mountain range. In arctic regions, foehn storms are *warm* winded storms that come out of the far north Arctic in winter. Explorer Peary wrote on pages 214 and 215 of his work, *Nearest the Pole*:

"I expected to hear later of our February foehn in other parts of Greenland, and I was not disappointed. Lieutenant Ryder was living for nine months at Scoresby Sound, on the coast of East Greenland, while we were at McCormick Bay. He was about four hundred and fifty geographical miles south of us. The maximum temperatures he recorded occurred in February and May.

He says (Petermanns Mittheilungen, XI, 1892, p. 256) that these high temperatures were due to severe foehn storms, one of which, in February suddenly, raised the thermometer to 50 F, 8 degrees higher than my instruments had recorded".

These warm Arctic winds have been happening since at least the days of Admiral Peary and Fridtjof Nansen in his 1892-1896 expedition.

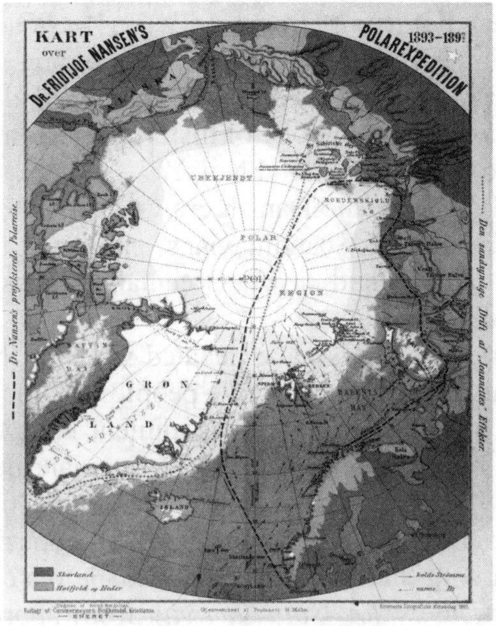

Fig. 68- A 19th century map showing the Nansen's expedition to the North pole. The voyage took just over three years, and the course of her drift changed many times, leading those on board to fear that they were trapped eddies or back waters of the main current. They did not get as close to the Pole as Nansen had hoped, the northernmost point reached was 85° 57' N, where they found themselves on October 16th 1895.

Therefore, this anomaly cannot be attributed to "global warming". What's the source of this warm, in the extreme north? Often the *dust,* carried by the wind, was unbearable. Some explorers, like Nansen, had to turn back due to the dust. Where could this dust come from in the extreme north, a land of ice and ocean? It was thought, it may be volcanic dust from an *unexplored* land to the north.

In 1904 Dr R. A. Harris of the US Coast and Geodetic Survey, published an article explaining why he believed that there must be a large body of *undiscovered land* or shallow water, in the polar basin northwest of Greenland. This will be discussed shortly.

A special mention deserves the *disappearance* of the Greenland Viking or Norse colony. The Greenland Norse colonized North America 500 years before Christopher Columbus "discovered" it, establishing farms in the sheltered fjords of southern Greenland, exploring Labrador and the Canadian Arctic, and setting up a short-lived outpost in Newfoundland.

The Norse colonies in Greenland were law-abiding, economically viable, fully integrated communities, numbering at their peak *five thousand* people. They lasted for four hundred and fifty years and by 1450, they *vanished*, posing one of history's most intriguing mysteries: What happened to the Greenland Norse?

In the early 1700's, the Danish crown sent a missionary here with orders to re-Christianize Scandinavians, presumed to have lapsed into paganism. Sailing up and down Greenland's coasts, he found only natives in kayaks.

For centuries, scholars have debated whether the Viking colonies were wiped out by marauding native peoples, by plagues from Europe, by inbreeding, by economic impoverishment brought by bishops sent from Rome or by abandonment by Scandinavian merchants' refusal to risk their ships in Greenland's increasingly iceberg-infested waters.

But historians have usually pinned most responsibility on the Norse themselves, arguing that they "failed" to adapt to a changing climate or a period of extremely *low solar activity* from approximately 1300 to 1850 a period of cooler global temperatures, referred to as the Little Ice Age (LIA).

The Norse settled Greenland from Iceland during a *warm period* around 1000 C.E. But even as a chilly era called the Little Ice Age set in, the story goes, they clung to raising livestock and church-building while squandering natural resources like soil and timber. Meanwhile, the seal-hunting, whale-eating native Inuit survived in the very same environment.

Over the last decade, however, *new excavations* across the North Atlantic have forced archaeologists to *revise* some of these long-held views. Now, through the use of modern scientific tools, new insights are emerging and scientists are pretty sure they have the answer: The Viking colonies simply *packed up and left*.

"When the climate deteriorated, and their way of life became more difficult, they did what people have done throughout the ages: They looked for a more opportune place to live," says Niels Lynnerup, a forensic anthropologist at the University of Copenhagen in Denmark who studies the Norse. Climate change was clearly driving the Norse, with their sheep and cattle farming traditions, to the edge of survival.

With the onset of the Little Ice Age (from 1300 to 1850), conditions deteriorated across the Norse lands, particularly for people living on marginal farmland in Iceland, northern Norway, and in the Western and Eastern settlements of Greenland. Scholars have wondered why the Norse failed to adapt, dropping agriculture in favour of hunting and fishing, like the Inuit. Turns out, they did up to a point.

An analysis of the bones of Norse buried at Brattahlid and other Norse sites, found that early settlers ate a diet consisting of 80 percent agricultural products and 20 percent seafood; from the 1300's the proportions reversed. But there were limits to their adaptations. Archaeological excavations indicate that the Norse never adopted the harpoons, kayaks, and fishing gear their Inuit neighbours used so successfully.

And while there are plenty of seal bones in Norse dumps, virtually no fish bones have been recovered, leading some to argue that they never took advantage of the ample fish resources in the streams and fjords, even in times of famine. This wouldn't make any sense at all, for a group of people facing starvation unless, the Norse *already knew* where to go for a more opportune place to live. So, why bother?

About 1350, an interesting event took place: the entire population of the Western settlement, about 1000 people, just disappeared. A Norwegian priest from the Eastern Settlement, who visited the Western settlement of the colony, did not find there a single living soul other than feral domesticated livestock. No corpses have been found. The Eastern Settlement lasted for another one hundred fifty years. In total around 3000 people vanished.

Archaeological sites show an *orderly* abandonment, not an apocalyptic end. "You don't find bodies in and around the ruins," says William Fitzhugh, director of the Arctic Studies Centre at the Smithsonian Institution in Washington. "People are being properly buried in church graveyards right up into the 15th century, so it doesn't look like they were wiped out by marauding Inuit or other big disruptions."

Another indication of an orderly retreat: no valuables such as crucifixes, chalices, or chandeliers at church sites, items often found in early medieval churches elsewhere in Scandinavia. "Nothing has ever been found of any real value, just everyday items," says forensic anthropologist Lynnerup. Furthermore, Dr. Lynnerup's genetic studies of modern Inuit from across Greenland has put another theory to rest:

that the Inuit absorbed the Norse. Their mitochondrial DNA (inherited from mothers only) show no European admixture. Archaeological evidence at medieval Inuit sites backs this up, suggesting contact between the two peoples was limited to minor barter.

The previous paragraphs lead to one question: if the Greenland Norse or the Viking colony just packed up and left when things got tough, as recent research shows; where did they go? Taking into consideration the *anomalies* encountered by polar explorers discussed before, the answer might be quite simple: they went somewhere northwards close to the North Pole, for greener pastures and *settled* over there.

In fact, Inuit oral traditions has that when the Norse were struggling to survive, searching parties were sent northwards. Then suddenly like a bombshell breaks upon the weary colony, the wonderful news: "We've found a polar paradise! Sunshine! Game! Grass! One moon's easy journey north! A short lap on the sea ice! Come!" What had they to wait for? A century had passed since the last ship sailed. The last man who had seen a real Norwegian had died. The homeland was but a myth. So, they "packed and, singing songs, departed," the native legend puts it, "suddenly to the northward." Never heard from them.

On the other hand, observers know that they are at the North magnetic pole when a dip needle points vertically into the ground. As the magnetic pole is approached the compass needle begins to dip down, until a horizontal force is too weak to overcome the friction in the compass's bearings. The needle becomes sluggish. Within 20 degrees of the magnetic pole, the directive force is so weak that the compass becomes insensitive and unresponsive.

However, it has been observed by many Arctic explorers who, after reaching high latitudes near to 90° were dumbfounded by the inexplicable action of the compass and its tendency to point vertically *upward*, when in fact it should point *south* as the magnetic pole has been left *behind.*

Because the geomagnetic lines have a *constant* direction, the only explanation for this phenomenon is that these explorers, were venturing on an Earth's surface that curves *inwards* and the compass needle instead to run parallel to the geomagnetic lines, pointed perpendicularly to these.

Another interesting anomaly observed in the Arctic is related with *tidal patterns* in the Arctic Ocean. In his 1904 Manual of Tides, Dr

Rollin A. Harris, a tidal mathematician for the US Coast and Geodetic Survey, gathered *evidence* from tidal patterns and sightings by Europeans and Inuit that all pointed to undiscovered lands near the Pole, a possibility grounded less in mythology and more in *science*.

The regular rise and fall of the ocean's waters are known as tides. Along coasts, the water slowly rises over the shore and then slowly falls back again. Everybody should know tides are caused mainly by the gravitational pull of the Moon and to a lesser extent the Sun. Because the Earth's surface is *not* uniform, tides do not follow the same patterns in all places. The shape of a seacoast and the shape of the ocean floor both make a difference in the range and frequency of the tides. Along a smooth, wide beach, the water can spread over a large area. The tidal range may be a few centimetres. In a confined area, such as a narrow, rocky inlet or bay, the tidal range could be many metres.

Tides are most easily observed and of greatest practical importance along seacoasts, where the amplitudes are exaggerated. Given a known portion of an ocean along with basic tides data coupled with physics background, one can *predict* an almost reliable tidal pattern anywhere, within the shores of this ocean.

This was precisely what Harris did, with tidal data gathered by Peary's 1908 polar expedition and other sources. Harris proceeded to elaborate a scientific report of tides to be used in future explorations and for the Navy. He based his studies in the *assumption* held in those days, that the mostly unknown arctic ocean, was an extended and *uninterrupted* deep polar basin.

Using tidal ranges data from different parts surrounding this ocean and the theory of equilibrium of tidal ranges, a mathematical device that governs the tides; Harris found that the observed values of tidal ranges *didn't* confirm the assumption of an uninterrupted deep basin.

Skipping the mathematical and technical terms, this means that if we imagine the Arctic Ocean like a bathtub filled 3/4 with water; a perturbation caused in one extreme will originate waves that reach the opposite extreme within a specific time and identifiable pattern.

However, if we introduce an *obstacle* in the middle of the bathtub, say a one feet diameter concrete cylinder, a few inches taller that the water surface, the same waves will be *deflected* by this obstacle, reaching the opposite extreme of the bathtub with a *different* timing, angle and shape. Harris found that the tidal hours and ranges where *not* propagated from the eastern Siberian coast to the Alaskan coast, *directly* across a deep an uninterrupted polar basin. Other observations he made, confirmed the same situation. The *discrepancies* were too significant and after a carefully examination, Harris concluded that no errors in the observations were made. The reader should recall that along seacoasts, tidal ranges are amplified.

Faced with this conclusive evidence, he envisioned the *necessity* for a tract of land, an archipelago or an area of shallow water in the Arctic Ocean. This will explain the fact tides, coming from the Atlantic, do not flow *straight across* the Arctic ocean, but follow closely the shore of Asia, cross at the eastern end and flow back to the west along the American shore.

This hypothetical landmass according to Harris, would be trapezoidal in form and containing nearly half a million square miles. Coincidentally, most of this hypothetical landmass is located in the Pole of Inaccessibility, mentioned before. Also, this is linked to one of the objectives of Nansen's Fram expedition.

Nansen in his remarkable endeavour, noticed that ice drifts were not parallel to the wind direction, but were consistently offset to the right, and suggested that it was due to the rotation of the Earth (Coriolis force). In order to explain this observation, another Norwegian, V.W. Ekman would develop a mathematical theory to describe wind driven surface currents but with no avail.

Nansen found this drift. The Transpolar Drift, the ocean current flowing from east to west in the Arctic Ocean, from the Siberian coast around the North Pole towards the Fram Strait between Greenland and Svalbard.

Bear in mind that *officially*, the Arctic Ocean is just a basin of 1 km average depth covered with water and surrounded by landmasses and islands. This definition is the same as the assumption of the Arctic Ocean, that Harris *disproved* through his scientific studies of tides. However, currently this fact is ignored by mainstream science or dismissed as a consequence of arctic currents, such as the Beaufort gyro.

Back to climate and focusing in the North Pole area, a key factor that must be explained is the warmth coming from there as reported by the early polar explorers. Bear in mind this is not a climatic weirdness blamed to hot springs or oceanic "warm" currents, let alone the so called "climate change".

This is a steady phenomenon covering hundreds of square miles. We know from the beginning of this chapter; the reasons why polar climate is bitter cold with temperatures sometimes well below zero. Anything warm coming from the tropical or temperate areas, whether by sea or airborne will inevitably be forced in the polar zones, to lose not only its thermal energy but also moisture (humidity) and thus it becomes dry and cold.

The poles are very dry, in fact, the South Pole is the driest place on Earth, surpassing the Sahara Desert.

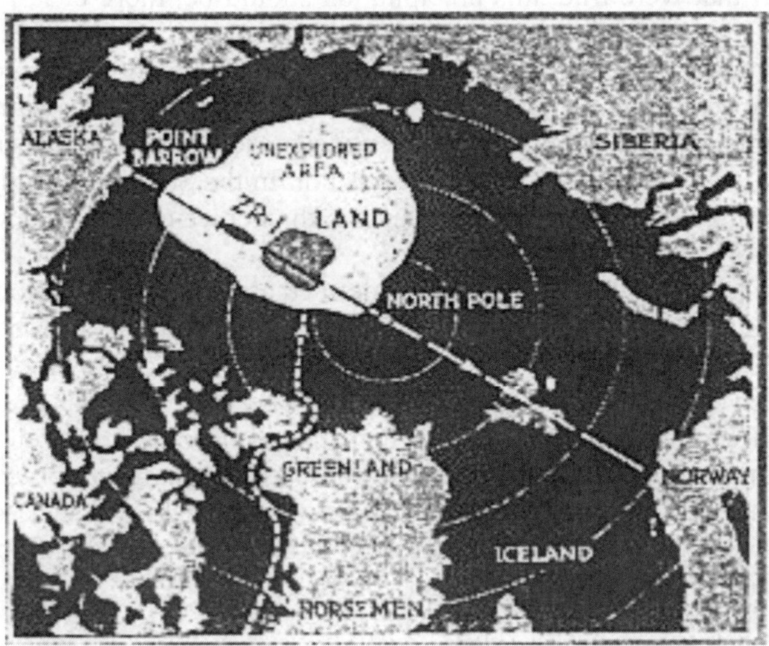

Fig. 69- This map shows the proposed transpolar air route of the ZR-1 (USS Shenandoah) from Alaska to Norway, in 1923. Cutting the distance to European and Asiatic capitals from 11,000 to 5000 miles, this route would pass across what may geologists believe to be an unexplored polar land on the opposite side of the Pole from Iceland. The curving dotted line indicates the possible route whit the "lost Norse colonies" may have followed to this mysterious arctic wonderland; not far from the “North pole of inaccessibility”.

This raises one apparently silly and perhaps overlooked question: where does all the *snow* from polar regions comes from? Because theoretically it shouldn’t be there any snow at all. It would be impossible snowing *without* moisture.

Besides evaporation in polar regions is too low to account for a significant snowfall/rainfall. However, there are an average of two kilometres thick, compacted snow caps covering Greenland and Antarctica. Something is missing in this picture.

The official explanation is that ice sheets, form in areas where snow that falls in winter, does not melt entirely over the summer. Over thousands of years, the layers of snow pile up into thick masses of ice, growing thicker and denser as the weight of new snow and ice layers compresses the older layers.

If that were true, this precipitation should be, more or less evenly distributed around the poles but instead, it is very *selective* because in the Northern Hemisphere the two kilometres of compacted snow cap only exists in Greenland and if we include the oceanic warm Gulf Stream this cap should be more widespread at least in Siberia and Northern Canada plus Alaska but it is not. In the Southern Hemisphere, East Antarctica has a more pronounced, thicker ice cap, than the western part.

Technically speaking, snow is defined as "solid precipitation which occurs in a variety of minute ice crystals at temperatures well below 0° C but as larger snowflakes at temperatures near 0° C". Snow forms when tiny ice crystals in clouds stick together to become snowflakes.

If enough crystals stick together, they'll become heavy enough to fall to the ground. Snowflakes that descend through *moist air* that is slightly *warmer* than 0 °C will melt around the edges and stick together to produce big flakes. Snowflakes that fall through cold, dry air produce powdery snow that does *not* stick together.

Therefore, snow is formed *only* when temperatures are low but *not* extremely low and there is *moisture* in the atmosphere in the form of tiny ice crystals. But in polar regions, air is *dry* and *too cold* to favour enough snowing. Unlike the tropics, where water vapor concentrations reach 40,000 ppm, there's little to no competition with water vapor concentrations at the poles.

Where does this moist and warm air comes from, which allowed the snowing that originated the thick ice caps we see today in Greenland and Antarctica? Because polar regions are practically *isolated* from the warm, moist winds and oceanic currents coming from the tropics. Therefore, the origin of this anomaly must be somewhere inside the polar regions and coming from *beneath* the surface. Invoking the so-called "climate change", to explain this anomaly, would be a farce.

In fact, the polar front *precludes* the access of warmer and moist winds from the Equator carried by the Farrell cell which is blocked by the Polar cell, mentioned before. In fact, the two air masses at 60° latitude, practically do not mix and form the polar front which *separates* the warm air from the cold air. Thus, the polar front is the *boundary* between warm tropical air masses and the colder polar air moving from the north. (The use of the word "front" is from military terminology; it is where opposing armies clash in battle.)

The polar jet stream aloft is located above the polar front and flows generally from west to east. The polar jet is strongest in the winter because of the greater temperature contrasts than during the summer. Above 60° latitude, the polar cell circulates cold, polar air equatorward. The air from the poles rises at 60° latitude, where the Polar cell and Ferrell cell meet and part of this air absorbs some thermal energy and returns to the poles, completing the Polar cell.

Therefore, almost no warm and moist air, coming from the tropics, can reach the poles. On the other hand, an often-overlooked fact is that the polar *cold* jet which is strongest in winter is often accompanied with warm *foehn* storms coming from the nearby North pole, precisely in winter. How can this be?

At this point is reasonable to figure out a plausible explanation of the observed anomalies in the polar regions. Apparently, they contradict the physical laws, mentioned at the beginning of this chapter.

In scientific research, it happens that, sometimes new data don't fit the current theory and this means either the theory must be revised or discarded but it also means that *something* hasn't been accounted for. This something is technically known as a *hidden variable*. For instance, during the 19th century, astronomers found that the orbit of Uranus was affected in such a way that didn't fit the Newton's Law of Gravitation.

The conclusion was that an unknown (hidden variable) celestial body, must be responsible for this odd behaviour of Uranus's orbit and when telescopes pointed where the calculations showed the likely position of this celestial body, Neptune was discovered. The hidden variable, in our case will be explained shortly.

There is also, the phenomenon of the *polar auroras*. The aurora (plural: aurorae/auroras) is a bright glow observed in the night sky, mostly in the polar zones. For this reason, some scientists call it a "polar aurora". In northern latitudes, it is known as the aurora borealis. Especially in Europe, it often appears as a reddish glow on the northern horizon, as if the Sun were rising from an unusual direction. The aurora borealis is also called the northern lights since it is only visible in the

North sky from the Northern Hemisphere. Its southern counterpart, aurora Australis, has similar properties. Auroras come in blue, green, red colours and twisting, diffuse or curtain-like features.

Officially, this phenomenon is caused apparently by collision of oncoming charged particles from the Sun (e.g. electrons), with atoms and ions in the Earth's upper atmosphere (at altitudes above 80 km). These charged particles are typically energized to levels between 1,000 and 15,000 electron-volts and as they collide with atoms of gases in the atmosphere, the atoms become energized. Shortly afterwards, the atoms emit their gained energy as light.

Variations in colour are due to the type of gas particles that are colliding. The most common auroral colour, a pale yellowish-green, is produced by oxygen molecules located about 60 miles above the Earth. Rare, all-red auroras are produced by high-altitude oxygen, at heights of up to 200 miles. Nitrogen produces blue or purplish-red aurora.

However, mainstream science would admit that this polar phenomenon, is still not fully understood. What is well known is auroras are lined up with the local direction of the magnetic field lines, suggesting that aurora is *shaped* by the Earth's magnetic field. Indeed, satellites show auroral electrons to be guided by magnetic field lines, spiralling around them while moving earthwards.

On the other hand, solar electromagnetic radiation (the so-called photons) and cosmic rays do reach upper atmosphere. The absorption of this short wavelength radiation, causes the atmospheric layer known as the *ionosphere* or thermosphere.

The ionosphere is part of Earth's upper atmosphere, between 80 and about 600 km where Extreme Ultraviolet (EUV) and x-ray solar radiation ionizes the atoms and molecules thus creating a layer of electrons. The ionosphere is important because it *reflects* and modifies radio waves used for communication and navigation.

The upper part of the mesosphere, and most of the ionosphere is where officially, auroras do occur in a band known as the *auroral zone*, which is typically 3° to 6° wide in latitude and between 10° and 20° from the geomagnetic poles. At all local times (or longitudes), most clearly seen at night against a dark sky.

However, few people know that the more beautiful the aurora, the greater the radiation. The North Pole (along the South Pole) is the part of the Earth that is bombarded with the most *cosmic radiation*. By its very nature, the North Pole is where cosmic particles from the Sun and elsewhere in the galaxy, are sucked into the Earth's atmosphere by the Earth's magnetic field. Radiation at the North Pole is two to four times higher than at the equator.

Furthermore, radiation also increases sharply at higher altitudes. This is why crew members of flights covering the polar or arctic routes, are at risk of developing certain types of cancer. Every time flight attendants and pilots, fly on the route close to the North Pole, they are exposed to as much radiation (0.1mSv) as if they were getting their chest X-rayed. Indeed, this cosmic radiation exposure exceeds that of many nuclear plant workers.

The trouble with the standard explanation of the origin of auroras, has to do with the misunderstanding of the concept of "solar wind" and its interaction with Earth's magnetosphere. The solar wind (Birkeland's current) is a stream of energized, charged particles, primarily electrons and protons, flowing outward from the Sun, through the Solar System at speeds as high as 900 km/s and at a temperature of 1 million °C.

This solar wind reaches Earth's magnetosphere which is the region of space surrounding Earth, where the dominant magnetic field is the magnetic field of Earth. In fact, the magnetosphere forms an obstacle in the path of the solar wind; causing it to be diverted around the magnetosphere, at a distance of about 70,000 km before it reaches that boundary, typically 12,000-15,000 km upstream, where a bow shock form.

The width of the magnetospheric *obstacle,* abreast of Earth; is typically 100,000 km and on the night side, a long "magnetotail" of stretched field lines extends to great distances. As the official explanation goes, a blast of charged particles from the Sun hits the Earth's magnetic field.

Then somehow it transfers energy and material into the magnetosphere, that excite charged particles (atoms and ions) in the Earth's upper atmosphere. These interactions cause oxygen and nitrogen molecules, to release photons of light and move along the magnetic field lines of the polar regions of the atmosphere, causing the aurora.

As explained before, auroras are *upper atmospheric* phenomena, reaching a maximum of 400 km of altitude, a tiny fraction of the magnetosphere. Solar wind is blocked or deflected some 15,000 km or more in *space.* Therefore, is *difficult* to transfer matter and energy from the boundaries of the magnetosphere that further, would reach the Earth's atmosphere. In fact, some particles from the solar wind can leak this shield but end up trapped in the Van Allen's belts, hundreds of miles from Earth. If some particles still manage to leak these belts, the amount would be too *small* to affect considerably the Earth's atmosphere and thus originate auroras.

Any transfer of energy from de solar wind, is very inhomogeneous: it happens only in specific locations and *not* always in the polar regions. In fact, 80 percent of the energy transfer happens in about 50 percent of the space.

A portion of this energy is transferred from the outer magnetosphere inwards by magnetohydrodynamic (MHD) waves, which propagate in the magnetospheric plasma along magnetic field lines into the upper atmosphere (ionosphere).

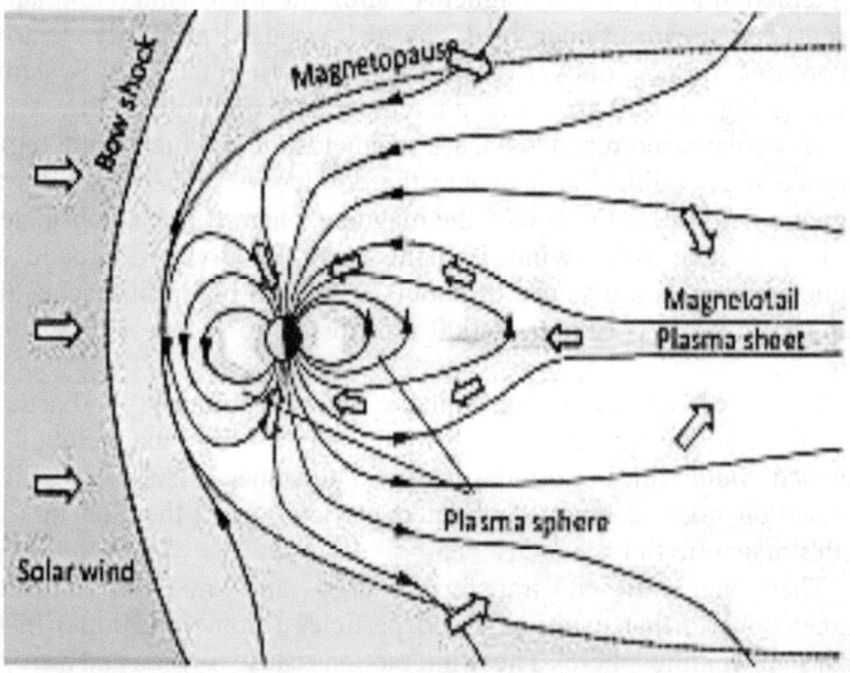

Fig. 70- Earth's magnetosphere acts as a barrier that deflects the "solar wind" (Birkeland's current), which is made of plasma. The boundary or barrier is known as the magnetopause and is located an average 90,000 km from Earth. Between magnetopause an Earth exists the Van Allen belt, which acts as a second barrier trapping "solar wind" particles, that might leak through the magnetosphere.

There, the energy is *dissipated* through frictional (Joule) heating, leaving practically *no* energy to light up auroras. Moreover, due to Sun's rotation, the solar wind is not blasting straight at us, but is swirling in a spiral way, with the swirls sweeping *across* us with less energy.

US and Russian space probes launched in the 1960's, *published* measurements of the particles more or less steadily emitted by the Sun, to have energies of no more than 0,5-1 keV. These energies are *too low* for auroral excitation. But today, mainstream science reassures us that the solar wind particles reach Earth with energies between 1 and 15 keV, as mentioned before. Why this deliberate discrepancy? Also, the energies of solar wind radiation have been measured to be about the same, *all* the way from the orbit of Venus out to the boundary of the geomagnetic field about 70,000 km from Earth.

Therefore, if the solar wind particles are indeed responsible for the auroras, as we are led to believe; these particles must be *locally accelerated.* This is, their energies would have to be suddenly increased somewhere in the neighbourhood of Earth, before colliding with its magnetosphere and there is no physical reason for that.

In fact, most of these energetic particles drift *around* the Earth so as to remain roughly in regions of *constant* magnetic field. It is known that the speed of solar wind has dwindled and so its energy, to its *weakest* state since recording began, researchers say. While researchers already knew that solar winds fluctuate in 11-year cycles, the current doldrums trump the declines seen over the past 50 years. With a weak solar wind's speed, how it is supposedly being *capable* to ignite auroras?

All of this is what makes *difficult* the standard explanation of auroral lights being caused entirely by solar wind. In fact, an alternate hypothesis was suggested (Hulburt, 1928) that the aurora is due to ultra-violet light of the Sun which produces *ions* and *electrons* in the high atmosphere of the Earth above 200 km or so. These diffuse along the magnetic lines of force, concentrate at the magnetic poles of the Earth, recombine and in some way (?) yield up their energy to form the aurora. The total aurora energy during a strong display is estimated to be 1015 erg sec^{-1}, which is in rough agreement with the energy of the upper spray of photoelectric ionization in the high atmosphere, indicated by wireless telegraphy.

On the other hand, as we will see shortly, an alternative cause to the "solar wind" that apparently causes the auroral lights, indeed travels *upwards* from Earth; not downwards from space as required by the official explanation. Also, it has the right energy to light up auroras. Very little is known about the work of Professor Neil Davis, author of

The Aurora Watcher's Handbook published in 1992, who used to be a geophysicist at the University of Alaska in Fairbanks. He found *evidence* that indicates that the auroral energy, must come from *inside* Earth.

Fig. 71- It is confirmed that aurora borealis and aurora australis occur together and are pulsating mirror images of each other, on a global scale. The pulsating forms varies in brightness at exactly the same times. This fact rules out "external factors" such as bundles of charged particles, bouncing back and forth, along geomagnetic lines.

What his research team discovered nearly half a century ago, was that pulsations in the auroras occurred exactly at the same time in the southern aurora as in the northern aurora, meaning this is a *synchronic* phenomenon. Professor Davis concluded that "Whatever the cause, it would seem that the triggering of the pulses must occur in the Earth's equatorial plane." No solar wind activity is involved at all".

Furthermore, a NASA study led by two scientists: Drs Rick Chappell and Barbara Giles titled: *Earth Weaves its Own Invisible Cloak*, published in 1997, shows that indeed "polar fountains" fill magnetosphere with ions.

After sending two satellites to collect data from the Artic and the Antarctic regions, they found that those probes measured *gases flowing upward* from the ionosphere into space.

At first glance, the particles that slammed into the ionosphere to paint the aurora borealis, also energized enough atoms to head spaceward. Also, the satellites *confirmed* through energy measurements of the lowest energy ions, that the source was somewhere between the ground and the ionosphere, *not* the solar wind. This vertical flow upward would explain a characteristic *shape* of auroras. Tall curtains of colour silently shift and alter across the night sky.

Polar fountains can explain why the intensity of the Van Allen radiation belt *increases* during an aurora, regardless of solar activity, as they simply "inject" particles into this radiation belt. This remarkable discovery, which should have made the scientific community to take a new look at decades-old *assumptions* about how the Earth interacts with the space environment, was ignored and dismissed as a mere scientific curiosity.

Indeed, as explained before, Earth's magnetosphere shields us from solar wind, deflecting it. That means that scientists do not need to look to the solar wind to fill the magnetosphere with energized particles. Polar fountains are the *real* cause of auroras. In fact, Earth does a good job of supplying all the materials.

A similar explanation was forwarded by English astronomer Edmund Halley, to explain an extraordinary aurora borealis display in 1716. After discussing certain magnetic aspects to this phenomenon, he suggested that it might be due to inherently luminous matter *inside* the Earth emerging through fissures in the North Pole.

Fig. 72- Depiction of the oxygen, helium, and hydrogen ions that gush into space from regions near the Earth's poles. The faint yellow gas shown above the north pole represents gas lost from Earth into space; the green gas is the aurora borealis-or plasma energy pouring back into the atmosphere.

The important question is *how* do those fountains of energized gas that are blowing from the North and South poles, can manage to cross Earth's crust and be able to reach the atmosphere? *What* causes them?

Undoubtedly, these polar fountains offer a hint that can explain the anomalies observed in the polar regions, discussed before.

On the other hand, recently it has been confirmed the rumour that auroras make *sounds*. Some people believed that the aurora raises the sound that accompanies the ripples and flow of the light (scientists said "no"- everyone else "yes"). If the aurora does make sound, this sound must have to be generated here on Earth's *surface* by some sort of electromagnetic-acoustic effect. Any noise generated by the aurora in the upper atmosphere, would take a long, long time to travel all the way to Earth, and the air up (ionosphere) is much too thin to carry sound. So, does the aurora make noises? Yes, it does.

The aurora really does make sounds and they appear to be generated in the air about 230 feet above the ground, according to a Finnish study published in 2012, that used an array of microphones to record and pinpoint the source of a weird surging hiss, during a magnetic storm.

This study was part of the project called Auroral Acoustics. "Our research proved that, during the occurrence of the northern lights, people can hear natural auroral sounds related to what they see," said engineer and acoustic researcher Unto K. Laine in his research posted by Aalto University, Finland.

The recorded, unamplified sounds can be similar to crackles or muffled bangs, which last for only a short period of time. Other people who have heard the auroral sounds have described them as distant noise and sputter. Because of these different descriptions, researchers suspect there are several mechanisms behind the formation of these auroral sounds. These sounds are so soft that one must listen very carefully, to hear them and to distinguish them from the ambient noise.

While the results *identified* the physical location in the *air* where the sounds came from, they didn't explain how the aurora triggered the hissing. Laine's findings prove that the sounds aren't illusions or hallucinations, nor created directly in the human brain by electromagnetic radiation, he wrote. "The result shows that the sound source at this particular case was real and, on the sky, not far away from the ground," he concluded.

No widely accepted physical model or reason for these sounds, officially exists. Mainstream science may either ignore this study or put forward an *ad hoc* hypothesis to hold the claim that this sound is made somehow, by solar wind. Definitely, something *unusual* is going on in the polar regions.

The hidden variable

Warm and moist winds, atypical dust, unexplainable animal migratory patterns, etc. According to the *definition* of polar regions, these things are inadmissible and should be found nowhere in the millions of square miles surrounding the poles unless...the *hidden variable* concept is put forward.

This means that in order to explain the anomalies discussed before, this warm moist air flow along the energized gas ejected upwards, which can be catalogued as *thermal emissions;* are coming from the *inside* of Earth through a sort of *polar opening* located somewhere around the geographic North Pole, more precisely, in the area where Harris placed his hypothetical landmass to explain the tidal anomalies. Similarly, this scenario repeats itself in the South Pole.

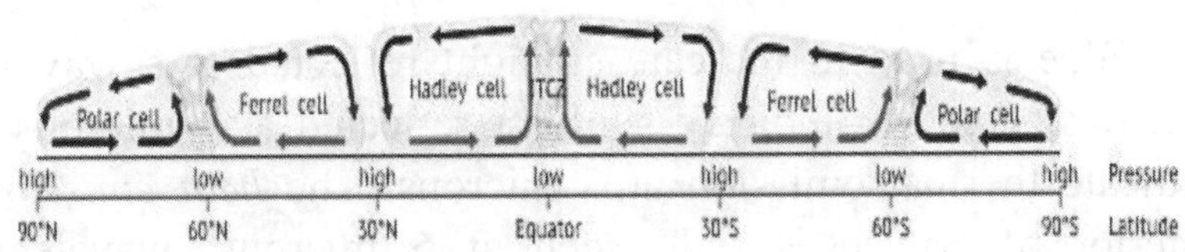

Fig. 73- The Polar cell is the littlest and coldest of the 3 types of the atmospheric circulation cells. This cell is located at each pole. At the pole the air is cold and dense, by consequence, it descends to the ground. Then it flows to a latitude of around 60° North to the Ferrell cell at the North pole and the same latitude South for the Ferrell cell at the South pole. While moving to this latitude the air start to warm, getting moisture and it rises in the atmosphere at around 60° latitude to go back to the pole, getting colder and losing moisture. It is impossible to have warm air currents in the polar regions, unless these are underground thermal emissions.

This idea might seem far-fetched from the conventional perspective, if we accept Earth as a solid body. But two entire chapters have been devoted to show that this concept, is *not* scientifically possible. First the standard solid Earth model is based on inconclusive evidence, has serious inconsistencies and it *lacks* backup from and independent source. Second a shell model with a diffuse core is more *consistent* with recent astrophysical observations and laboratory experiments, than the unrealistic chaotic-accretion model of the solid Earth's origin.

It is noteworthy to mention that Earth is not the only planet to have this sort of *thermal emissions* originating in the poles. This false colour composite image of Saturn is constructed from data collected in the near-infrared wavelengths of light:

Fig. 74- Different degrees of thermal emission showed in red on the southern hemisphere of Saturn. During the Voyager mission in the early 80s, it was possible to measure Saturn's emitted power in the infrared wavelengths. It was found to be symmetrical and pretty constant in latitude, while the incoming sunlight is not. This rules out the Sun as the main heat source to warm the planet, in a such way.

Blue is used to indicate sunlight reflected at a wavelength of 2 microns, green to indicate sunlight reflected at 3 microns and red to indicate thermal emission at 5 microns. The *heat emission* from the interior of Saturn is only seen at 5 microns' wavelength in the spectrometer data, and thus appears red. The dark spots and banded features in the image are clouds and small storms that outline the deeper weather systems and circulation patterns of the planet. They are illuminated from underneath by Saturn's thermal emission, and thus appear in silhouette.

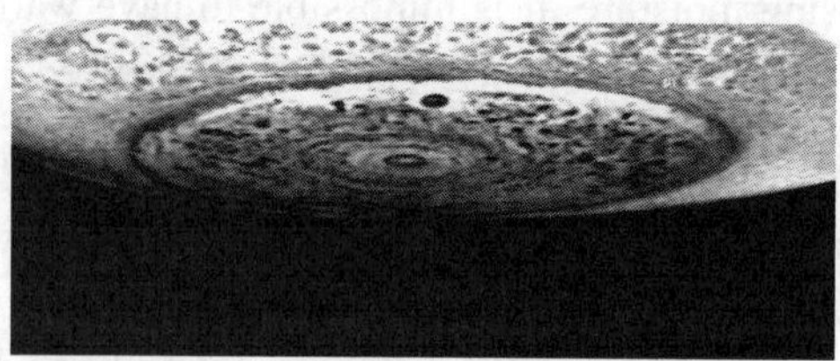

Fig. 75- Perfectly delimited thermal emission encircling Saturn's South Pole. The upper atmosphere (thermosphere) is hottest at Saturn's poles, where these host auroras, which are similar to the northern and southern lights on Earth. Astronomers using flawed computer models, believe electric currents heat up the thermosphere at the poles.

It was discovered that the hot spot temperature contrast in the stratosphere is broader than seen in the troposphere and weakens with increased altitude. One of the key issues is the peculiarly high temperature of Saturn's thermosphere, a great deal hotter than expected from solar heating alone.

In astronomers' jargon, this is known as Saturn's "energy crisis". Apparently, the solution put forward for this crisis, are auroras originated in Saturn. These auroras connect the thermosphere to the magnetosphere, the bubble of magnetic field that surrounds Saturn. The interaction between the two "creates" electric currents, which in turn heat up the thermosphere at the poles.

You will notice the south pole of Saturn is nearly entirely red which indicates, that any heat shown on the poles is strictly thermal emission from the planet itself, *not* reflections from the Sun, let alone "electrical currents" modelled with *imaginary* resistivity parameters; flowing within Saturn's upper atmosphere.

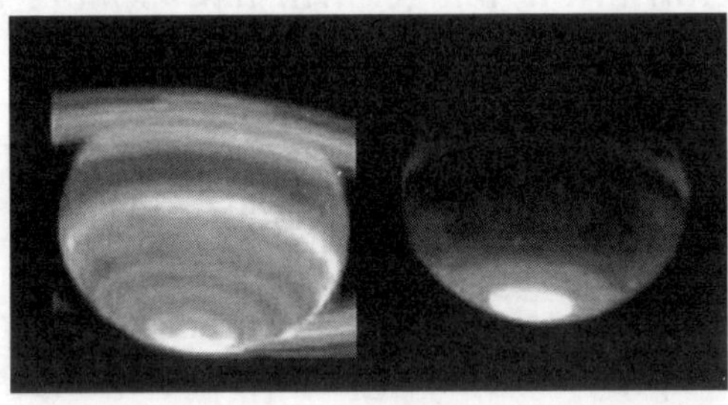

Fig. 76- These side-by-side false-colour images show Saturn's heat emission. The prominent hot spot at the bottom of each image is at Saturn's south pole. The warming of the southern hemisphere was expected, as Saturn was just past southern summer solstice, but the abrupt changes in temperature with latitude were not expected.

Now, put what we have just learned in context of this infrared image of Saturn, which shows a bright hotspot at the extreme South pole. We just can find out that any polar heat is generated by the *planet itself* and is not a reflection of solar emissions.

In 2016 NASA's Juno spacecraft executed several flybys about 4 200 km (2500 miles) above Jupiter's swirling clouds. It was Juno's first flyby with instruments switch on and it sent back first-ever images of Jupiter's North pole.

What Juno saw was storm systems and weather activity, unlike anything previously seen on any of our Solar System's gas-giant planets.

Along with JunoCam snapping pictures during the flyby, all eight of Juno's science instruments were energized and collecting data.

The Jovian Infrared Auroral Mapper (JI-RAM), supplied by the Italian Space Agency, acquired some remarkable images of Jupiter at its North and South polar regions in infrared wavelengths. "JIRAM is getting under Jupiter's skin, giving us our first infrared close-ups of the planet," said Alberto Adriani, JIRAM co-investigator from Istituto di Astrofisica e Planetologia Spaziali, Rome. He also added:

"*These first infrared views of Jupiter's North and South poles are revealing warm and hot spots (italics by the author) that have never been seen before. And while we knew that the first ever infrared views of Jupiter's south pole could reveal the planet's southern aurora, we were amazed to see it for the first time. No other instruments, both from Earth or space, have been able to see the southern aurora. Now, with JIRAM, we see that it appears to be very bright and well structured. The high level of detail in the images will tell us more about the aurora's morphology and dynamics.*"

Among the more unique data sets collected by Juno during its first scientific sweep of Jupiter, was that acquired by the mission's Radio/Plasma Wave Experiment (Waves), which recorded ghostly-sounding transmissions *emanating* from above the planet. These radio emissions from Jupiter have been known about since the 1950's but had never been analyzed from such a close vantage point. The researcher concluded:

"*Jupiter is talking to us in a way only gas-giant world can. Waves detected the signature emissions of the energetic particles that generate the massive auroras which encircle Jupiter's north pole. These emissions are the strongest in the Solar System. Now we are going to try to figure out where the electrons come from that are generating them.*"

The likelihood is that scientists will try to pin down this phenomenon, either to the solar wind or a stream of particles coming from Jupiter's moons. In reality, the cause are Jupiter's *polar fountains*, similar to the ones found on our planet.

Back to Earth, one might ask what about the data obtained by the artificial satellites orbiting our planet? There are specially designed ones, to collect data of the poles by being put into *polar orbit*. Let there be no more speculation as to the orbit of polar satellites. NASA explains themselves and even shows in multiple video segments, how the polar orbit of their ICESAT - Ice, Cloud and Land Elevation Satellite intentionally *circumvents* the polar regions of the planet. Never crossing above them.

According to the author of *Deep Black - Space Espionage and National Security*, William Burrows, the very first polar orbiting satellites that the United States attempted to launch, those fell out of orbit over the poles. This is because they attempted to launch them close to the geographic poles and when the satellites entered the region of the polar openings, the satellites were unable to *maintain* their orbits due to the unanticipated gravity profile of the polar hole and subsequently crashed.

In fact, the variables momentum of the satellite and gravity, must be perfectly *balanced* for an orbit to happen. A low-density rock structure or deficit of mass on the surface of Earth, would cause what is known in geophysics a *negativ*e gravitational anomaly or less pull.

This would *destabilize* the satellite's orbit on the spot. The government quickly learned that to achieve a somewhat polar orbit, the path must be slightly off-centre from the pole so as not to dip into the area of the polar openings.

If one was to speculate, it would presumable the satellites more or less traverse the *rim* of the polar opening before it begins to dip in, so the hole is still a way inside the no-fly-zone, further limiting the actual size of this hole. And what about the innumerable photographic evidence from space showing the polar regions? Well, this is a tricky question indeed.

Because the issue is *who* is providing those images and *what* is going to be shown to the public. In other words, we are dealing here with *censorship*. This is not a conspiracy theory. Take Google Earth for instance, there are plenty of images that have been deliberately *altered* or modified with patches, pixelating or blurring, concealing information as discussed in chapter one. Therefore, polar apertures along other things, won't be shown.

The same applies to many images coming from probes sent to planets, moons and comets exploration. This means without a shadow of a doubt, that there are things that ordinary citizens are *not* allowed to know.

Antarctica, a special polar region

Antarctica, the coldest, driest and windiest place in the world. Is the southern-most continent on the Earth covering 13.9 million km^2 and the continent that we know (apparently) the least about geographically.

Two factors make it difficult to study the geography and other subjects in Antarctica:

One, the cold temperatures and strong winds, along with the 24-hour period of darkness during the long Antarctic winter, make it a very difficult place to work and collect any scientific data.

Two, less than 3% of Antarctica is ice-free. In fact, ice-free Antarctica can be roughly divided into two biogeographic regions: the Antarctic Peninsula, and the continental coast of Queen Maud Land.

Farther inland from the coast, in the mountain and glacier zone that encircles the central ice sheet, the climate becomes colder and drier. All monthly average temperatures are below -5°C, there is no rainfall, and precipitation is only the water equivalent of 10 cm annually (EPA 2001).

Several parts of Antarctica are not covered by any glacier ice at all because it does not snow. This area is considered Antarctic Desert. Being a frozen and windy wasteland, Antarctica still was in those days as it is today, an object of special interest among the nations.

Nonetheless, the coldest place on Earth does contain small patches of land with large freshwater lakes marked by significantly *warmer* temperatures. For numerous decades, this *anomaly* has been observed by perplexed explorers and scientists.

The notable US polar explorer Richard E. Byrd, used a cryptic phrase regarding Antarctica as: “That enchanting continent in the sky, land of everlasting mystery” The exploring expedition organized by back then, US Rear Admiral Richard E. Byrd in 1928, may be considered the first of the mechanical age of exploration in Antarctica. The program was the first of its kind to utilize the airplane, aerial camera, snowmobile and massive communications resources. Although Sir Hubert Wilkins, on November 6, 1928; was the first to fly an airplane in Antarctica, he preceded Byrd by only ten weeks. (Byrd first flew on January 15, 1929).

However, Byrd's flights, made with three planes (Ford monoplane, Fokker Universal and a Fairchild monoplane), were much more significant than Wilkins since they were made in higher latitudes and were tied in with ground surveys.

Great Britain expanded its territory in 1933 by claiming the Australian sector, which to this day is the largest claim in Antarctica.

This could potentially influence Norwegian whaling, and voices were heard expressing the desirability of a Norwegian occupation. But Norway's relations with Great Britain weighed heavily. The superpower did not recognize the Norwegian fishing zone in northern Norway, and the Norwegian government was reluctant to exacerbate this conflict by also occupying land in Antarctica. The issue was therefore set aside in 1934.

During those times, another country that showed keen interest in Antarctica was Germany. In December 1938, Adolf Hoel leader of NSIU, *Norges Svalbard-og Ishavsundersøkelser*, the forerunner of the Norwegian Polar Institute; had been in Berlin to give a lecture about Norwegian polar research.

He recalled that the Germans had contacted NSIU with enquiries about all literature concerning occupation of Arctic and Antarctic regions, and purely by coincidence, he heard about a secret German expedition to Antarctica.

The objective was to occupy areas that many considered Norwegian. Also, in mid-November 1938, when preparations for an Antarctic expedition where in full development, the notable Antarctic explorer Richard E. Byrd came on board by invitation of the *Polar-Schifffahrtsgesellschaft*, to Hamburg to lecture them on what to expect.

There Byrd led in front of 82 persons present, an Antarctic film (title: With Byrd to the South Pole). 54 of these persons were members of the ship's crew and came to the training and prepare for this Antarctic expedition. Byrd, being possibly the most well-informed polar explorer ever, divulged his vast knowledge and details of his exploits to the Germans. He was asked to participate in this expedition. Despite his great expertise with polar explorations, Byrd declined the invitation.

On his way home, Hoel communicated his concerns to Jens Bull, chief of staff at the Ministry of Foreign Affairs. The government reacted with indignation and immediately arranged for a secret meeting on 5 January 1939. During the meeting, Hoel expressed his opinion, which was that if Germany annexed this territory it would create a political scandal. Germany annexed the territory after the *Schwabenland* expedition returned home in August 1939. The government published a decree, establishing a German Antarctic sector under the name *Neu-Schwabenland* and refused to ratify the Norwegian annexation.

On a winter night of December 17th in 1938, the German research ship *Schwabenland* cast off from a dock in Hamburg, destination Antarctica and spent nearly a month over there. Aboard, accompanying the ship's handpicked crew, was some of Germany's top airmen, technicians, oceanographers, biologists, meteorologist and Earth scientists, all members of the first *Deutsche Antarktische Expedition* of 1938-1939.

The people who made up the costly Antarctic expedition were under secret orders not to divulge their mission, the purpose of which little is known today. Apparently, that expedition would have had to do with developing whaling activity and to map uncharted territory, along recording meteorological and oceanographic data. It was suggested another reason was, the search for energy sources such as coal as it would be needed for the looming war. Anyway, they knew what to look for.

The German Antarctic Expedition 1938/39 was a unique and special event in the history of polar research. It was not a traditional scientific expedition only, but also a characteristic although lesser known momentum of the Third Reich's power and political ambitions, on the distant South Atlantic waters.

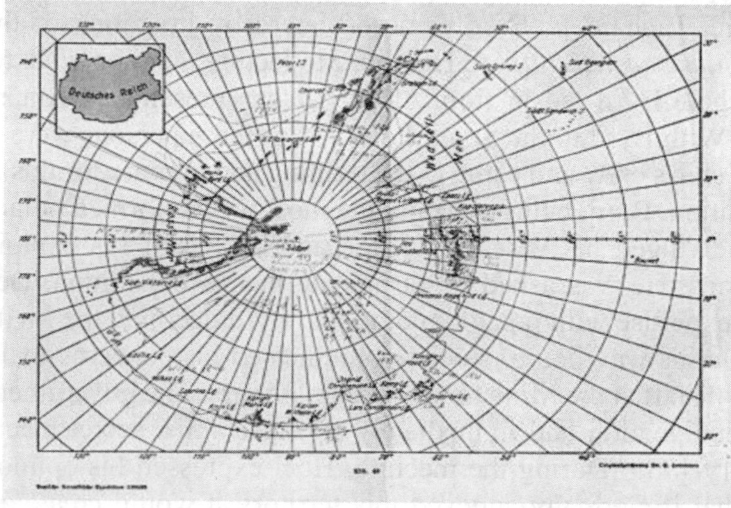

Fig.77- Map issued by the *Deutsches Antarktische Expedition 1938-39*, showing the German annexed territory of New Schwabenland, in Eastern Antarctica. This expedition can be seen undoubtedly, as model for systematic use of aircraft for scientific investigations from the air.

This expedition visited the western part of what is now known as Dronning Maud Land or Neu-Schwabenland and was organized and launched by the Kriegsmarine, the German naval high command together with the assistance of Lufthansa and their ship: *Schwabenland.*

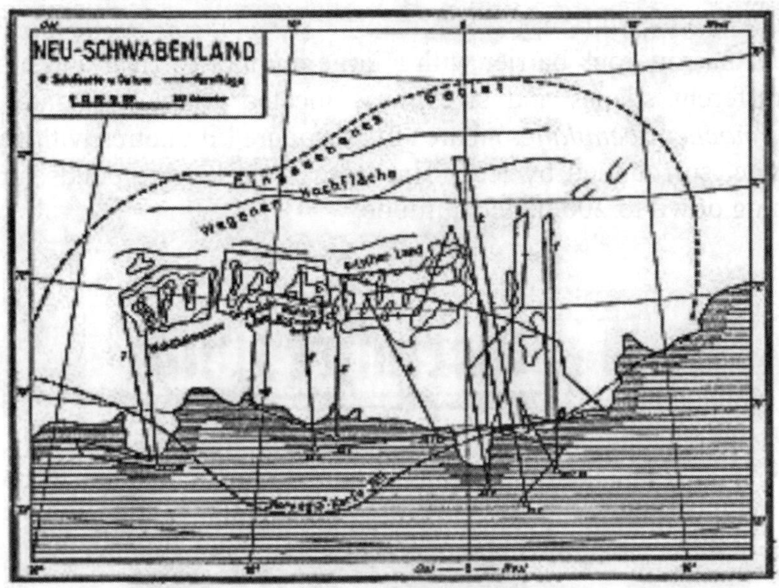

Fig. 78- German expedition map of the Neu Schwabenland region of Antarctica, marking the paths of the seven flights (top of map is directly south). Everyone envisioned a flat, largely featureless, ice-covered wasteland punctuated by large mountain ranges. They were wrong.

The expedition leader was Captain Alfred Ritscher of the German Navy, who possessed the unique qualifications of having both polar and flying experience; in fact, several of the scientific and flying personnel also had some polar experience.

The *Schwabenland* carried two Dornier Wal (Whale) 10-ton seaplanes, the Boreas (North Wind) and Passat (Trade Winds). These were perfect aircraft for polar work. While flying for up to 15 hours, a Wal could carry nearly 4.5 metric tons of fuel, personnel, and equipment over hundreds of kilometres.

Until the *Deutsche Antarktische Expedition* of 1938, it was generally assumed that Antarctica was all snow-lying "island", covered all over by thick ice, save for a few volcanic protrusions, only one of which was still active: Mount Erebus in the Ross area.

For the first time, *ice-free* areas with lakes, signs of vegetation (mostly lichen and moss) and cave inlets were discovered inland 73° latitude South. On February 3, 1939, the crew of the Dornier-Wal *Boreas* plane, made a special discovery while returning from a photographic reconnaissance flight in the *Wohlthat* massif to the floating base of the expedition.

A narrow rock barrier with a large number of clear and blue lakes of different shapes and sizes, was located which was later named *Schirmacher Seenplatte;* an area of 35 square kilometres with wild *mild* climate, surrounded by ice. The Boreas circled the ponds, gradually coming down to 200 meters altitude.

Fig. 79- The discoveries in Antarctica by the Germans in 1939, of clear ground with a vast mountain range and open lakes, with water that remained unfrozen; were on the cover of the magazine for the German government economic policy. The mild climate would allow a plan for territorial determination, to secure German population settlements.

The temperature began to rise to just below freezing (0° Celsius). After a careful study of the water below, to ensure that there were no dangerous ice or rock protrusions, the landing site was selected.

The Boreas skimmed the placid surface of the water, then splashed down, making red waterspouts on either side of its grey hull. After cutting the engines, the flying boat glided slowly across the pond and came to rest, gently nudging the two-foot-thick ice shelf which surrounded this strange body of open water. making red waterspouts on either side of its grey hull.

The weather was so pleasant, in fact, that the pilot was able to shed his sealskin flying suit and remain quite comfortable. The discovery of the *Schirmacher-Seengruppe* of warm water lakes in the Antarctic ice desert was like finding an oasis in the Sahara.

Why is this relevant? It goes to show that *Neu-Schwabenland* was much more than a territory claimed for whale hunting purposes only. Huge areas of *mild climate* ideal for human settlement in an otherwise *frozen* land, would suit the German government.

Since it appears in an old-school globe, this could indicate that, at some point, it was in fact considered a German state, like Hesse, Thuringia, Bavaria or any other of Germany's 16 states and taught to German students.

Of course, the results of this scientific research were used not only for civilian but military goals such as a reconnaissance mission to further set up a naval base for instance. Otherwise, Germany's interests in Antarctica wouldn't have bothered the US government, to worry about a potential threat to the solidarity and defence of the Western Hemisphere, should a second world war occur.

As we can see from the above paragraphs a major confrontation between American and German forces seemed to be a very real and present danger in Antarctica in 1939. However, in retrospect we find that, though the German presence in the "land of everlasting mystery" was the publicized reason for Admiral Byrd's hurry-up expedition, at no time did Admiral Byrd or those under his command make any attempt to observe what the German expedition was up to.

In fact, President Franklin D. Roosevelt established the U.S. Antarctic Service and designated Admiral Byrd as commanding officer. For the first time, Congress authorized partial funding for one of Byrd's expeditions.

Previous Antarctic expeditions were a privately-financed affair. The United States Antarctic Service Expedition (aka Byrd Antarctic Expedition III), mapped a thousand miles of coastline between the two new established bases.

Scientific studies were carried out in geology, biology, cosmic rays, and other disciplines and recommendations were made on types of cold weather clothing and gear. US flags and benchmarks were left throughout the newly explored territory, which was unofficially claimed for the United States. The venture ended in 1941 due to the Second World War.

Despite this expedition was highly publicized by the media and followed by many during its preparation, almost *no* public information was released *after* it finished. Just vague statements about the expedition being successful, regarding scientific observations.

However, the *Deutsche Antarktische Expedition* did publish the scientific research and several books were written for the public in Germany. Captain Ritscher and his two Dornier Wal flying boats (*Boreas* and *Passat*) flew extensively and produced more than 11,000 photographs that covered an area of over 250,000 square miles.

Even the maps made during this venture, proved to be a useful tool in future international Antarctic expeditions. Nonetheless, is possible that certain information could have been kept classified.

Is it possible that these "secret orders" contained the real reason for Admiral Byrd's expedition? And was the fact that the Germans were also conducting a "secret" expedition in Antarctica make the Admiral's real reason for putting together the US expedition a matter of up-most importance? Could both expeditions have been carrying secret orders to explore unknown "lands beyond the pole?"

Were the Americans and Germans involved, in 1939, in a race to be the first to gain entrance, and explore the legendary lands inside our Earth?

In 1946, the US Navy launches Operation Highjump, a military expedition to the South Pole organized by Admiral Richard E. Byrd and led by Admiral Richard Cruzen. They assembled officially a training mission including: 4,700 sailors, 13 ships and 33 aircraft along with a cohort of scientists and at the time Operation Highjump, was also the *largest* operation of its kind to Antarctica. Operation Highjump established the Little America base and started aerial surveys of the southernmost continent.

The expedition discovered, as the Germans did in 1939, that Antarctica had areas that were *ice free*. Admiral Cruzen spoke of several "oases" with ice-free land and warm water. Lieutenant Commander David Bunger and his men flew inland, west from the Queen Mary Land coast where they were amazed to see "a land of blue and green lakes and brown hills in an otherwise limitless expanse of ice."

They landed their 'flying boat' on a lake and tested the water, which was about 30°C. How this could be possible? On February 19, 1947, during a 1,700-mile flight across the South Pole, Admiral Byrd was perplexed by the *lack* of snow and ice in a supposed frozen wasteland wherein he recorded in his diary the following:

"1000 Hours - "We are crossing over the small mountain range and still proceeding northward as best as can be ascertained. Beyond the mountain range is what appears to be a valley with a small river or stream running through the centre portion. There should be no green valley below! Something is definitely wrong and abnormal here! We should be over ice and snow! To the portside are great forests growing on the mountain slopes. Our navigation instruments are still spinning, the gyroscope is oscillating back and forth!"

The term oasis was adopted in the popular press of the West during the US Navy Operation Highjump on February 11th, 1947. It may be worth quoting a few lines from Byrd (1947):

"Early in February the crew of the flying boat commanded by Lt. Cmdr. David E. Bunger, found themselves above a land-scape which made them question their own eyes a land of blue and green lakes and brown hills in an otherwise limitless expanse of ice. This so-called oasis was by far the most important as far as public interest was concerned the single geographical discovery of the expedition".

Bunger Hills, also known as Bunger Lakes or Bunger Oasis, is a coastal range on the Knox Coast in Wilkes Land in Antarctica, consisting of a group of moderately low, rounded coastal hills, overlain by morainic drift and notably ice free throughout the year, lying south of the Highjump Archipelago.

The Bunger Hills are surrounded by glaciers. On the southeast the Bunger Hills is bordered by the steep slopes of the Antarctic ice sheet, on the south and west by outlet glaciers, and on the north by Shackleton Ice Shelf, which separates the area from the open sea.

The ice-free area measures 450 km^2 (174 sq mi) according to some sources even 750 or 942 km ^{2} (290 or 364 sq mi), though these latter values include a marine area not covered by continental ice or the Shackleton Ice Shelf.

The topography is characterized by rugged hills, and there are many freshwater and salt lakes. The largest and deepest lake, Algae Lake is 25 km (16 mi) long and up to 137 meters (449 ft) deep. A sample of the lake was obtained and on analysis it was found that the water was brackish, indicating that the lakes were connected to the sea.

The leader of Operation Highjump, Admiral Richard E. Byrd stated that the Bunger Hills was '...one of the most remarkable regions on Earth. An island suitable for life had been found in a universe of death"

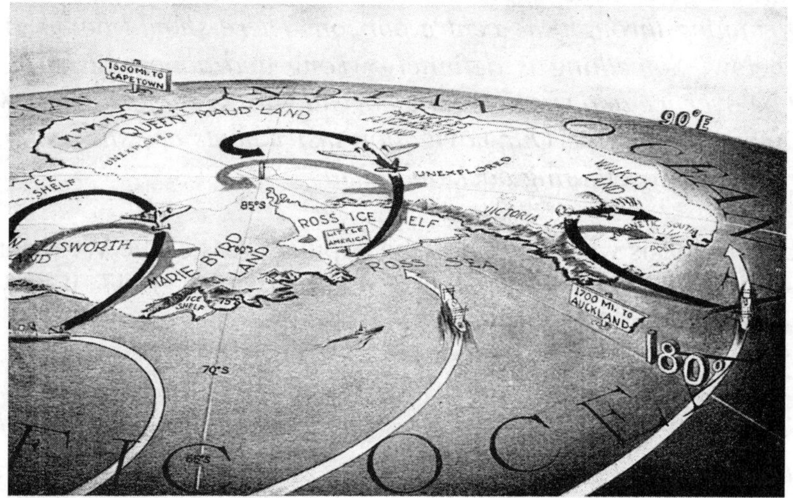

South Polar continent sends one group to eastward and another to westward. Central group bases on Ross Shelf.

Fig. 80- Operation Highjump technically termed the Antarctic Developments Project 1947, was billed as the US Navy's expedition into the "Deep South" concerned primarily with testing standard naval ships and equipment under frigid conditions and training personnel. Under the slogan "Men Against Ice – All Hands" it sold an image of Antarctica being "cruel and rugged, covered by an ice cap colder than the Arctic, by an average of 40 degrees and swept by pounding gales and blinding snowstorms."

The reasoning behind the minute amount of ice and mild climate in the area, is still relatively unknown and remains under intense debate amongst scientists today. For example, a scientific paper (Sledzinski, 1999) describing the area reads:

"The station is situated in a very specific Antarctic "oasis" where the nunatak stones appear above the ice, where there blow very strong

winds, there is very high solar radiation and very high temperature gradient, big differences of the day and night temperature; in summer time the day temperature is usually above zero and the humidity is very low".

It is interesting the remark of the *high* temperature gradient. However, this statement is flawed in regards of the "very high solar radiation". Solar radiation or insolation is all of the light and energy that comes from the Sun. As it was explained before, because the Earth is round, the frigid polar regions *never* get a high sun with a considerable thermal energy.

Therefore, is *impossible* to have "very high solar radiation" in the Bunger Oasis area. Either the author of this paper doesn't know the cause of this unusual warmth or this argument is a *diversion* from the real cause. Due to the considerable area covered and geological studies carried out, heat sources such as underground volcanic activity has been ruled out.

More recently, a new kind of *anomaly* arose in the southern continent. Being Antarctica a hotbed for scientific research, from time to time new discoveries are being made.

Since March 2016, researchers have been *puzzling* over an event in Antarctica, where cosmic rays did *burst out* from the Earth, and were detected by NASA's Antarctic Impulsive Transient Antenna (ANITA), a balloon-borne antenna hovering about 40km over the southern continent, leading physicists to proposed several hypotheses for these "upward going".

ANITA searches for the signals of high-energy particles from space, including the *controversial* lightweight, ghostly particles called neutrinos. Those neutrinos can interact within Antarctica's ice, producing radio waves that are picked up by ANITA's antennas.

The surprise came when it detected *high energy* particles coming off the icy ground instead coming from space.

Calculations show that these particles must have had energies more than 0.5 exa-electron volts or 70,000 times higher than the energy achieved with the most powerful *particle accelerator* in the world.

At the beginning, it was thought that cosmic rays entered the North Pole region, crossed Earth's diameter and surfaced in Antarctica. But high-energy neutrinos, as well as other high-energy particles, have considerable cross-sections which means that they'll almost always crash into something soon after zipping into the Earth and never make it out the other side.

When the ANITA events were detected, the main hypotheses were an astrophysical explanation (like an intense neutrino source), a systematics error (like not accounting for something in the detector), or physics beyond the Standard Model.

Right now, researchers just know that whatever this particle is, it interacts very weakly with other particles, or else it would have never survived the trip through the planet's dense mass. In reality, a crack or opening in the South pole would make the transit of this "particle," easier.

Standard physics is at *a loss* to explain this phenomenon. A new particle about 500 times as massive as the proton, has been hypothesized that might explain this anomaly. Based in only two observations, scientists admit such limited statistics make it hard to really pin down the cause of the anomalous events.

However, this anomaly brings similarities with the "polar fountains" mentioned before, in the sense that it is coming out from inside Earth and thus, shares a common origin which will be discussed later. This is a solid evidence that something new has really arrived.

The reader should notice that for some reason, the frozen and unhospitable wasteland of Antarctica, has become a reserved or *restricted* area in several regards. Since the 1940's, the US had gradually kept a keen interest into developing policies to be carried out in Antarctica to ensure its jurisdiction.

For the United States Antarctic policy is set by the interagency Antarctic Policy Group (APG), which was established in 1965 by President Johnson on advice of then acting Secretary of State George Ball.

The APG is composed of the Secretary of State, who chairs the group, the Secretary of Defence and the Director of the National Science Foundation (NSF).

Responsibilities are divided roughly as follows: State formulates overall United States Antarctic policy, NSF manages and funds the entire United States Antarctic Program (USAP) and designates the senior United States representative in Antarctica and Department of Defence plans and executes logistical support for the USAP.

The NSF is the *responsible* for representation (photos, maps, etc.) of the polar regions. In addition to its strategic placement of stations, the

United States exerts its influence in other ways over parties to the Antarctic Treaty, to ensure that they operate within the framework of the treaty.

The interdependence of the New Zealand and United States programs on Ross Island is one example. This relationship, the closest and most cordial working relationship among parties to the Antarctic Treaty, has endured and thrived since 1953.

In the 1970's, US officials also have expressed an interest in better regulation of what they regard as a new threat to the Antarctic environment: private expeditions and tourism. The National Science Foundation (NSF), which exercises a *monopoly* over all official American trips into the Antarctic region, takes a rather dim view of the upswing in *private* activities on the ice, for one thing, visits to US stations by tour groups can be disruptive.

The obligation to brief and escort visitors takes away from the already limited time available for research. US officials also are concerned about the obvious dangers associated with travel in the remote and forbidding region, as well as the difficulties of rescuing stranded explorers and tourists.

As any avid reader of *National Geographic* knows, private explorers continue to follow in the footsteps of Scott, Amundsen and other heroes by testing their mettle in the Antarctic continent or surrounding seas. The same spirit of adventure also has attracted an increasing number of tourists to the region, starting with the first commercial venture in 1966. At the end of 1988, at least five operators had entered the Antarctic tourism market, offering trips ranging from cruises around the Antarctic peninsula to skiing expeditions on the continent. Some airline companies saw an opportunity to make business.

In the 1970's, Antarctic overflights or flightseeing were a new and exciting breakthrough in airborne tourism tough it wasn't exempt of risks. After operating for just two and half years in 1979, an Air New Zealand aircraft crashed into Mount Erebus in the Ross Dependency, with no survivors. This ended what had been a popular tourist experience.

Air New Zealand business competitor in the same area, the Australian airline Qantas, cancelled its flights in February 1980 to resume these in 1994. Antarctic flights cover *only* the coastal areas and a few miles inland of Antarctica. Touristic and commercial flights don't fly *across* Antarctica nor the South Pole.

Unveiling the clue that points out at the real cause of magnetic fields

As discussed in chapter two, the current explanation of how Earth's magnetic field originates, has *no* scientific basis. Dynamo Theory was based upon the *wrong* assumption of a molten iron core within Earth and the logical fallacy of *oversimplification* applied to this core, to create a magnetic field. Scientists know that magnetic bars exist and have a magnetic field associated to them. In the laboratory under normal conditions, they observe wires carrying electrical current, generating magnetic fields. Then scientists observe the effects of the Earth's magnetic field and wonder, how it is caused.

All they know is that this magnetic field, comes from inside Earth and nothing else. So, scientists rely in the only two known observable magnetic phenomena, where their origins are visible. Magnetized bars and electrical currents. After scrutiny, it is decided that due to high temperatures inside Earth, magnetized material, demagnetizes and ceases to exist. This leaves apparently one option: induced electric origin trough wires. Scientists start to think about an *analogue* electrical mechanism inside Earth that can explain its magnetic field, though there might be *other* causes as well.

The complicate, shaky and ultimately *unfeasible* Dynamo Theory is created. However, two objections arise: first the alleged (unobservable) fluidic metallic currents inside Earth's core, would be *different* from a solid wire under laboratory conditions and more variables (fluid mechanics, pressure, density, etc.) are at play. Second and more important: is this mechanism a reflection of a stable and efficient *Gestaltung* working feature of Nature?

To put it in this way: we *observe* a rainbow and wonder how or what causes it. Then we *observe* a dispersion of light through a prism with the same colour pattern of a rainbow and here it is: a rainbow is caused by drops of water (rain) that wen sunlight hits them at the right angle, these droplets act as prisms dispersing light in colours. Droplets of water and a prism have the same *Gestaltung* when light crosses them, organizing and shaping the rainbow's colours.

Is there a *Gestaltung* for a magnetic field of any celestial body or Earth for that matter? Yes, indeed there is. Earth and all *rotating* celestial bodies do have magnetic momentum *associated* with their angular momenta. In Earth's case this association is what *causes* its magnetic field. How? To answer this question, it is necessary to abandon the *conventional* scientific ideas and go in depth with wat *really* causes magnetism.

The only form of magnetism known until the beginning of the 19th century was ferromagnetism. It was then known that certain materials, when "magnetized", would attract certain other materials. The only materials attracted by a magnet were those that could become magnetized themselves. Since only certain materials exhibited magnetic properties, scientists concluded that magnetism was an "inherent" property of materials. Then in 1820, Danish physicist Hans Christian Ørsted, (1777-1851) studying the relatively new field of *electrical currents* discovered that moving charges produce magnetic effects (Ørsted effect).

A current traveling through a loop of wire "creates" a magnetic field along the axis of the loop. The direction of the field inside the loop can be found by curling the fingers of the right hand in the direction of the current through the loop; the thumb then points in the direction of the magnetic field. With this discovery, magnetism appeared to occur in two different manners: *Ferromagnetism* depending on the material and *Electromagnetism* caused by the so-called "electric currents".

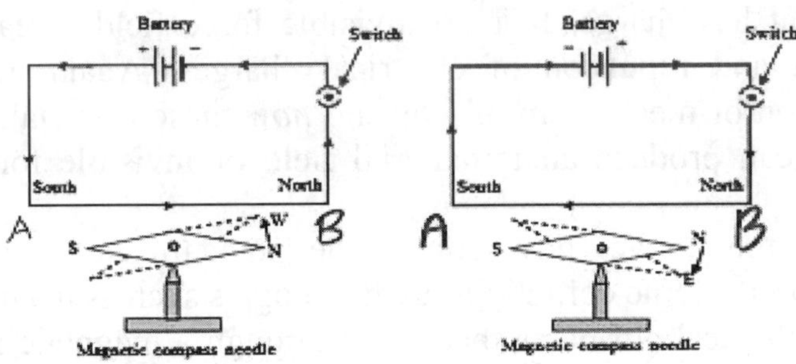

Fig. 81- Experimental setup showing the Ørsted effect. When the switch is pressed and current is allowed to pass in the wire from A to B (from south to north then the North pole (N) of the needle is deflect towards the west. Apparently, the current (or the moving charges) produces a magnetic field. When the direction of the current in the wire is reversed by reversing the terminal of the battery, the North pole of the compass needle is deflected towards the east.

It is important to discuss here tough briefly, the importance of the Ørsted effect and some basic tenets of Electromagnetism, meaning the *real framework* of electromagnetic phenomena and their dynamics; not the current conventional and *superficial* theories about electricity and magnetism. The problem here is that Electromagnetism involves action at distance trough *space*.

In order to make sense of this phenomenon, the official explanation found in any conventional textbook of Physics, avoids to include the *space* between the wire and the needle. You can read that a moving material "particle" or electron, inside the wire, can "create" a magnetic field. There is no explanation or mechanism of *why* and *how* these moving material particles, can produce an immaterial field trough space.

In fact, Physics is *unable* to explain the most basic questions, like WHY is there inertia, gravitation, and HOW does space carry, transfer, and propagate gravitational forces, and the tremendous forces and energies of electro-magnetic interactions and radiation.

Modern physics gets rid of such questions by proudly declaring the absence of *physical explanations* and understanding, thus implying the divinity of its postulates and of its mathematically derived findings.

The same problem is found when science defines charge also known as electric charge, electrical charge, or electrostatic charge and symbolized *q*. Is a characteristic of a unit of matter that expresses the extent to which it has more or fewer electrons than protons.

According to this concept, an electrical field will surround any object that has charge. It is an invisible force field "created" by the attraction and repulsion of electrical charges. Again, there is NO explanation or mechanism of *why* and *how* these material, ponderable charges, can produce an immaterial field or invisible force, through space.

Said the above, now comes the standard definition of what *magnetism* is. Some definitions are tautologies such as magnetism is the class of physical attributes that occur through a magnetic field, which allows objects to attract or repel each other.

Another definition is a force of Nature produced by moving electric charges and this concept is used as a basis to "explain" the origin of Earth's magnetic field.

The problem with this definition is that mainstream science assumes moving electric charges "produce" a magnetic field when in fact, both are correlated having a *common origin* not recognized by "modern physics". Correlation doesn't necessarily mean causation. As we will see shortly, a dynamic space or *vacuum* is what produces the magnetic field and electromagnetic phenomena in general.

The experimental setup of the Ørsted effect is a *system* composed of battery + wire + needle + space. In order to *fully* understand and explain the magnetic behaviour of a needle at distance from a wire carrying an electric current, all *four* key components of the system *must* be included. However, as explained before, mainstream science *ignores* space and pins down electricity and magnetism on material "particles" either moving or at rest.

The electron, considered a "fundamental particle" as it is known today, was "discovered" nearly eight decades *after* Ørsted published his experiments. In between, an extraordinary advance was achieved by notable and talented physicists in the 19th century; to unveil the foundations of how Electromagnetism *really* works.

One of those physicists was the outstanding Michel Faraday perhaps the best *experimental physicist* ever, who made landmark discoveries in both chemistry and electromagnetism, had notions regarding matter and force altogether *different* from most scientists. Faraday did consider the *space* between the needle and the wire in the Ørsted effect and in electromagnetic phenomena in general.

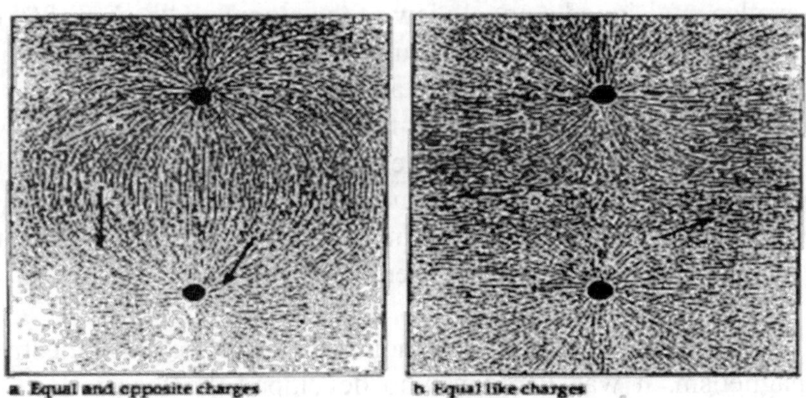

Fig.82- After observing how iron filings arrange themselves around a bar magnet, Faraday called the curved paths of the filings, lines of force, which endow space with *physical properties*. The arrows (vectors) indicate the magnitude and direction of the force. The same applies for electric charges. How can electric charges exert such pushes and pulls without coming in contact with each other? In fact, how does one charge "know" that another is around?

That's why he introduced the concept of *field,* a three-dimensional causal nexus surrounding the flowing electric currents and magnets.

What is the nature of space where fields operate? Is it empty or structured; dynamic or static? Is there a subtle medium of the Universe, an *aether* that underlies matter, energy and fields? These questions were always in Faraday's inquisitive mind.

He maintained that if the aether exists, it should have other functions than simply the transmission of radiations. In 1844, Faraday proposed that lines of force are *real physical entities* of disturbances in the aether propagating over distance. Here is the key to understand the cause of electromagnetic phenomena: a *disturbance* in a specific medium filling space.

This disturbance o cause is an initial *imbalance* of forces in space, an imbalance in the energetic structure of the vacuum energy. The effect is the phenomenological response trying to *restore* this balance. No mediating particles, "virtual photons" nor mathematical abstractions devoid of physicality, are required at all.

For example, lines of magnetic force, and their convergence or divergence from their respective poles, appear in the lineup of iron filings when sprinkled on a piece of paper placed over a magnet. Faraday further conjectured that lines of magnetic force formed *closed circuits*. Faraday also envisioned a novel view of the atom as a "center of force", with each atom extending throughout the whole of space, with infinite range, such as the electric and gravitational force.

Thus, Faraday considered that forces carried by the aether were more *fundamental* than matter. Furthermore, since he *observed* that lines of force tended to contract longitudinally and expand laterally, he modeled them as tubes with a fluid inside, rotating axially and creating centrifugal bulging, making an ensemble of *dynamic vortices* as in a fluid.

Though Ørsted was first to demonstrate a link between electricity and magnetism, it was Faraday who developed the theory of *why* a moving magnet generates an electric current in a nearby wire. This phenomenon is called induction. In 1831, Faraday first reported his discovery of *electromagnetic induction*, arguing that electric currents were induced in a changing magnetic field, or when a conductor "cut" what he called magnetic "lines of force."

These lines of force, made *visible* by iron filings in the space around magnets, formed the backbone of Faraday's theory. They demonstrated *physical activity* in the space around current-carrying wire, and around magnets. As Faraday's theory of the field evolved, he increasingly insisted that the space *around* conductors was an *active component* of the electromagnetic energetic system. However, "modern physics" dismisses this approach. Space is "empty" and non-dynamic.

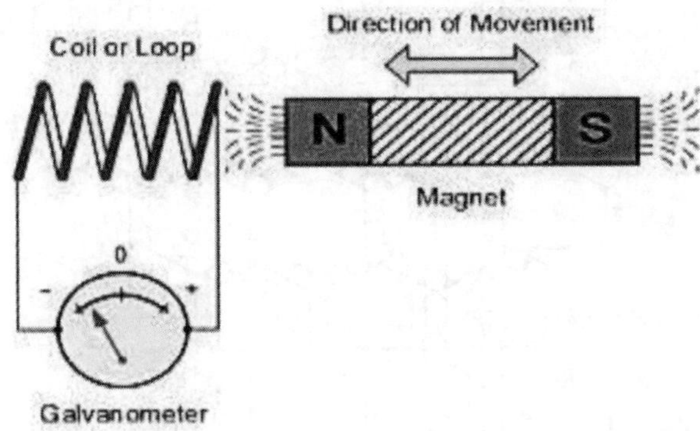

Fig. 83- Electromagnetic induction by a moving magnet. Faraday noticed that when he moved a permanent magnet in and out of a coil or a single loop of wire, it induced an electromotive force or emf. In other words, a voltage and therefore a current was produced. The induced voltage in the coil did not depend on whether the magnet was moved toward the coil or if the coil was moved toward the magnet. Only a *disturbance* in the medium, or space, produced by either the movement of the coil or the magnet, was required. This relative motion creates *rotational stress* in the surrounding vacuum (aether) while dragging it along, producing a current in the loop.

Remarkable Scottish physicist and truly, an unsurpassed genius James Clerk Maxwell, read Faraday's *Experimental Researches* in three volumes and was struck by the lines of force concept. Also, Faraday turned to Maxwell for a theory of his experimental discoveries, who devised a *physical conception* of the electromagnetic field, Maxwell is credited with having brought electricity, magnetism, and optical phenomena, together into one unified theory. However, Maxwell's *real* achievements are in fact totally alien to "modern physics."

In Parts I and II of his 1861 paper, *"On Physical Lines of Force"* Maxwell explained magnetic force and electromagnetic induction *hydrodynamically* in the context of an all-pervading sea of tiny *aether vortices* in which electric particles circulate around the edge of the vortices. These vortices bond together by aligning along their mutual *rotation axes* to form solenoidal vortex rings which constitute the prevailing *magnetic* lines of force.

Stability is established on the basis of a *balance* between tension acting along the lines of force and centrifugal pressure acting sideways from them. This Maxwell's sea of aether vortices is called a *granular* model of space.

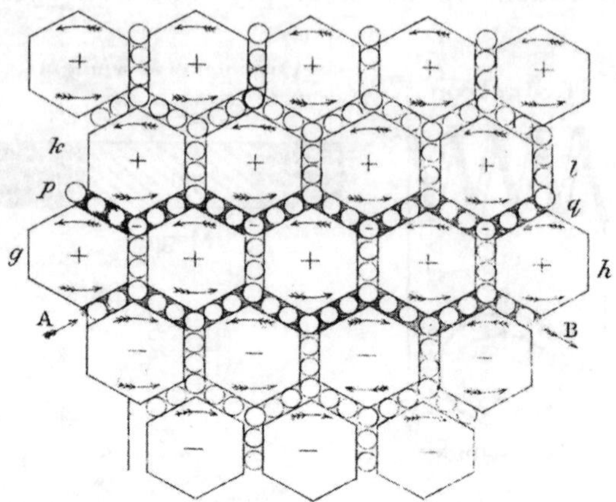

Fig. 84- Maxwell's diagram from the *Philosophical Magazine* of 1861 showing the aether's rotating vortices represented by hexagons and the idle wheels between them. He identified the idle wheels with electric "particles" which, if they were free to move, would induce an electric current as in a conductor; already permeated and surrounded by the aether.

Remarkably, this physical model for the aether could account for *all* known phenomena of electromagnetism. As an example of induction, consider the effect of embedding a second wire in the magnetic field **H** of a wire carrying a current (I). If the current is steady, there is no change in the current in the second wire. If, however, the current *changes*, a rotational impulse is *communicated* through the intervening idle wheels and vortices and a reverse current is *induced* in the second wire with it correspondent electric field **E**. (Fig. 85).

Both of the generated electric and magnetic fields are created *simultaneously* (and also changing at the same instance) by their common source; time-variable electric current. It is nonsensical to state that an electric field causes or precedes a magnetic field and vice versa. Electric and magnetic fields are *inseparable*. Both are synchronous when a disturbance in the medium filling space (aether), occurs.

Maxwell always emphasized the importance of *rotation* and *centrifugal forces* in his Theory of Electromagnetism. He mentions Lord Kelvin in relation to identifying the rotatory nature of *magnetism*. Maxwell identified the cause of magnetic repulsion in terms of the centrifugal pressure arising in a sea of molecular vortices.

He identified the mechanism for the force on a current carrying wire, and also for motionally induced EMF, in terms of differential centrifugal pressure in this sea of molecular vortices. The centrifugal force bit was *crucial* for explaining magnetic repulsion, yet both centrifugal force and aether are stringently *denied* by "modern physics,"

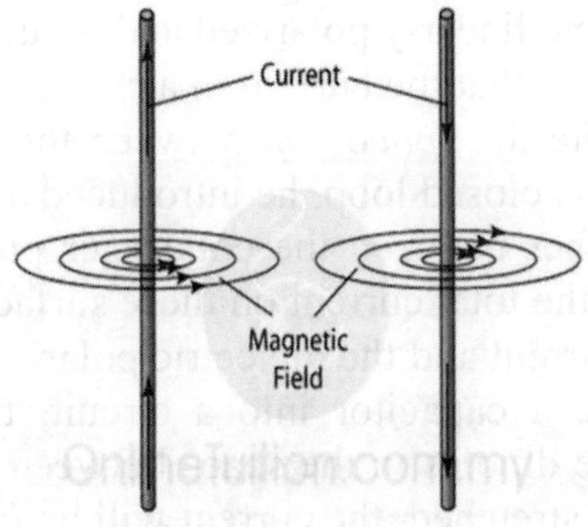

Fig. 85- Two independent wires or conductors separated trough space or vacuum containing aether. A current (I) is flowing steadily in one wire while the second wire is static and neutral. This current is associated with a magnetic field **H**. A change in the current will create an internal stress in the vacuum medium around, which will rotate in the direction parallel to the motion's axis. So, it will create rotational stress in the surrounding vacuum medium while dragging it along, producing a current in the second wire with opposite direction. A disturbance mediated through space or vacuum, (aether) induces a current.

Maxwell's physical model of the aether can be better understood if we replace his molecular vortices with *rotating* electric "particles" (+) and (-) dipoles, each of which consists somehow of an "electron" in a mutual circular orbit with a "positron" so to speak. Such a vortex will then double for both an electric dipole and a magnetic dipole. This is the *dielectric* nature of space. The rotating electron-positron dipole, by its very nature is the prototype dielectric.

It contains a single positive display and a single negative display. It possesses self-restoring transverse elasticity and it can be linearly *polarized* when subjected to an external electric field. Therefore, the very presence of space implicates it is *filled* with a polarized field comprising two opposing forces and motions: the *dielectric*, with counter spatial and contractive tendencies, and the *magnetic*, with spatial and expansile attributes. Electron-positron dipoles are like keys.

When we turn these keys by applying a *torque*, we get energy, mass, and angular momentum.

Maxwell's considered the existence of a dielectric aether and became of great importance when he examined the electrical current flowing in a *condenser (capacitor) circuit*. A dielectric medium such as vacuum, impedes electric current due to the fact that the constituent aether-dipoles become linearly polarized and induce a back EMF.

Maxwell was sure that the current in a condenser circuit must form a closed loop, despite the *space gap* between the condenser plates. In order to complete this closed loop, he introduced a *polarization current* in the dielectric aether between the condenser plates, normally filled with air. He treated the total current on these surfaces to be equal to the sum of the circuit current and the dielectric polarization current.

If we introduce a capacitor into a circuit, the aether flow will linearly stretch these dipoles in the space between the capacitor plates. When they are fully stretched, the current will be *blocked* from flowing and the magnetic field will collapse. As the current collapses, the aether that flows out of the dipoles in the magnetic field will now flow into the dipoles between the capacitor plates.

The stored *energy* in the circuit now takes on the form of a linear stretching of the dipoles between the plates of the capacitor. This kind of linear stored energy in the *vacuum*, not in the capacitor's plates, has the effect of *pushing* the current back again in the opposite direction when the power source is removed.

Based on extensive *experimental research* made by Ampère, Ohm, Gauss, Faraday, Helmholtz and others including his own; Maxwell developed a theory that would encompass *all* the known facts of electricity and magnetism.

It is of extreme importance to know the *original* purpose of Maxwell's theory, because its foundation is totally dismissed by "modern physics." Only some equations and concepts, conveniently remain. In Maxwell's own words:

"The theory I propose may therefore be called a theory of the Electromagnetic Field, because it has to do with the space in the neighbourhood of the electric and magnetic bodies, and it may be called a Dynamical Theory, because it assumes that in that space there is matter in motion, by which the observed electromagnetic phenomena are produced".

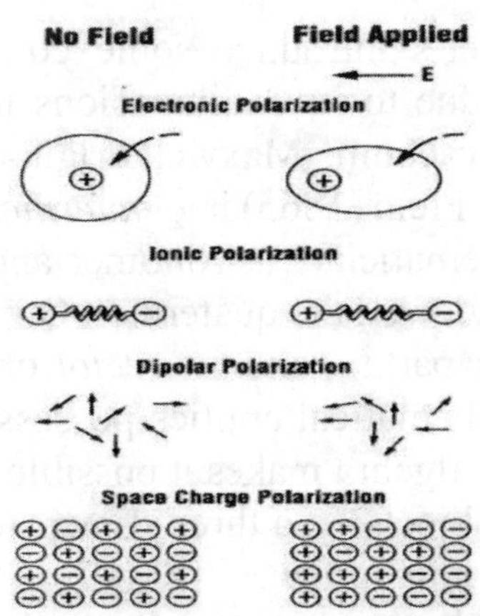

Fig. 86- All space is filled with a dense electrically neutral sea of a sort of rotating "electron-positron" dipoles, a subtle substance that constitutes the aether, which also pervades all matter. A rotating electron-positron dipole is the primary physical unit of Electromagnetism. It contains a single positive display and a single negative display. A dense sea of "electron-positron" dipoles would be electrically neutral unless a disturbance in the medium is applied, causing *polarization* either in space (electric or magnetic field) or within material bodies (charge). Bodies have a specific inductive electric capacity, mediated by the transmission of force between the aether dipoles surrounding the electrified or charged bodies (Faraday, 1837).

Central to his theory was the *elasticity* of the medium (aether) through which electromagnetic phenomena are to be propagated including light, leading to the concept of the displacement current as a necessary part of the theoretical apparatus. Equally important is the formalization of the processes of electromagnetic induction that can produce currents, these can result in dissipative energy loss in resistances and chemical dissociation in electrolytes. As an omnipresent *inertial medium,* aether is the ground for *all* observed physical spatial phenomena, affecting properties of all bodies at any given point.

In developing his theory, Maxwell contrasts his approach with that of other investigators such as Helmholtz and Weber, who deduced the phenomenon of induction from their mechanical actions. Maxwell proposes to follow the opposite agenda: to deduce the mechanical actions from the laws of induction contained in *space.*

To achieve this, he needs the general equations of the electromagnetic field, which consist of *20 equations* with 20 variables. However, current Physics and Engineering courses only teach 4. What did happen with the rest? What facts are missing and why?

To begin with, let's introduce some concepts that are used in Maxwell's equations due to its 3 dimensions nature and the *physical properties* of space or vacuum. Maxwell originally described his theory of the Electromagnetic Field (1865) in *Quaternions* that were discovered in 1843 by Irish mathematician, astronomer and physicist Sir William Rowan Hamilton. He viewed the quaternion Q = a+bi+cj+dk as the sum of *two* parts: the *scalar* part (a) and the *vector* or tensor part (bi+cj+dk).

These unique mathematical entities possess distinct multiplication principles. Quaternion algebra makes it possible to describe the rotation of three-dimensional objects on a three-dimensional space filled with a medium, plus a *scalar component*.

This medium (aether) is the abode of the scalar component in Maxwell's original equations and it is *non-local*. In general, it contains the *virtual form* (potential) of any electromagnetic manifestation, which is actualized only when all its material parts have taken up their appropriate places and are influenced by *energy*.

This scalar component represents the *fourth* component of the electromagnetic 4-vector, is *related* to the properties of the medium in which an electromagnetic wave propagates. It holds *active* properties, conveys *information* and it does not "propagate" but appears *at once* all over the Universe (non-locality). Furthermore, it doesn't coincide with neither matter, field nor the fictitious "four-dimensional space-time".

Quaternion algebra can be translated in the *Navier-Stokes* equation being a model of a fluid. A *scalar* is a physical quantity that is completely described by its magnitude; examples of scalars are volume, density, speed, energy, mass, and time. A quantity described by multiple scalars, such as having both direction and magnitude, is called a *vector*. Curl and divergence are fundamental concepts in vector calculus that play a pivotal role in understanding the dynamics of vector fields, offering insights into *rotational* and *flux* properties, respectively.

The *curl* measures the tendency of a vector field to rotate about a point, providing a vector that describes the rotation's axis and magnitude, while *divergence* assesses a field's tendency to converge or diverge from a given point, revealing sources or sinks within the *field*.

Grasping these principles is essential as key tools for analyzing physical phenomena ranging from fluid dynamics to *electromagnetism*. After Maxwell died in 1879, Quaternions were *abandoned* and replaced by vectors. A key feature of Maxwell's original 20 equations, the *scalar potential* of the vacuum ψ, was dismissed and almost forgotten.

Maxwell is most famous in connection with a set of four equations which bear his name, but these equations have been totally removed from the *physical context* within which Maxwell was working, and outside of that physical context the full meaning of these equations is *lost*.

In the truncated Maxwell's equations lists of nowadays, there are two curl equations. One of these curl equations is referred to as Faraday's law although it only deals with time varying electromagnetic induction, whereas the *original* Faraday's law had *wide*r application to the convective kind as well.

There was another curl equation in Maxwell's original list of 1865, but it does not appear in modern sets of "Maxwell's Equations". This very important curl equation:

$$\text{curl } \mathbf{A} = \mu\mathbf{H}$$

which relates to the *fly-wheel* nature of the magnetic field. Maxwell identified Faraday's "electrotonic state" with a vector **A** which he called the *electromagnetic momentum*.

The vector **A** relates to the magnetic intensity **H**, through this curl equation. **A** is the momentum of free electricity per unit volume, and so to all intents and purposes it is the same thing as the vector **J** that is used to denote electric current density.

The coefficient of magnetic induction μ is closely related to the mass density of the *medium* for the propagation of light, and it would appear to play the role of "moment of inertia" in the magnetic field in space.

According to Maxwell in 1861, the electrotonic state corresponds to "the impulse which would act on the axle of a wheel in a machine if the actual velocity were suddenly given to the driving wheel, the machine being previously at rest."

This shows that space has dynamic properties. However, this original equation was played down nowadays in favour of the much *less informative* equation:

$$\text{div } \mathbf{B} = 0$$

which is obtained by taking the divergence of the original curl equation containing **A**. Obviously, with this sort of distortion "modern physics" *ignores* magnetism at its fundamental level. The vector **A** represents the aether field momentum.

The involvement of the electromagnetic momentum **A**, in the displacement mechanism, flowing at the same speed as electric current as established by the 1855 Weber-Kohlrausch *experiment*, indicates that electromagnetic waves must be interwoven with an *aethereal* electric current, swirling from vortex to vortex.

It is a virtually unknown fact in "modern physics" circles that the expression that bears the misnomer "the Lorentz force" is actually of the *original* set of Maxwell's Equations.

The Lorentz force existed when Lorentz was a young boy and it appears as equation in an earlier paper by Maxwell entitled '*On Physical Lines of Force*' in 1861.The so-called Lorentz force is the solution to Faraday's law of electromagnetic induction and in modern vector format it is written as:

$$\mathbf{E} = \mu o \mathbf{v} \times \mathbf{H} + \partial \mathbf{A}/\partial t - \text{grad}\psi$$

where $\mu o \mathbf{v} \times \mathbf{H}$ can be expressed as $\mathbf{vXB}$ (a Coriolis inertial force). In this equation, **E** is the electric field. The first and third terms on the right-hand side this equation are the *pressure* and *tension* respectively.

The centrifugal force, $\mu o \mathbf{v} \times \mathbf{H}$, presses outwards in the equatorial plane of the tiny vortices as they strive to dilate, where **v** is the circumferential speed of an individual vortex relative to the medium and **H** is its vorticity, that being the basis of the concept of *magnetic intensity*, while μo is a measure of the density of the vortex sea. This force acts sideways on wire that is carrying electric current in a magnetic field, and it is also the *convective component* in electromagnetic induction.

The $-\text{grad}\psi$ term is the electrostatic force, where ψ is the *scalar potential*, while the $\partial A/\partial t$ term in the middle is the force that is involved in time-varying electromagnetic induction, that being the time derivative of the electromagnetic momentum **A**, where **A** is nowadays known as the often-neglected *magnetic vector potential*. The vector **A** actually represents the momentum density of the electric fluid, or aether, and it would appear to represent electric current at the most fundamental level.

When Oliver Heaviside and Josiah W. Gibbs *corrupted* Maxwell's original equations expressed in form of Quaternions in 1884, they put them into partial time derivative vectorial format. The Maxwell's original equations were considered to be defined in the aether. Thus, Maxwell used *total time derivatives*.

Instead of using the original Lorentz force, Heaviside and Gibbs chose to use a partial time derivative version of Faraday's law of electromagnetic induction, and in doing so they *lost* the convective **vXB** or dynamic component of the Lorentz force and eliminated a key component of electromagnetism: the *scalar potential.* The nowadays mediocre version of the Lorentz force as addendum to the four canonical Maxwell's equations is:

$$\mathbf{F} = q\left(\mathbf{E} + \mathbf{v} \times \mathbf{B}\right).$$

These four Maxwell's equations cannot bear only his name. In reality these are the *Maxwell-Heaviside-Gibbs* equations and together these are sold by the academic establishment fraudulently, as a "complete description" of the production and interrelation of electric and magnetic fields, within an "empty space".

This statement is outrageously *false* and *misleading*. In 1959, Israeli physicist Yakir Aharonov and US physicist David Bohm, theoretically predicted that a relative phase shift can exist even when the electron beams pass only through spaces *free* of **E** and **B**. This effect was later called the *Aharonov-Bohm effect*.

They confirmed this effect to potentials ψ and **A**, which "modern physics" considered to have no physical meaning in classical physics; which obviously is not true.

The Aharonov–Bohm effect or AB, sometimes called the Ehrenberg–Siday–Aharonov–Bohm effect, is wrongly labelled a quantum mechanical phenomenon because it was already *present* implicitly in classical electromagnetism.

An electrically charged particle is affected by an electromagnetic field (**E**, **B**), despite being confined to a region in which both the direction of magnetic field **B** and electric field **E** are *zero*. But, how can it be?

The underlying mechanism is the coupling of the *electromagnetic potential* with the complex phase of a charged particle's wavefunction, and the Aharonov–Bohm effect is accordingly illustrated by interference experiments. The Aharonov–Bohm effect illustrates the *physicality* of the *aethereal medium* or electromagnetic potentials, ψ and **A**, whereas previously it was possible to argue that only the electromagnetic fields,

E and **B**, were physical and that the electromagnetic potentials, ψ and **A**, were non-existent or purely "mathematical constructs".

This is false. Electromagnetic potentials are *real* and capable of affecting matter and were the cornerstone of the *original* Maxwell's equations, based on Faraday's lines of force that reside in *space* not in material objects. Maxwell incorporated potentials (magnetic vector potential and electric scalar potential as physical agents), because he viewed electromagnetism as perturbations in a universal continuum constituted by the *aether.* Thus, potentials are physically more fundamental than fields.

This electromagnetic potential ψ was (and still is) located in the medium that fills *space* or aether, before "modern physics" *concealed* their existence as physical fields, embedded in an actual material medium. In fact, despite its extreme importance, the AB effect is presented, in most of the textbooks dealing with electromagnetism and quantum physics, mainly from a theoretical point of view, (Shadowitz, 1988; Feynman et al., 1965; Felsager, 1998), while experimental aspects are *disregarded* for NOT supporting an "empty space."

In his Theory of Electromagnetism, Maxwell makes NO explicit mention of the concept of electric charge. Maxwell talks about "free electricity" and "electrification".

By free electricity, it would appear that he is talking about a *fluid-like* aethereal substance that corresponds to the vitreous fluid of Franklin, Watson, and DuFay, and it would appear that when Maxwell is talking about the density of free electricity that he is talking about a quantity which corresponds very closely to the modern concept of electric charge.

Charge would therefore appear to be aether pressure and aether tension. Such a *hydrodynamical* approach to charge enables us to explain *how* net charge enters an electric circuit when it is switched on, and also *why* the linear polarization process in a dielectric, result in a net charge. These things *cannot* be explained using the modern idea that charge is a "fixed property" that is attached to a particle such as the modern concept of electron, a "fundamental particle."

On the other hand, as Maxwell showed, magnetism *does not* exist as a standalone entity but is part of the *intertwined* spectrum of electromagnetic phenomena and can only be *induced* via field interactions, NOT by independent "charges in motion" such as electrons or "virtual photons".

This is key in understanding the true origin of Earth's magnetic field and of the celestial bodies in general. Magnetism originates in the *dielectric* (space) and, like matter, represents one of its discharge modalities.

Magnetism in all cases is the only aether modality to possess *spatial dimensions* and thus space itself can only be an attribute of magnetism. Magnetism represents the *polarization* of dielectricity when in contact with matter or induced by electricity. To say magnetism is spatial is to say that as the polarized field expands out of counterspace it creates open (but not empty) space.

At the end of the 19th century, another *corruption* came to twist Maxwell's original and workable electrodynamic approach. In 1897, the electron was "discovered". This event needs a brief discussion to understand how "modern physics" displaced the *field* or space filled with a medium, in favour of matter or particles, as an attempt to either explain or understand the interaction between electromagnetism and matter.

Dismissing the *outstanding* work of Faraday and Maxwell on the fundamentals of electromagnetism, several mediocre, second-rate physicists attempted to explain electromagnetic phenomena by action-at-a-distance forces between electrical particles. As an example of this approach to electrodynamics consider the work of Wilhelm Weber.

In the 1870s he attempted to construct an electrical theory of matter, based on *discrete* electrical particles, suggesting that each of the identical ponderable atoms combining to form chemical elements was a neutral system consisting of a negative, but highly massive, central particle with a positive satellite of much smaller mass. Sounds familiar?

The suggestion of the name 'electron' (amber in Greek) for a hypothetical *small unit* of electric charge is generally attributed to George Johnstone Stoney. He was able, by the 1870s, to comprehend a unit of electric charge being exchanged in *electro-chemical* processes (electrolysis) involving molecules of water.

The constituents of these molecules were hypothetically the same atoms that were so vividly invoked in the Kinetic Theory of Gases and, particularly, through information on viscosity, where sizes of atoms, their number, and their mean free path were being estimated numerically. The regularities in the phenomena of electrolysis need not have implied the atomic character of electricity. The *inference* from Faraday's law to the "atomicity of charge" required a prior conviction in the atomic structure of mater and was clearly drawn by Helmholtz in 1881. During the 1890's the concept of electron was one of the main topics in scientific research, involving *cathode rays*.

These rays were detected in the discharge of electricity through gases at very low pressure. Their main observable manifestation was that they gave rise to a *fluorescent* spot when they *collided* with certain substances. The function of the cathode ray tube is to convert an electrical signal into a visual display.

Early experimenters were simply trying to figure out what was happening inside the tubes and were mesmerized by the different coloured lights they could produce. In the 1850s, Julius Plücker, a physicist and mathematician at the University of Bonn, and his glassblower colleague Heinrich Geissler observed that the green phosphorescence on the glass of a high-vacuum tube was *magnetic.* When they placed a magnet near the cathode, the light spread out in a pattern similar to iron filings around a magnet.

The purely scientific question: What are cathode rays? Many German physicists believed that visible cathode rays resulted from an interaction with the *aether*, a colourless, weightless substance that enveloped all of space.

French and British scientists, meanwhile, were beginning to argue that cathode rays were electrified subatomic particles. However, there were *conflicting* pieces of evidence about the nature of cathode rays.

On the one hand, the fact that they could not be deflected by an electric field and that they could pass through metals that were impenetrable to particles of atomic size suggested that they were waves in the aether.

On the other hand, the fact that their trajectory was influenced by a magnetic field and that they were "carriers of electricity" supported their representation as charged particles. In other words, cathode rays showed, depending of the experimental set up, a sort of wave-particle *duality*. Sounds familiar?

Cathode rays were studied in a *discharge tube*. It is a cylindrical hard glass tube that is fitted with two metallic electrodes (anode and cathode) connected to a high voltage source. At normal atmospheric pressure, air is almost an *insulator*. An electric field of 3 million volts is required to make it conduct electricity.

In a discharge tube, the air was subjected to a very low pressure (~0.0001 atm) maintained by a vacuum pump and high voltage (~10,000 volts) to generate the cathode rays. William Crookes saw in 1874 that cathode rays could turn a paddle in their path (they have momentum). By 1896 it was known that they could be deflected in a magnetic field, and that they cast strong shadows, suggesting that they travelled in straight lines from the cathode.

In 1893 Heinrich Hertz had shown that they could pass through thin metallic films. This topic was pursued by his student Philipp Lenard. Such rays became known as Lenard rays. In 1895 Jean Perrin had demonstrated that the rays carried an electric charge.

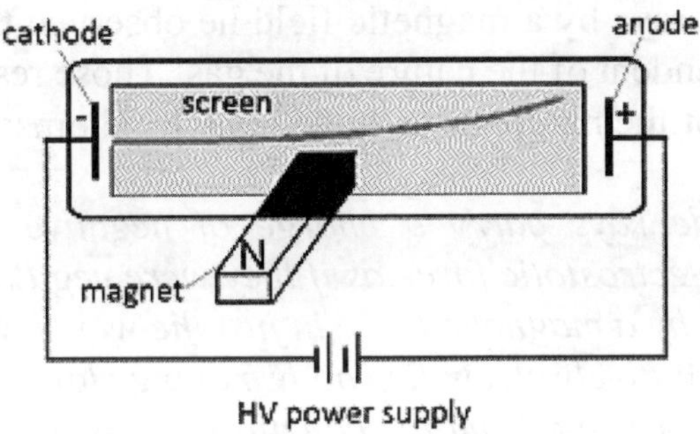

Fig. 87- A discharge tube contain gas at very low pressure in a glass tube which has a high potential difference applied across it. It is believed that the emission of light from a discharge tube, happens because the voltage applied to the tube is so high that it "pulls electrons out" of some of the gas atoms in the tube, causing electromagnetic radiation. Cathode rays get deflected when placed in an electric field and magnetic field. The direction of deflection shows that they are negatively "charged." Cathode rays and electric current are essentially the same thing: polarized aether flow caused by a *disturbance* in the medium (high potential difference).

The contribution of British physicist J.J. Thomson in 1897 to the acceptance of the corpuscle (electron) hypothesis was closely tied with the experimental and theoretical investigation of cathode rays.

It is important to understand that Thompson in reality, presented his "discovery" as a conclusion, not as a *fact* obvious to everyone. In order to identify an event or a process as the discovery of an unobservable entity, one needs a criterion (or a set of criteria) that would enable one to say that such a discovery has taken place.

The discovery of the electron wasn't like the discovery of penicillin, the American continent or nuclear fission. Pretty much straightforward and observable events. Gravity as such cannot be seen or touched, but its effects from massive objects are straightforward.

Not so with the electron because it was *never* observed and isolated directly. It was *computed* from the raw experimental data, in order to make sense of the observations and reach a conclusion.

In his research, Thomson argued that cathode rays consisted in extremely small, sub-atomic particles. He inferred their small size from their ability to pass through thin sheets of metals (which was already known). Moreover, the unexpectedly small value of their mass to charge ratio (m/e) supported further that *inference.*

By that time, he had managed to deflect the rays by means of an electric field, but 'only when the vacuum was a good one.' Furthermore, by deflecting the rays by a magnetic field he observed that 'the path of the rays is independent of the nature of the gas' Those results threw light on the question of the nature of cathode rays. In Thomson's words:

"*The cathode rays carry a charge of negative electricity, are deflected by an electrostatic force as if they were negatively electrified, and are acted on by a magnetic force in just the way in which this force would act on a negatively electrified body moving along the path of these rays, I can see no escape from the conclusion that they are charges of negative electricity carried by particles of matter*."

There is an escape. Thomson was trained in the belief that all phenomena could be described ultimately in terms of *matter* in motion and assuming an *inert* vacuum (when in fact it is dynamic). This is unacceptable in the Maxwellian system, where an "isolated charge" is a set of *polarized* composite rotating electron-positron dipoles, due to a disturbance in the medium or vacuum (aether).

Thus, unfortunately and for the worse, Thomson ended up being a mediocre physicist, *misinterpreting* the deflection of the rays by electrically charged plates and magnets, as evidence of "bodies much smaller than atoms" (electrons) that he calculated as having a very large value for the charge-to-mass ratio. Later he estimated the value of the charge itself. Having established that cathode rays were "charged material particles", Thomson then inquired further into their nature.

The following question arises: What are these particles? are they atoms, or molecules, or matter in a still finer state of subdivision? His mass to charge (e/m) measurements aimed at elucidating this question. The charge-to-mass (e/m) ratio of the particles (cathode rays) can be measured by observing their motion in an applied magnetic field.

Thomson repeated his measurement of e/m many times with different metals for cathodes and also different gases.

Having reached the same value for e/m every time, it was concluded that a "fundamental particle" having a negative charge *e* and a mass 2000 times less than the lightest atom known by then (hydrogen ion in electrochemistry), existed in all atoms.

Thomson also conjectured that the corpuscle was a universal constituent of all matter. The mass to charge ratio of cathode rays depended neither on the chemical composition of the gas within the cathode ray tube, nor on the material of the tube's electrodes.

This shouldn't come as a surprise because *aether* is a universal substance filling space and pervading *everything* and *everyone*. Thus, traces of aether will be found *everywhere*. It should be emphasized that in 1897 Thomson did not guess correctly one of the key properties of the electron, the value of its charge: 'The smallness of m/e may be due to the smallness of m or the largeness of e, or to a combination of these two' He preferred the last option.

Also, the charge and mass of the electron are only two of the *bizarre* properties now associated with electrons (quantum numbers, position-momentum, wave-particle duality, indeterminate etc.) which were literally inconceivable in Thomson's time. It is not at all clear why the theoretical and experimental detection of just two of these properties constitutes the "discovery" of the electron.

If one insists on using the expression "the discovery of the electron", one would have to use it as meaning the complex process that led to the consolidation of the *belief* that "electrons" denote real entities if so. And here is the *problem*, a huge one because the concept of electron since its inception at the end of the 19th century, has no room in Maxwell's Theory of Electromagnetism discussed early.

Both are mutually exclusive, which constitutes a *hindrance* in the progress of science. A central aspect of the research program initiated by Maxwell, was that it avoided microscopic considerations altogether and focused instead on *macroscopic* variables (e.g. field intensities). This macroscopic approach proved *successful* in unifying electricity, magnetism and light (optics) in a coherent theoretical framework and *predicted* the existence of electromagnetic waves, with other frequencies than visible light.

This was *confirmed* experimentally by German physicist Heinrich Hertz in November 1886, when he became the first person to transmit and receive controlled radio waves. However, some scientists believed that Maxwell left several details unattended. One in particular was to provide an understanding of electrical conduction. It was in response to these details that British physicist Joseph Larmor, started to develop a theory whose aim was to explain the *interaction* between aether and matter.

In the first stage in that development, Larmor explained that in order for a medium to be able to sustain electric displacement it must have *rotational elasticity.* In the original formulation of his theory conductors were conceived as regions in the aether with zero elasticity, since Larmor had assumed that the electrostatic energy is null inside a conductor.

Conduction currents were regarded, in Maxwellian fashion, as mere *epiphenomena* of underlying field processes and were represented by the *circulation* of the magnetic field in the medium *encompassing* the conductor; which is pretty much right.

Nonetheless, being influenced by other physicists claiming that the "discrete structure" of electricity was an independently established fact in electrochemistry, that did not follow from Larmor's theory; he changed course and had to add these discrete or molecular *charged entities*. Currents were now *wrongly* identified with the transfer of free charges ('monads'), which were also the "cause" of magnetic phenomena.

Those charges had the ontological status of independent entities and ceased to be epiphenomena of the *field* (when in fact they are). Later on, Larmor changed the name of these monads for electrons. The problem here is that Larmor makes the *mistake* of changing a cause (field) for an effect (electric current) in order to "explain" the interaction between electromagnetism or aether and matter. What this moron did was to put the cart before the horse, *misleading* the course of electromagnetism related research such as geomagnetism, for more than a century and fooling many scientists to this day.

An electric charge *cannot* be independent of the field because it is a manifestation of the hydrodynamic *polarization* of the field, due to a disturbance in it. Physics textbooks have been preaching this misconception, over and over through generations.

An electric charge (measured in Coulombs) is a measure of the rate of inflow or outflow of the aether, into or out of sinks or sources, and so it becomes of relevance in the irrotational Coulomb force. The vorticity of the aether is connected with the quantity that we call spin, and it becomes of relevance in the angular $\partial \mathbf{A}/\partial t$. Rather than representing a "charged particle", the so-called electron is the fleeting *effect* of the *interaction* between the dielectric and magnetic fields at the atomic level.

In contrast to Maxwellian theory which *did not* attribute independent existence to charges at all, in Larmor's theory the electron acquired independent reality and so did other "charged particles" subsequently.

Furthermore, the macroscopic approach to electromagnetism was *jettisoned* and microphysics was launched. Macroscopic charges were considered constituted by an "excess of particles" whose charges have a determined sign; an electric current would be a true stream of these corpuscles.

Thus, electron now was defined as a negatively charged subatomic particle. It can be either free (not attached to any atom) or bound to the nucleus of an atom. In electrical conductors either wires or plasma, current flow results from the "movement of electrons" from atom to atom individually and from negative to positive electric poles in general.

Whereas Maxwell wisely used a *Gestalt* approach, considering the wholeness in the study of electromagnetic phenomena, "modern physics" starting with Larmor and other physicists; used a *reductionistic* approach later fully developed in Quantum Mechanics.

This difference refers to the observation that the *whole* is greater than the sum of the parts; furthermore, the whole is *qualitatively* different. A system composed of many constituent interacting parts has properties that we often *cannot* anticipate from knowledge of the properties of the individual parts, as the reductionistic approach emphasizes *ad nauseam*.

Following this microphysical and *misleading* paradigm, the Universe can now be explained studying how its "building blocks" behave. The pseudoscience of "particle physics" was born. Since then, scientists have persisted in erroneously ascribing electromagnetic phenomena to *imaginary* particles like electrons and photons, even engaging in a prolonged and futile search for the so-called 'God particle'. They *failed* to recognize as Maxwell did, that light and electricity are utterly mass-free.

For example, a *misconception* which most physicists acquire in their formative years is that the *photoelectric effect* requires the quantization of the electromagnetic field for its explanation; using energy packets called photons. However, it was long suspected that the so-called photon, *cannot* have an independent existence (fundamental particle) and rule interactions between light and matter.

Since the mid-1960s, we have known that the photoelectric effect does not imply the existence of photons (Lamb & Scully, 1968). From the photoelectric effect alone, it is actually *ambiguous* whether it is the electronic levels or the impinging radiation that should be quantized. Physicists are *unable* to explain: what exactly are light quanta?

Fortunately, more than a century later after a mediocre physicist published his "explanation" of the photoelectric effect invoking photons in 1905, Indian electrical engineer Dhiraj Sinha has stirred the scientific pot with a bold new study *Electrodynamic excitation of electrons* published in *Annals of Physics* (2025). His research implies that photons cannot be as uniquely quantum as we've always believed: they actually emerge from *classical electromagnetism* instead.

According to Sinha, the *key equation* for understanding light was already tucked away in Maxwell's original 19th-century equations. This challenges long-accepted ideas about the nature of photons and reignites an old debate in physics. Light's particle-like behavior emerges simply from the quantization of *magnetic flux*, thus no quantum quackery required and goodbye to the notorious "wave-particle duality," forever.

In general, we shall see that all of the experimental photoelectric phenomena are described by a theory in which the electromagnetic field is treated *classically* while only the matter is treated quantum mechanically, according to Planck's law.

All particle-based conceptions of such dynamic phenomena are impossible and solely a product of the defunct atomistic (corpuscular) perspective. The field, as conceived by modern science, is an *abstract* particle-based nexus derived from *erroneous* notions of space and time, (mathematical space-time), which because it does not include an aethereal medium, *cannot* explain how charge, gravitational and electromagnetic forces, or distant actions are mediated, except of course by *imaginary* "particle exchanges".

For decades, physicists hunted in vain for new "elementary particles" that would have explained why Nature looks the way it does and *failed.* In private, many physicists admit they do not believe the particles they are paid to search for exist. The purpose of "modern physics" through its theories during the 20th century, was to *conceal* the existence of physical fields as an actual *material media,* as these were experienced by Faraday, Maxwell and Tesla. Force action of these media is *replaced* in these theories by the *mathematics* of kinematical non-material quantities, where the dynamical cause-effect is *absent*. As long as physicists can perform a calculation, that allows them to predict some results, they are happy. Don't ask. Shut up and calculate.

In fact, these kinematical quantities describe (but cannot explain, so you better don't ask) the physical material world, which works on the principle of balance and changes of pressure of the densities of the material bodies and of the *material force fields* surrounding these bodies.

Without the recognition of the existence of physical fields grounded in the *aether,* as actual physical substance, all mathematical descriptions of physical reality are merely the *kinematical* numbers of the ratios of the lengths and times on paper.

So, at the end what in reality are these notorious cathode rays? In fact, cathode rays and electric current are essentially the same thing. They are pure *pressurized aether flow*. The term 'cathode ray' is used when the electric current jumps the gap in an electric circuit. Cathode rays will move directly into the *electron-positron sea,* usually as part of an electric circuit, and they will linearly *polarize* the electron-positron *dipoles*.

It would seem that cathode rays are one of the least understood phenomena of all, and they have been *confused* for "beams of electrons" or independent (isolated) particles with "intrinsic charge". Cathode rays possess *energy* which will be a combination of aether pressure and velocity as per Bernoulli's principle. This would explain their propagation in straight line and turning a paddle in their path. It is believed nowadays that electric current, and hence cathode rays, are a stream of "charged particles". In the case of cathode rays, these are believed specifically to be electrons. This uninformed belief follows because the aether flow often drives a stream of charged (polarized) particles along with it (drag).

Despite the rejection of physical fields by "modern physics" and enthroning the corpuscular electron as a sort of demigod; no much time went by without *inconsistencies* arising.

In the 1920's, besides its charge and mass, another electron's property was found when Dutch physicists Sam Goudsmit and George Uhlenbeck, were the two theorists who first suggested electron *spin* publicly, to explain certain *anomalously* split lines in the spectrum of atomic hydrogen, the so-called forbidden transitions.

A few years earlier, the Stern-Gerlach experiment was performed, where a beam of silver atoms directed through a nonuniform magnetic field would be forced into *two* beams.

This experiment apparently confirmed the quantization of electron spin into two orientations, independent of its orbital angular momentum but at the time when it was done, it was not interpreted in those terms. Both experimental situations suggest just two possible states for this angular momentum and following the pattern of quantized angular momentum, this requires an angular momentum quantum number of ½ and - ½. However, electrons intrinsic spin and its associated magnetic moment are considered to be *paradoxical* and *problematical* from the Standard Model standpoint.

The electrons spin is accepted because it provides a fourth quantum number (½) for atomic electrons and helps *explain* known spectroscopic frequency data. But the spin is paradoxical because the electron seems point-like, with a size no larger than 10^{-18} meters. To spin, an object must have a *radius*. The electrons magnetic moment is also accepted, because its physical effects are detected experimentally. But it is also paradoxical because a *magnetic* moment would normally be created by a *current* moving in a finite-sized loop or by a point charge moving in a finite-sized closed circulating path. However, the electron is experimentally found to be point-like.

And *no* closed circulating path of a point-like charge that would produce the electrons magnetic moment, has *ever* been observed. So, the question is: *where* does the electron's charge, spin, mass and magnetic momentum comes from? *What* shapes and organizes the physical characteristics, observed in the electron?

Bear in mind that the Standard Model doesn't even attempt to predict where this intrinsic internal angular momentum comes from. It's just a property of electrons, taken as a given or *intrinsic*. Asking "why do electrons have spin?" is like asking "why are electrons charged?" Pretty much the same applies to mass in general.

Scientists have simply accepted the existence of inertia as a given. It is an "intrinsic" property of material bodies. Indeed, the word intrinsic is used in such cases, as a wildcard. If the alleged "Higgs particle" is the giver of mass to all other particles, what then gives the Higgs itself its mass? Or it is intrinsic?

From a mechanistic-reductionist point of view, where isolated point of masses or charges interact with each other within an "empty" space; it seems that there is no answer to what underlies this "intrinsic" property. However, in reality, the answer lies in the surroundings or *environment.* In other words, material spatial force fields.

In the electron's case, it is the often-called *quantum vacuum field* or quantum plenum, an euphemism for an aetheric medium. The problem is that empty space is *not nothingness.* Physicists are gradually coming to realize that there is a level of existence *beneath* the realm of reality of photons, electrons, quarks, electric-magnetic fields and all quantum interactions. For instance, the jittering of subatomic particles technically known as *Zitterbewegung,* predicted by the Dirac Equation, considered the most important equation of Quantum Mechanics.

Zitterbewegung of a free electron has a frequency of order 10^{21} Hz and an amplitude of order 10^{-13} m, which makes difficult measurement with current technology. But *why* would a single electron in free space, with *no* forces acting on it will spontaneously move back and forth at the speed of light? What causes it? This seems unphysical.

Some physicists believe zitterbewegung is not a real physical phenomenon but just an artefact of an incorrect single-particle interpretation of the Dirac Equation. Other physicists believe zitterbewegung is a real physical phenomenon, with experimentally observable effects, as per recent published papers. Zitterbewegung of a free particle is an *unresolved* puzzle in the foundations of the of Quantum Mechanics.

A quantum fluctuation is the temporary appearance of energetic particles in our observable reality, out of "empty" space, as allowed by Heisenberg's uncertainty principle. The uncertainty principle states that for certain pairs of related variables such as position/momentum or energy/time, it is impossible to have a precisely determined value of each member of the pair at the same time. But, this still *doesn't* satisfactorily explain why tiny particles (or fields) can pop in and out of existence from an apparently "empty" space devoid of material "substance".

Nowadays, there is a growing awareness of some sort of *space medium* with which energy and mass particles interact in a very peculiar manner. This *interaction* is what gives the *observed* properties of the electron and matter in general (spin, charge, inertia etc.). This substrate is a *subtle* medium which is also known as *aether* and it is permeating all space and everything within. The term *vacuum* is a covert or another way of saying aether. Indeed, the assumed "empty" space has measurable physical properties that imply the existence of a *subtle substance* herein called the "aether."

Matter can be considered a dynamic state of the aether. A rudimentary *consciousness* may be rooted in this deep substratum of existence. Aether was the foundation of Faraday's experiments, Maxwell's Theory of Electromagnetism and several of Tesla's scientific breakthroughs. The *air* we breathe is filled with aether. It is the *essence* of the Universe.

However, "modern physics" *rejects* the existence of aether mainly because of an obscure and mediocre physicist with poor academic record, working as a third-class clerk in a patent's office in Switzerland; wrote an infamous paper published in 1905, featuring the following statement:

"The introduction of the luminiferous aether will prove to be superfluous inasmuch as the view here to be developed will not require an absolute space provided with special properties, nor assign a velocity vector to a point of the empty space in which electromagnetic processes take place"

Clearly the author of these lines dismisses the field, a *dynamic continuum*, originating in the aether and consists of aether, filling space; which manifests in the modalities of dielectricity, electromagnetism, mass and gravity. He *washes away* the experimental and theoretical work of notable physicists and chemists such as Hooke, Huygens, Lavoisier, Young, Laplace, Fresnel, Faraday, Maxwell, Mendeleyev, Tesla and others; in just one go because he did feel like. This act of vandalism has polluted, disfigured and *derailed* the course of physics in a distorted and unnatural way, that persists to this day.

The other reason in rejecting the aether, is the propagandized "null result" of the notorious and most *misunderstood* experiment in the history of science: the *Michelson-Morley* experiment, performed in 1887, designed to test what sort of structure or form, the aether possessed on a cosmic scale in relation to celestial bodies, particularly the Earth. Instead of electricity or magnetism, light was used trough an *interferometer*, an optical instrument. In the 19th century *two forms* of aether were proposed. The first, formulated by French physicist

Augustin-Jean Fresnel in 1818, held the aether to be essentially stationary within the Solar System in which case there should be a detectable aether wind of some 30 km/s at the Earth's surface, in its orbital speed around the Sun. The second, formulated in 1844 by Irish physicist George Stokes, where the aether is dragged by Earth in which case, there should be little or *no measurable* aether speed at the Earth's surface.

Mainstream scientists recoil at the minimum insinuation of the aether word. The term aether is today like a verbal obscenity, the unspeakable "no-word" that no professional physicist, shall be heard to utter on pain of being branded a deranged crackpot. This attitude originated since the 1880's when the notable Michelson-Morley experiment (MME) to detect the "aether wind" mentioned in chapter one, was performed. It was *wrongly* concluded that the aether didn't exist because apparently, no aether drift (wind) was observed. However, the non-existence of the aether was not easily accepted by the scientific community.

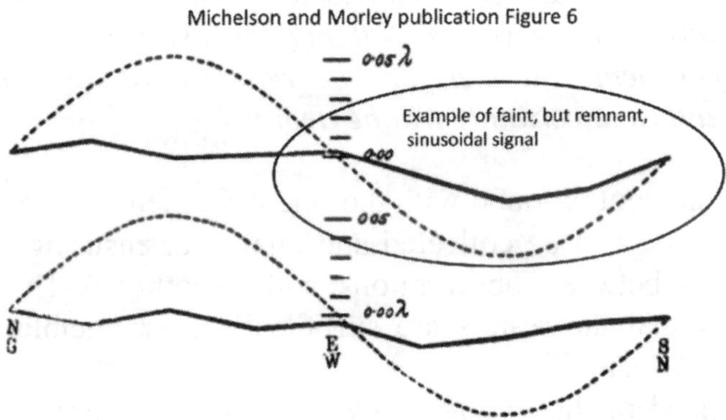

Fig. 88- Graphic version of the Michelson and Morley's experiment results. The upper solid line is the curve for their observations at noon, and the lower solid line is that for their evening observations. The dotted curves represent one-eighth of the theoretical displacements, showing a <u>lower</u> value of aether drift than expected. They concluded: "the aether is at rest with regard to the Earth's surface." Michelson and Morley confirmed that Stokes' model of aether drag is correct.

Based on this *wrong* conclusion, Irish physicist George Fitzgerald suggested that the aether can be saved by *assuming* that fast-moving bodies shorten their dimension in the direction of motion. This is the alleged "length contraction". The aether could also be saved by the alleged "time dilation", derived by Dutch physicist Hendrik A. Lorentz. This effect would cause time to run slower in fast moving vehicles.

To top it off, French physicist Henry Poincare in 1901 and Lorentz in 1904 introduced the ideas of the "universal constancy" of the speed of light, of the "time dilation", of the "length contraction" and of the "non-constancy of mass" of moving bodies. Instead of saving the aether, these *bizarre*, *illogical* and *non-physical* conceptions, were used in the Special Theory of Relativity (1905) to abolish it.

In their paper of 1887, besides stating that: "The experimental trial of the first (Fresnel) hypothesis forms the subject of the present paper". This explains why the observed values didn't match the expectation, because the aether was dragged as Stokes predicted (second hypothesis).

Also, the Michelson-Morley team made an important qualification:

"In what precedes, only the orbital motion of the Earth is considered. If this is combined with the motion of the Solar System, concerning which but little is known with certainty, the result would have to be modified; and it is just possible that the resultant velocity at the time of observations was small though the chances are much against it. The experiment will therefore be repeated at intervals of three months, and thus uncertainty will be avoided."

This statement is sound with the scientific method. It asks for *more* experimental data to be collected and analyzed, ensuring there are no discrepancies between observations and hypothesis. Then the team would think that the results are sufficiently reproducible, to reach a definitive conclusion.

The speed of the *interferometer* through the entrained (dragged) aether (detected by an observer resting in the aether) should be the *sum* of the velocity of Earth's rotation, Earth's orbital velocity, the velocity of the Solar System around the center of the Milky Way, and of our galaxy in the Universe; making it *not* immediate to know exactly, *what* value for the fringe shift to expect in the MME.

Unfortunately, this repetition (which later was to prove to be very important) was never carried out and the *warning* stated in the Michelson-Morley paper, was *neglected* by mainstream science.

Over the years their partial "aether wind" result was *downgraded* to a "null" result by many scientists. This famous "null result" is the one often quoted in most physics textbooks. However, *contrary* to the official version of that no traces of aether were ever found, the values obtained of Earth's speed relative to aether where:

Midday readings: = 6.22+/-1.86 km/s
Evening readings: = 6.80+/-4.98 km/s

With an average of some 6.5 km/s well *outside* of experimental error, which is quite different from zero. Simply because something is smaller than expected doesn't mean that it doesn't exist. Michelson never *questioned* the existence of the aether, but only the extent to which it is entrained by the Earth's motion. This is evident from the title of one of Michelson's papers: *"The Relative Motion of the Earth and the Luminiferous Aether"*. So, after all *there was* aether involved in the *experiment*. But mainstream science would lead us to believe that *no* aether was detected.

The 20th century saw a growing *aversion* to the concept of an aether; inexplicably, the evidence of aether detection first in 1887 and most conclusively in 1925-1926 by US physicist Dayton Miller, was completely *ignored*. Its existence was either dismissed or simply denied and generations of students have been falsely taught, that light propagation requires no medium.

The official story, based upon just only *one* sloppy experiment carried out in 1887, was and still is that this experiment didn't find any traces of aether. Thus, space is "empty". From the scientific method standpoint, a single experiment reporting a finding *cannot* be considered valid let alone the basis for a theory; unless it can be *replicated* by independent researchers and corroborate whether the alleged finding is true or not. No replication no science.

However, although Michelson never did carry out the recommendations contained in his concluding remarks, others did and it is their experiments (performed during the 20th century) using far more sensitive apparatus than the one used by Michelson and Morley, that accumulated definitive evidence and *confirmed* the existence of aether.

The experiments of Georges Sagnac, Dayton Miller, Roland deWitte, Maurice Allais, Héctor Múnera and others; clearly and consistently showed an *aether drift effect*. Mainstream science denies or ignores this endeavour, dismissing it as pseudoscience. Why?

The aether or vacuum has a structure *Vakuumgliederung* with real physical properties that imply a substantive-dynamic medium. The vacuum has an elastic modulus, a stress tensor, a sheer tensor, a magnetic permeability coefficient, magnetic susceptibility, and a characteristic electromagnetic wave impedance of 377 Ohms, among other *measurable* physical properties.

The quantum vacuum became known as a *quantum plenum*, a "neo-aether". It might also provide unifying explanations and deeper insights into physical phenomena. For example, the nature of electron or any particle spin may be seen as a *vortex* in the aether. Electric fields may be regarded as a *polarization* of the aether.

Matter itself may be seen as a dynamic modification of the aether at the foundation of Quantum Field Theory (QFM). This "neo-aether" is dynamic and without a preferred reference frame. In other words, it is *absolute*. It has both micro and macro-fluctuations with resulting in-homogeneities that can create local observable differences.

Aether IS the real fabric of space and all matter is simply "undulations in the fabric of space." All radiation particles, all energy particles, all mass particles, and all electromagnetic energy fields *absorb/consume* aether. That's why science prodigy Nikola Tesla through his experiments, stated that there is no energy in matter other than that received from the *environment*. The conversion matter-energy would be a subset. Space is NOT "nothing," the mere location of bodies and forces, but a *medium* capable of supporting the strains of electric and magnetic forces.

In fact, the *energies* of the world are not localized in the particles of matter from which these forces arise but rather are to be found in the *space* or environment, surrounding them. However, the property most relevant is the all-important ability to *conduct* quanta of electromagnetic radiation. This defines the aether as being *luminiferous*, making it the vital light-conducting medium of the Universe. Therefore, an empty space or void *doesn't* exist at all. A "void" in conventional science, is a volume of space that does not contain matter but gravitational or electromagnetic fields. However, it contains *something else*. At the beginning of twentieth century, studies of radioactivity began showing that the empty vacuum of space had spectroscopic *structure* similar to that of ordinary quantum solids and fluids.

In 1930, British electrical engineer and physicist P.M.A. Dirac (1902-1984) formulated the existence of an aether described as a "sea of negative energy particles" or the *Dirac sea* but it was highly criticized. No wonder as anything that smells to aether is *rejected* by mainstream science. The Dirac sea, a direct consequence of his famous equation, made the physics community to greet it with alarm and outrage.

This was because the equation gave twice as many states as they thought it should have. They expected a Ψ (the fourth-component Dirac spinor wavefunction) with *two* components; but this equation gave *four*. After the discovery of the positron, it was realized that its *four* solutions call for electrons and positrons of positive energy, and electrons and positrons of *negative energy*, known as the Dirac sea. However, any reference to a universal substance that "undetectably" filled space sounded too much like an aether.

Since all negative-energy states have lower energy than any positive-energy state, Dirac wondered why there were *any* filled positive states, since according to Hamilton's law, all entities tend to seek the lowest-energy state.

He suggested that all of the negative energy states must be filled, like the filled electron shells in the Pauli exclusion scheme. Then, unless a "vacancy" occurred, positive energy particles would "float" on the surface of the negative-energy "sea" and stay positive.

Dirac was right. In order to *properly* explain and understand quantum electrodynamics, we need to have a sea of tiny *aether vortices*, and in order to have a sea of tiny aether vortices, we need to have *sources* and *sinks* in the aether.

These sources and sinks are what we call electric particles, and it is a dense 'Electric Sea' of positive (source) and negative (sink) particles that causes the fundamental aethereal based forces to manifest themselves in the particular guise of electromagnetism and quantum electrodynamics.

However, the Dirac equation was a direct *threat* to the reigning paradigm. As Dirac noted, physicists had always arbitrarily *ignored* the negative energy solutions, that in one way or another point to the *aether*. Dirac argued that aether had been prematurely rejected.

Recognizing the *impossibility* of an absolute void, he wrote that physicists *must* make 'profound alterations' in their theoretical conception of the vacuum. Based on emerging and irrefutable *evidence* in electrodynamics, Dirac protested, 'we are forced to have an aether.'

Dirac's request fell on deaf ears, to this day. This is because in the technical framework of quantum field theory, the vacuum is not a part of *physical reality*. Only a mathematical requirement. However, subsequent studies with large particle accelerators, have now led us to understand that space is more like a piece of window glass than ideal Newtonian emptiness. It is filled with 'stuff' that is normally transparent but can be made *visible* by hitting it sufficiently hard to knock out a part.

Because the Standard Model has, so far, *failed* to predict the masses of the elementary particles; there have been numerous efforts to develop *alternative* theories (e.g. Skyrme, 1962; Martin 2005; El Naschie, 2008). One of the most successful is the "lattice model" that allows to predict the mass and spin of all the long-lived mesons, baryons and leptons, including the electron-muon and tau neutrinos.

Israeli physicist Menahem Simhony, based on the *Dirac sea*, developed Electron Positron Lattice (EPOLA) model concerning the cubic lattice structure and dynamic behaviour of the aether, which would have satisfied Faraday's search for a dielectric aether and the mechanistic basis for the Maxwell original twenty equations, published in 1873.

Maxwell's *original* works are pioneering and of enormous value which pointed us in the right direction as the outstanding achievements of Nikola Tesla proved. Any shortcomings within these works pale into insignificance when compared with the *errors* that followed in Maxwell's wake. A series of *derailments* culminated with the mad world of Relativity where two clocks can both go slower than each other, and where electromagnetic waves can propagate in a pure vacuum or "empty space," without the need for any *physical* displacement mechanism. Since 1983, the situation has degenerated even further when the speed of light is now a defined quantity rather than a *measured* quantity

Once we have grasped the modern meaning of this "neo aether" or quantum plenum, which *pervades* everything, it is easy to understand the concept of *torsion fields* related to it and associated to rotational bodies and its link to the origin of magnetism in stars and planets.

Torsion fields, spiralling toroidal flow of aether

Bear in mind that in Western science, the physical implications of *rotation* or spinning present in observable bodies are barely noticed, despite of the fact that rotation or spin is a *widespread* feature in the Universe at macro and microscopical levels. The idea that a spinning body such as a wheel or a planet, can *affect* physical reality beyond its known angular momentum and centrifugal force, might sound unthinkable but this doesn't mean necessarily, it is impossible.

It was an accepted idea that electricity and magnetism where separate and mutually exclusive until an *experiment*, performed by Danish physicist Hans Christian Øersted in 1820, proved otherwise something unthinkable by the standards of those days. In the same way, the variety of physical effects of spinning bodies has been proven. The first researcher who experimentally detected the *unusual* effects associated with rotation, was a professor of the Russian Physical-Chemical society, N.P. Mishkin in 1906, who conducted a series of experiments using scales.

Torsion fields is the name associated to the effects originated by *rotational bodies*. Over the course of latter decades, tens of *unexplainable* microscopic and macroscopic effects in natural sciences and especially in physics and biology; have been revealed and investigated in different countries, that couldn't be explained by the conventional framework of the Standard Model. Should be emphasized that a large part of these phenomena where demonstrated by objects having *spin* or angular momentum.

Take for example a gyroscope. Despite its utter simplicity, it is after all, nothing more than a wheel on an axle: it remains the most fascinating and mysterious device ever created. Spin up the wheel, place it on a pedestal, and there it stays, pointed at…? A key concept to understand the behavior of spinning bodies is *inertia*.

Inertia is a fundamental physics concept that describes the tendency of an object to *resist changes* in its state of motion. The force of inertia is directly related to the *mass* of the object: the more massive an object, the greater its inertia. This means that heavier objects are more resistant to changes in their motion than lighter ones. This is pretty much straight forward for an object, moving linearly at the same speed. The problem arises when a body with a given mass is *spinning*.

The gyroscope also bends to inertia's will, but in confounding ways. Touch it, and the gyro opposes you by veering in unexpected directions. If it is spinning extremely rapidly, the gyroscope remains rigidly locked in the direction it has been set. In ballistic missiles its sights fixed on a specific target hence the term *inertial* guidance systems. If a rocket veers off the gyro's fixed course, a sensor detects the error, and a servomechanism realigns the missile with the gyroscope axis. Isaac Newton would argue that the gyroscope is pointed in a fixed direction relative to absolute space, what physicists term an *inertial reference frame*, indeed, the ultimate inertial reference frame. Think of it as an invisible reference grid somehow etched into the fabric of the Universe. However, since "modern physics" abolished the concept of absolute space along the aether in 1905, and considers centrifugal force as *fictitious*, whereas for the great masters of physics from the past, it was a real causal agent. Physicists today still don't know how a gyroscope stays pointed in a fixed direction.

The problem lies in that researchers have not considered the action of *inertial forces* produced by the mass elements and center mass of the spinning rotor that create internal resistance and precession torques. Resistance torque is established through the actions of *centrifugal* and *Coriolis* forces. Precession torque is established through the actions of common inertial forces and a change in angular momentum. These internal torques act simultaneously and interdependently on two axes and represent the fundamental principles of gyroscope theory.

These gyroscopes internal torques are *real* active physical components, and the well-known change in the angular momentum, that is, the non-primary acting component, which does not play the first role in gyroscope properties. The results of new studies make it clear why the gyroscope theory is far from perfection.

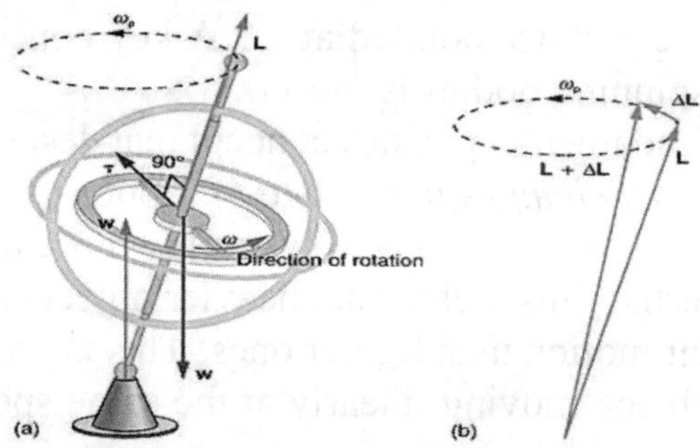

Fig. 89- a) Basic representation of a spinning gyroscope on a stand. **W** is its downward weight (mg) and also the upward normal reaction of the stand, ω is the angular velocity and **L** is the angular momentum (Iω). The torque **T** experienced by the gyroscope is in a direction perpendicular to its angular momentum about its own rotation axis. b) Torque is the rate of change of angular momentum, as a result the angular momentum vector **L** of the gyroscope begins to "follow" the torque; so, the gyroscope itself begins to move around in a circle ω_p around the z axis (precession).

Experimental evidence suggests that when a torque is applied to a spinning gyroscope such that the torque axis is perpendicular to the spin axis, then an *induced torque* will be generated in the gyroscope that is mutually perpendicular to both the spin axis, and to the applied torque axis. This induced gyroscopic torque exhibits the *three way* mutually perpendicular characteristics of the motion of a charged particle in a magnetic field.

Applied mathematics textbooks *do not* however recognize the existence of induced gyroscopic torque as a *distinct* fundamental force in its own right. Textbooks assume that when a spinning gyroscope appears to be defying gravity, that this can be "fully explained" without having to recognize the existence of any *additional* forces beyond downward Newtonian gravity (**W**) and upward normal reaction of a surface (-**W**). As research shows, this is not the case. This induced torque has the power to *deflect* the effects of gravity sideways.

On the other hand, the mathematical formulation of Newton's law of gravity is based on the existence of an *inertial frame* of *reference*, in which the background stars appear to be fixed. This implies the existence of some kind of *aethereal medium* with which motion is measured to be relative to. The very concepts of position, velocity, and acceleration, imply the existence of particles moving in that aethereal medium.

As Maxwell pointed out, space is filled with tiny aethereal vortices that press against each other with centrifugal force while striving to dilate and he referred this sea of vortices as *luminiferous medium*. These same tiny aethereal vortices are responsible for the *inertial forces*.

As the luminiferous medium flows through the interstitial spaces between the atoms and molecules of all moving bodies, as like water flowing through a basket, this will generate a *gyroscopic interaction* akin to the principles lying behind Maxwell's explanation for Ampère's Circuital Law. This interaction leads to the formation of *vortex rings,* concentric on the line of motion, and centered on the moving body. This is the foundation of *torsion fields*.

A theoretical approach of torsion fields was put forward by French mathematician Elie Cartan, to improve the flawed General Relativity, which *only* considered the mass of the objects. In fact, the equations of General Relativity, which describe what happens in an accelerating frame, were written *without* considering the *spin* of elementary particles. Therefore, when a spinning particle accelerates, its behaviour is not correctly described by these equations. Cartan suggested that "space-time" torsion could be used as the macroscopic manifestation of the intrinsic angular momentum of the matter. Cartan was the first to point to the possibility of the existence of *fields* generated by *spin* polarity and angular momentum density of *rotating* objects, showing a close interconnection between gravitation and torsion and some unorthodox Maxwell's equation solutions.

Albeit Cartan proposed that spiralling torsion fields could not move, (i.e. they would remain static,) and could only exist within a space far smaller than the atom. Cartan's theory, however, was proposed before the discovery of the electron spin and this was perhaps, one of the reasons his ideas went essentially unnoticed for some decades.

It was probably not until the work of Kibble and Sciama, who laid down the foundations of the description of the gravitational field that introduces an additional (rotational) degree of freedom to the (imaginary) "space-time" fabric, that the role of "space-time" torsion in modern physics was appreciated. This is perhaps, the only acknowledge of a variant of torsion fields in Western science.

However, the *main experimental work* on torsion fields came mostly from Russia. Besides academic interest, the aim was to investigate the *anomalies* demonstrated by gyroscopes and inertial systems; which were interfering with the Russian aerospace program.

Probably the first researcher to establish that the behaviour of gyroscopic systems, *cannot* be explained in the framework of Newton's classical laws of mechanics; was notable Russian astrophysicist Nikolai Alexandrovich Kozyrev (1908-1983).

This discrepancy is because Newton's laws *do not* distinguish between the spinning and the non-rotating object, that represented the state of mechanical knowledge at Newton's time. Because Newton did not distinguish between rotation and non-rotation, General Relativity (GR) did not *distinguish* between the so-called inert and "gravitational mass." The fact that rotation *affects* the mechanical properties of objects places Newton's Laws as a special case and *invalidates* a geometrical (relativistic) interpretation of space.

In fact, the central principle of GR is known as the *Equivalence Principle*: It arbitrarily states that accelerations and gravitational fields are equivalent. Apparently, there is no experiment that can *distinguish* one from the other. However, it is often neglected that in rotational bodies, an angular momentum vector is aligned *antiparallel* to the local gravitational field, which violates the Equivalence Principle.

This assertion (Equivalence Principle) was concocted upon the *alleged* equivalence principle and the experimental results of the

Hungarian physicist Loránd Eötvös, around the turn of the twentieth century, which verified the equivalence of inertial and gravitational mass to within about one part in 10^8.

These are generically referred to as Eötvös experiments, designed to investigate if there is any violation, no matter how small; of the proportionality between inertial and gravitational mass. He used a *torsion balance* which compared the gravitational force on two objects composed of different substances. The results were null. No mechanical changes were observed.

However, these experiments were carried out in fixed, *non-rotational* set ups giving null results; whereas gyroscopes and *rotational physics* do *affect* the mechanical properties of objects, as the experiments performed by Kozyrev did show. These experiments carried out by Eötvös, are an example of a *poor* research design and data analysis, encouraging false-positive findings.

Planets, stars and galaxies all *spin* and therefore do *accelerate* and can affect their surroundings. In other words, these are *non-inertial* frames of reference. In Nature, non-inertial frames of reference are the *real* ones. With the exception of the absolute inertial frame of reference (aether), inertial frames of reference *do not* exist.

Inertial frames of reference were only *hypothesized* so mathematicians in the 18th century, could simplify the calculations over a very small region/integral. In fact, they even always noted that their *assumptions* of fixed inputs (like gravity) were for simplification, and should never be considered static.

Kozyrev is in many ways a forerunner, the father of today's efforts to re-interpret physics in a way that does not contradict *intuitive* understanding. The western scientists' efforts to reconcile the inherent *contradictions* of the Standard Model of Physics, firmly based in GR and Quantum Mechanics, have brought *less* than satisfying results. Kozyrev has measured spin or torsion field effects at a time when Western science, was busy smashing atoms into ever smaller fractions. He investigated time and the aether before most of us in the West, ever thought of *questioning* the workability of our modern interpretations of the Universe.

However, the Western world is largely uneducated about Kozyrev. In the early 1950's, Kozyrev conducted an extended series of *experiments* with gyroscopes and found that the gyroscope evidences *variations* in weight, which varies as a function of *angular velocity* and the *direction* of rotation. Left rotation doesn't show any change in weight. Kozyrev showed that the gyroscopes shed more torsion waves when shaken or spun, so the aetheric energy that sustains the *inertial mass* of the top was shed *back* into the background *sea* of the aether. The momentary deficit of aether energy, accounted for the very slight weight *drop* due to a tiny loss of inertial mass.

This should not come as surprise because other researchers, have concluded that inertial mass is deriving from an accelerated body's *interaction* with the Zero-Point Field (ZPF), also known as aether. (Haisch, Rueda, Puthoff. 1994).

This sort of phenomena with gyroscopes, were *independently confirmed* outside Russia by Japanese scientists Hideo Hayasaka and Sakae Takeuchi in a 1989 paper, published in the journal *Physical Review Letters*. In fact, very little has been published in the scientific literature on the generation of *gyroscope weight-reduction*. A problem exists for most theorists because it is necessary to go *beyond* standard textbook theory, wherein the gyroscope pivot is always kept fixed.

The Japanese team experimented with three gyroscopes. Each weight change of three spinning mechanical gyroscopes whose rotor's masses were 140, 175, and 176 g was measured in inertial rotations without systematic errors. The experiments showed that the weight *changes* on rotations around the vertical axis are completely *asymmetrical*: The right rotations (spin vector pointing downwards) causes weight decrease of the order of milligrams (weight), proportional to the frequency of rotations at 3,000 – 13,000 rpm.

However, the left rotations do not cause any change in weight. This confirms the right-handed nature of the phenomenon, as Kozyrev had noticed before. The *fact* of a physical phenomenon, that appears only in one direction of rotation is rather unprecedented. It bluntly violates de *dogma* of symmetry and "invariance" of the Standard Model of Physics.

Nature *distinguish* between the left and the right, indicating that there is a *dynamically* preferred way of matching-up the points of space at different times; that are independent of matter and its relative motion. Nature does have a *preferred* coordinate system (the inertial frame) which *contradicts* Special Relativity. Other *independent* researchers around the world, arrived at the *same* conclusions. (De Palma, 1977).

Classic mechanics *does not* provide explanation for certain lifting and/or lowering effects observed in spinning gyroscopes. Researchers point out that a possible interpretation relies upon the hypothesis of velocity/acceleration fields associated with *vortices* of cosmic plenum (aether) and relevant "velocity circulations".

In an analogy, the magnetic field created by a coil conducting continuous electrical current might also be viewed as a special "gyroscopic behavior" of the plenum involved. When the sense of the spin and rotation are the same the acceleration is towards the observer, and when not, away. Thus, one can observe both weight gain and weight loss. The *rotation* is a critical factor. The unjustified rejection of *changes* of gyroscopes' mechanical properties such as weight is a token of the dominant academic mentality, according to which what has *no* clear explanation within the accepted scientific paradigms, either cannot exist or can somehow be proved by means of that which is already known.

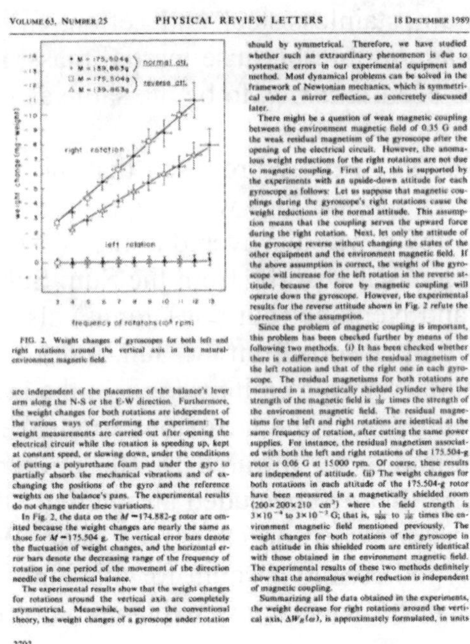

VOLUME 63, NUMBER 25 — PHYSICAL REVIEW LETTERS — 18 DECEMBER 1989

FIG. 2. Weight changes of gyroscopes for both left and right rotations around the vertical axis in the natural-environment magnetic field.

are independent of the placement of the balance's lever arm along the N-S or the E-W direction. Furthermore, the weight changes for both rotations are independent of the various ways of performing the experiment: The weight measurements are carried out after opening the electrical circuit while the rotation is speeding up, kept at constant speed, or slowing down, under the conditions of putting a polyurethane foam pad under the gyro to partially absorb the mechanical vibrations and of exchanging the positions of the gyro and the reference weights on the balance's pans. The experimental results do not change under these variations.

In Fig. 2, the data on the $M = 174.882$-g rotor are omitted because the weight changes are nearly the same as those for $M = 175.504$ g. The vertical error bars denote the fluctuation of weight changes, and the horizontal error bars denote the decreasing range of the frequency of rotation in one period of the movement of the direction needle of the chemical balance.

The experimental results show that the weight changes for rotations around the vertical axis are completely asymmetrical. Meanwhile, based on the conventional theory, the weight changes of a gyroscope under rotation should by symmetrical. Therefore, we have studied whether such an extraordinary phenomenon is due to systematic errors in our experimental equipment and method. Most dynamical problems can be solved in the framework of Newtonian mechanics, which is symmetrical under a mirror reflection, as concretely discussed later.

There might be a question of weak magnetic coupling between the environment magnetic field of 0.35 G and the weak residual magnetism of the gyroscope after the opening of the electrical circuit. However, the anomalous weight reductions for the right rotations are not due to magnetic coupling. First of all, this is supported by the experiments with an upside-down attitude for each gyroscope as follows: Let us suppose that magnetic couplings during the gyroscope's right rotations cause the weight reductions in the normal attitude. This assumption means that the coupling serves the upward force during the right rotation. Next, let only the attitude of the gyroscope reverse without changing the states of the other equipment and the environment magnetic field. If the above assumption is correct, the weight of the gyroscope will increase for the left rotation in the reverse attitude, because the force by magnetic coupling will operate down the gyroscope. However, the experimental results for the reverse attitude shown in Fig. 2 refute the correctness of the assumption.

Since the problem of magnetic coupling is important, this problem has been checked further by means of the following two methods. (i) It has been checked whether there is a difference between the residual magnetism of the left rotation and that of the right one in each gyroscope. The residual magnetisms for both rotations are measured in a magnetically shielded cylinder where the strength of the magnetic field is $\frac{1}{100}$ times the strength of the environment magnetic field. The residual magnetisms for the left and right rotations are identical at the same frequency of rotation, after cutting the same power supplies. For instance, the residual magnetism associated with both the left and right rotations of the 175.504-g rotor is 0.06 G at 15000 rpm. Of course, these results are independent of attitude. (ii) The weight changes for both rotations in each attitude of the 175.504-g rotor have been measured in a magnetically shielded room ($200 \times 200 \times 210$ cm^3) where the field strength is 3×10^{-4} to 3×10^{-3} G; that is, $\frac{1}{1000}$ to $\frac{1}{100}$ times the environment magnetic field mentioned previously. The weight changes for both rotations of the gyroscope in each attitude in this shielded room are entirely identical with those obtained in the environment magnetic field. The experimental results of these two methods definitely show that the anomalous weight reduction is independent of magnetic coupling.

Summarizing all the data obtained in the experiments, the weight decrease for right rotations around the vertical axis, $\Delta W_R(\omega)$, is approximately formulated, in units

2702

Fig. 90- Excerpt of the Hideo Hayasaka and Sakae Takeuchi 1989 paper, showing the graph of the *weight changes* of gyroscopes, where aetheric energy that sustains the inertial mass of these, was shed back into the background sea of the aether (aka Dirac Sea) during their experiments. The upper slanted lines show mass variation with right rotation. The bottom horizontal line shows no mass variation with left rotation. Gyroscopes and rotational physics, do affect the *mechanical properties* of objects. This means that accelerations and gravitational fields are NOT equivalent, which invalidates the basis of General Relativity and therefore, mainstream Cosmology.

There is more: when a spinning laser gyroscope is placed near a super-cooled rotating ring, the gyroscope accelerates a bit in the same direction as the ring, and scientists aren't sure why. The anomalous acceleration was discovered in 2007, by Austrian physicist Martin Tajmar at the Space Propulsion group at the Austrian Institute of Technology in Seibersdorf, Austria. So far, the effect has only been observed in this one laboratory, but it has similar connotations with the ones explained before. Since then, scientists have been looking for an explanation for the so-called *Tajmar effect*.

It is suggested that the gyroscope's observed acceleration stems from a change in its inertial mass and an attempt to conserve momentum with respect to a super cooled rotating ring. Tajmar's experiments used rings that were cooled to 5°K and made of a variety of materials, such as niobium, aluminum, stainless steel, and Teflon. Given the above, it is clear that these *anomalies* are because inertial mass, has not been well understood and has been *assumed* to be the same as gravitational mass (the Equivalence Principle, EP).

During the 1940's, Kozyrev proposed that the *rotation* of stars is *related* to their energy *output*. This would apply for planets and moons as well. In fact, in 1969 it was discovered in the West, that Jupiter emits *twice* as much energy as it absorbs from the Sun. Subsequently, it was found that Saturn and Neptune (but not Uranus) also emit about twice as much energy as they absorb from the Sun. Indeed, all of the planets and moons in our Solar System emit *strongly* in the infrared.

This infrared emission is the heat from atmospheres and surfaces, which peaks in the mid to far-infrared 15-100 microns (a micron is 1 millionth of a meter). It was concluded an internal energy source was responsible for this *anomaly* or excess of heat. Western scientists proposed several hypotheses to account for this unusual energy output but, half a century later, a satisfactory explanation of this phenomenon still is pending.

It is a pity that Kozyrev's work in astrophysics and other subjects, is practically ignored in the West while in Russia is confirmed and recognized. Dr. A. P. Levich of Moscow University (Levich, 1996), who states: "N.A. Kozyrev, an outstanding astronomer and natural scientist, enriched the dynamic picture of the world by introducing a new entity, possessing 'active properties,' and coinciding with neither matter, nor field, nor space-time in its usual understanding."Levich's report describes experiments by dozens of Russian scientists who *verified*

Kozyrev's predictions using torsion balances, resistance bridges, photocells, piezoelectric elements, mercury thermometers, chemical reactions, and changes of mass and density. Levich describes *experiments* which validate Kozyrev's ideas in astronomy, electromagnetism, gravity and *subtle energy* (torsion fields).

Kozyrev proved beyond any doubt that such an energy source *does exist*. More than half a century passed since he developed his theories, and yet it is not easy for mainstream scientists, to reconcile his findings with the present scientific paradigms. His torsion field experiments established the aether as an energy source that and transformed our understanding of the Universe, at least in Russia.

In fact, Kozyrev is a fine example of a what a *true* scientist must be. He is a genuine descendant of the Scientific Revolution. Working on its own and sticking to the *scientific method*, tied to the laboratory or the astronomical observatory, he brings *real* and testable scientific knowledge to the table.

Kozyrev never was a *theorist*, who speculates about how the Universe ought to work, per alleged mathematical principles along untestable concepts/hypothesis and on top of that, believing it is the "truth" with vague or null references to *physical reality*.

Furthermore, several theorists in the West have been propped up and idolatrized by the western media as it is commonly seen today. These "scientists" are constructs, a made-up persona, who never bothered to perform real *experimental research*, not just regurgitate the same old trash and add some patches; to give the appearance that something new is achieved. A flood of propaganda for them is started by articles in journals and newspapers, by textbooks and by lecture tours.

And to top it off, a material with *nil* scientific substance is being sold to the gullible public as "science". This is indoctrination. It is no wonder that in Russia, such *exaggerations* and *distortions* of scientific reality and its proponents, are taken to be characteristic of Western reporting on science.

This new entity, with 'active properties' is known as *torsion field.* It is produced by the *spin* of a mass, whether it is an electron or as large as a star, in an analogous way as electric charge and mass "produce" electromagnetic and gravitational fields.

Electromagnetic fields such as light, radio, and magnetism are *vector* fields in that they possess both magnitude and direction of propagation. Gravitational fields are also vector fields with both magnitude and direction. Torsion fields are *scalar*; they only have the property of magnitude. Torsion fields operate *independently* of direction. These fields are also known as *subtle energies*.

They also have the extraordinary property of holographically operating throughout the entire Universe, without respect to *time* nor *distance*. These torsion waves are called *non-hertzian* because they do not correspond to the classical theory of Hertz and Maxwell; describing the behaviour of waves such as light, radio and microwaves. Equations can predict torsion behaviour and inventions and practical devices, have been made using torsion fields.

The torsion field have been shown to couple with consciousness. Contrary to electromagnetism, in torsion fields energy of the *same* polarity is *attracted* to itself and opposite polarities are repelled. Thus, the popular Law of Attraction is scientifically sound. With the addition of torsion-field research, we can see that the spinning objects can *harness* naturally spiralling torsion waves in its *environment*, which gave it an additional supply of energy, loosing inertial mass.

All spinning objects “grab-hold” of the Universe's *aetheric energetic fabric* and can have “extra” energy, that doesn’t come from common sources such as combustion or sunlight. This explains, for example, the “anomalous” energy output observed in Jupiter and other planets, mentioned before.

Torsion fields demonstrate *axial* symmetry, unlike electromagnetic and gravitational fields which demonstrate central symmetry. Unlike electromagnetic waves, torsion waves transmit information and under specific circumstances, energy.

They propagate through physical media without interacting in the traditional sense, with the media. But propagating torsion fields have been shown by many experimenters, to *alter* the spin state of *physical media*. Thus, torsion fields can be *detected* by various types of devices.

Torsion fields cannot be shielded by most materials, but they can be shielded by materials having certain spin-structures. There exist both left and right torsion fields, depending on the classical spin orientation or rotational orientation. Unbeknownst by conventional science, torsion fields are the *real* cause of the so-called *violation of parity* (lack of left-right symmetry) in the Standard Model of Physics.

During the middle of the 1950’s, the misnomer of "Law of Conservation of Parity" that claimed "universal forces are *not* aware of "left/right" or "positive/negative charge" or "N/S magnetic polarity"; was put to *test* by US-Chinese physicist Chien-Shiung Wu, who experimented it in low-energy electron movements not dominated by strong outside influences and micro/macro threshold forces.

In other words, the most "natural", unperturbed and stable state. A cobalt variety, will leak electrons under certain conditions and she could prove that these charged "particles", were repelled *only* by the South pole of a magnet instead of both, as predicted by the Law of Conservation of Parity.

For the first time, we knew that electromagnetic states both building blocks and supporting forces of material Universe, do indeed slope from "left-right" and from "positive-negative" and from "N-S". This parallels the phenomenon of *chirality,* discussed in chapter one and it is *connected* with the right-handed *loss* of inertial mass, observed in spinning gyroscopes.

However, perhaps by a pervading sexism in the scientific community, the Nobel prize went *not* to Chien-Shiung Wu but to Tsung Dao Lee and Cheng Ning Yang, male physicists who had speculated on the possibility without coming to firm conclusions. The Chinese principle of Yin and Yang, which states that all things exist as inseparable and contradictory opposites; for example, female-male, dark-light and old-young has proved to be scientifically sound, so to speak.

Torsion fields are generated not only by a single spinning particle, but also by an ensemble of particles. This situation is similar to that demonstrated by electricity, where we often encounter the collective electric fields generated by an ensemble of electrical charges such as atomic nuclei, atoms, charged bodies, etc. Thus, *any* nuclear spin polarized target is the *source* of a torsion field. This fact has been repeatedly *observed* and *verified* by numerous research groups. Since analogous spins attract and opposite spins repel, the interaction of a spin-polarized particle with a spin polarized target nucleus, results in the appearance of "anomalous forces" which *depend* on mutual spin *orientation* of the particle and the target.

For example: in 1977, US physicists A.C. Tam and W. Happer showed *experimentally* that two circularly polarized laser beams attract or repel, depending on the mutual orientation of their circular polarization. They discovered and verified that if the direction of rotation is similar, then these beams attract, and if the rotation of their respective polarizations is opposite, then they repel. These results, which have been repeated and experimentally verified many times, *violate* a number of rules embodied in the Standard Model of Electrodynamics and could *not* adequately be explained by mainstream science at the time this work was originally being conducted. In the mid-80's, the US physicist A.D. Krisch and his group, were experimentally investigating the interactions between spin-polarized protons and a spin-polarized proton target. It was found that the observed process of spin-spin interaction could *not* be *described* in the framework of the quark model developed by Gell Mann and Fermi Lab.

The experimental results *did not* conform with the Standard Model of Chromodynamics (SMC) and could *not* be explained in terms of *any* other generally accepted models of physics. Sceptics would have a hard one in trying to "explain" this situation, without the help of *torsion fields*. However, torsion fields are considered "pseudoscience" by many scientists in the West.

Every electromagnetic (EM) or electrostatic field is accompanied by or contains a *torsion component*, meaning that *all* organic and inorganic objects have their own signature torsion fields. The torsion or *scalar component* can be separated from the EM field. Around 1900, Nikola Tesla was the first to accomplish it, experimenting with two spiral shaped wires.

He fed these with alternating currents in exact *opposite* phase, which made the net result a *nil* electromagnetic field, the zero field. Even though the two opposite electromagnetic fields *eliminated* each other, Tesla managed to *demonstrate* that these spiraled wires, were capable of *sending* energy over very long distances.

This is because strong torsion fields are generated by high electrical potentials and by devices, having organized circular or *spiral* electromagnetic processes. Tesla discovered a totally *new* form of energy. Incredibly, these scalar waves did not *lose* any energy when carried over a *distance*, as we see happening with normal electromagnetic waves, mainly because scalar waves can travel only *longitudinally* not transversally. Also, these scalar waves don't follow the *inverse square* rule of gravity and electromagnetic fields.

As explained earlier, Maxwell in his *original* 20 mathematical equations concerning Electromagnetism, published in 1865, established the theoretical existence of Tesla waves (longitudinal or scalar waves) through the *scalar potential* (Ψ). After his death, mediocre physicists assumed that several of these equations were "meaningless" and wrote out the scalar component. Many sceptics of course, would dismiss scalar waves as pseudoscience. Only the well-known (transverse) Hertzian waves can be derived from the truncated Maxwell's field equations, whereas the calculation of longitudinal SWs gives zero as a result. This is a *flaw* of the field theory because SWs exist for ALL particle waves (e.g., as plasma wave, as photon or cosmic radiation).

In fact, through the use of high-energy directed weapons, also known as Star Wars or the Strategic Defense Initiative (SDI); the military are well aware of the existence of *scalar waves* (SW_S) aka longitudinal waves (LWs). These waves proved much more difficult than initially thought; this was mostly because of the wave's *subtle densities*, fluctuating frequencies, and orthogonal directional flow. As a result, the *truncation* of Maxwell's equations was upheld, dealing only with of transverse (Hertzian) waves. Nevertheless, SW/LW in free space are quite real.

Besides Nikola Tesla, empirical work carried out by electrical engineers (e.g., Eric Dollard, Konstantin Meyl, Thomas Imlauer, and Jean-Louis Naudin, to name only some) has clearly demonstrated SW/LWs' existence *experimentally*.

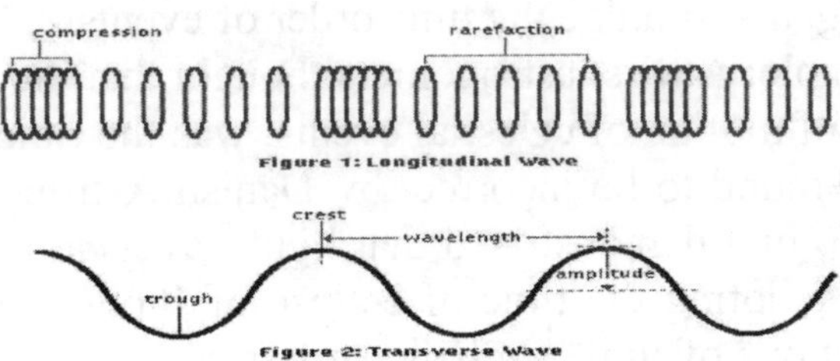

Fig. 91- Unlike the transverse waves of electromagnetism, which move up and down perpendicularly to the direction of they are heading, Tesla waves vibrate in line with the direction of propagation. Tesla waves have magnitude but no direction (no vectors), since they are imagined to be the result of two electromagnetic waves that are 180 degrees out of phase with one another, which leads to both signals being canceled out. This results in a kind of 'pressure wave' or scalar wave emanating from the potential Ψ, which radiates in all directions at once. They are a carrier of energy and information.

To test the torsional waves coming from rotating celestial bodies such as stars, from the mid 1950's to the late 1970's, Kozyrev with V.V.

Nasonov conducted *astronomical observations* using a receiving system of an entirely new and novel variety. When his telescope was directed at a selected star, the detector (designed by both scientists) positioned within the telescope, registered the incoming signal even if the main mirror of the telescope, was shielded by dense metal screens. This strategy relied on the notion that electromagnetic waves (in the form of light) embody and intrinsic (scalar) component which cannot be shielded by dense metallic screens.

When the telescope was directed not at the visible position but at the *true position* of the star, calculated from sidereal tables and ephemeris charts, the detector then registered an incoming signal that was much stronger than the one received by targeting the apparent visible location. The detection of the true positions of different stars using this technique, could only be explained as a function of the detection of radiation emitted by the star, which exhibited transport velocities many times *greater* than the speed of light *c*.

This fact is extremely important because it makes useless, worthless the conception of "relativity of simultaneity," an outcome of Special Relativity. It says that inertial observers in relative motion disagree on the *timing* of events at *different* places. If one observer thinks that two events are simultaneous, another might not. Apparently, the issue lies on using *light signals* to judge the time order of events.

For example: At first astronomers thought that the time in the laws or equations of motion of celestial events, was the time of *observation*. But this was found to be incorrect by Danish Astronomer Ole Rømer who, in 1676, first demonstrated that light travels at a *finite* speed. He achieved it by noting the time of eclipse of Jupiter's first satellite for different positions of Jupiter relative to Earth.

When the periods of revolution of Jupiter's satellites were found by observation of their eclipses, transits, and occultations when Jupiter was near opposition, the results were used to *predict* these phenomena. When Jupiter was situated otherwise, *errors* ranging up to a quarter of an hour were found.

Rømer gave the correct explanation, namely that the time in the equations of dynamics (event) is not the time of observation, but the time when the event under discussion actually happened, and that in a visual observation the time of observation is *later* because it takes light a finite time to travel from the object observed to the observer. The *delay* corresponds to a time of passage across the Earth's orbit of about 16 minutes. When the times were corrected to allow for this effect the anomalies or errors disappeared,

The need for invent Special Relativity with its truly *distorted* notions of relative space and time, arose from two facts about light: first, that it has a finite velocity (apparently, unmatched and unsurpassed by anything), and second, that this velocity is independent of the motion of the observer. If light had travelled with an infinite velocity we could have identified the time of the event with the time of observation, and there would have been no further trouble. And this is precisely the matter: radiation in form of torsion fields and gravity, propagates at *superluminal speeds*, millions of times greater than the speed of light *c*.

Therefore, any distant events detected by different observers using these radiations, will be simultaneous *regardles*s of the type of the frame of reference, as Kozyrev and other researchers proved. This implies that space and time are *absolute*. Kozyrev also found that the detector registered an incoming signal when the telescope was directed at a position symmetrical to the visible position of a star, relative to its true position. This fact was interpreted as a detection of the future positions of stars. However, very little of these results are known in the West.

Torsion fields play an important role in the *Solar System*. A spinning source that releases energy in any form; such as the Sun, a planet or the centre of the galaxy and/or a spinning source that has more than one form of movement occurring at the same time, such as a planet that is rotating on its axis and revolving around the Sun at the same time; a *dynamic* torsion field is produced.

This phenomenon allows torsion waves to propagate through space instead of simply staying in a single "static" spot. A type of energy or force, is associated with spinning sources which dynamically radiate torsion field waves. Thus, these little known torsion fields, like gravity or electromagnetism are capable of moving from one place to another in the Universe.

It has been clearly demonstrated that the Lagrangian of a *spinning* source which is dynamically *propagating* radiation contains many terms with constants which do *not* depend on *G* or *h.* These facts are however, either *ignored* or *neglected* by mainstream scientists. Therefore, their apparent understanding of how the Solar System works is *incomplete* and consequently, *anomalies* obviously will arise. In fact, there is a puzzling anomaly caused by planets that *modify* the trajectory and velocity of probes, which *don't fit* the explanations of the Standard

Model. In the early 1960's, scientists developed the gravity-assist method, where a spacecraft would conduct a flyby of a major body, in order to increase its speed.

Many notable missions have used this technique, including the Pioneer, Voyager, Galileo, Cassini, and New Horizons missions. In the course of many of these flybys, scientists have noted an *anomaly* where the *increase* in the spacecraft's speed did not accord with orbital *models*.

This has come to be known as the "flyby anomaly", which has endured despite decades of study and resisted all previous attempts at explanation. To address this, a team of researchers from the University Institute of Multidisciplinary Mathematics at the Universitat Politecnica de Valencia in Spain, have developed a *new* orbital model based on the maneuvers conducted by the Juno probe, when it arrived Jupiter in 2016.

Their model took into account the tidal forces exerted by the Sun and by Jupiter's larger satellites Io, Europa, Ganymede and Callisto and also the contributions of the known zonal harmonics. They also accounted for Jupiter's multipolar fields, which are the result of the planet oblate shape; since these play a far more important role than tidal forces, as Juno reaches perijove (the point in the orbit of a satellite of Jupiter nearest the planet's centre). In the end, they determined that an *anomaly* is also be present during the Juno flybys of Jupiter. They also noted a significant radial component in this anomaly, one which decayed the farther the probe got from the centre of Jupiter.

As one researcher commented:

"Our conclusion is that an anomalous acceleration is also acting upon the Juno spacecraft in the vicinity of the perijove (in this case, the asymptotic velocity is not a useful concept because the trajectory is closed). This acceleration is almost one hundred times larger than the typical anomalous accelerations responsible for the anomaly in the case of the Earth flybys.

This was already expected in connection with Anderson et al.'s initial intuition that the effect increases with the angular rotational velocity of the planet (a period of 9.8 hours for Jupiter vs the 24 hours of the Earth), the radius of the planet and probably its mass."

The team highlights the *rotation* of Jupiter as a *key* determinant in the anomaly. Also, they determined that this anomaly appears to be dependent on the ratio between the spacecraft's radial velocity and the speed of light, and that this decreases very fast as the craft's altitude over Jupiter's clouds changes. These issues were *not* predicted by the standard Newtonian or General Relativity theories, so these flyby anomalies are the result of an *unknown* effect that has been overlooked.

Bear in mind that in this case, the Standard Model takes into account only the mass and gravity of celestial objects; ignoring its *spin* either at macro or microlevels as Eli Cartan along with other scientists, have pointed out.

However, a recent research conducted in the West, is suggesting a *new* non-standard force called Topological Torsion Current (TTC), put forward by Portuguese physicist Mario J. Pinheiro from the University of Lisbon in 2015, which can *explain*, in a simple and direct way, the *anomalous* acceleration detected in space crafts during close planetary flybys.

It points to the existence of a yet *undiscovered* relationship between linear momentum of the probe and *angular motion* of the celestial body, through the agency of a vector potential or force, as it still is defined in the Standard Model, but it is really a *torsion wave*. The calculations show that indeed, the anomalous velocity variation of the probe, is *independent* of the mass of the planet (no gravity), remaining dependent on its radius, *angular velocity*, and the orbital inclination. These results coincide with the Lagrangian of a spinning source, mentioned before

Pinheiro's TTC is a reliable approach to tackle problems in the frame of gravitational and electromagnetic rotating systems. This would be the first time that the West acknowledges, that the *rotation* of a celestial body is *related* to an energy output, since the Kozyrev's research in the 1940's, on the same subject.

On the other hand, it is now generally accepted that the space surrounding the Earth and perhaps the entire Galaxy has "right-handed spin," meaning that energy will be influenced to spin clockwise as it travels through the physical vacuum or *aether*. This would explain the "something" that was slowing down NASA's Pioneer 10 & 11 spacecraft on their journey out of the Solar System.

Another anomaly in the Solar System is that one associated *only* with total solar eclipses and is known as the "Allais effect ", in which minor changes in the behaviour of paraconical, Foucault or even simple pendulums, do occur as the Sun passes behind the Moon and *cannot* be explained by the Standard Model.

In 1954, French economist and physicist Maurice Allais, was experimenting with pendulums over the course of a 30-day period, during which he observed and recorded their movements.

During this time, an eclipse of the Sun occurred, under which conditions Allais noticed that the movement of the pendulum appeared to *quicken*; although the effect was very minor, it was nonetheless *noticeable*, and has been *reproduced* several times over the years, despite there being controversy over *what*, precisely, causes the so-called Allais effect.

In 2004, *The Economist* reported on a study by Dutch physicist Chris Duif with the Delft University of Technology in the Netherlands, who was convinced after further *research* of his own; that the phenomenon known as the Allais effect is not only *real*, but that it could still qualify as being *unexplained*.

Many explanations have been proposed, from experimental *errors* and atmospheric/temperature tricks to alterations in the *inertial mass* of pendulums, that might be allowed after squeezing Quantum Theory, hoping to get any reasonable explanation.

None of these can satisfactorily explain this phenomenon, because it is caused by a *variable* not accounted for, by any conventional theory. The usual explanation is that the anomalous torque may originate from the so-called "gravitomagnetism" of the motion of the Sun and the Moon. In other words, the pendulum anomalies during total solar eclipses, might provide preliminary evidence which show the existence of gravitomagnetism; which it is *assumed* like the electromagnetic field, in this case gravity is "retarded".

Trouble is that this gravitomagnetism from the Moon and Sun, would *decrease* with the inverse of the square, weakening. Also, electromagnetic waves travel at the speed of light *c* whereas gravity propagates instantaneously, as it will be explained in chapter five.

Torsion waves are the *real* cause of the Allais effect. In normal circumstances, torsion waves coming from the Sun which don't follow the inverse of the square rule and thus are *not* weakened, reach Earth's surface where a pendulum is swinging. These waves do *dampen* the pendulum oscillations that are normally observed and recorded, following the conventional science rules.

Indeed, the pendulum in normal conditions is working "under pressure" of torsion waves. When a total eclipse of the Sun occurs in the location where the pendulum is swinging, the considerable size and mass of the Moon *blocks* the oncoming torsion waves from the Sun, preventing these to *reach* the pendulum. Therefore, the dampening effect is *eliminated* and the pendulum oscillations quickens.

Given the above, the scientific community in the West still dismisses the concept of *torsion fields* and the Russian work on this subject as "pseudoscience". One of its current leading scientists, Dr Gennady Shipov is catalogued as a crackpot. Thus, the public is being deliberately *misinformed.*

However, the Soviets during the Cold War in the 1970's, under the leadership of Premier Leonid Brezhnev following the policy of his predecessor Nikita Khrushchev "to crush the West with nonconventional weapons, at any cost"; used this "pseudoscience" as a weapon to deliver a headache to the US in regards of an "unusual" weather patterns and several aerospace "accidents" still classified.

The UK got its share as well, with the "enigmatic" Bala earthquake that shook most of North Wales on the night of 23 January 1974. With a magnitude of 3.5 it represents the size of earthquake to be expected in the UK with a return period of about one year.

Nevertheless, the prominent atmospheric lights observed at the time of the shock and the report of a mysterious bang and rumbling, led to speculation that an aircraft had crashed, and search-and rescue teams were deployed. The following day, newspaper reports spoke of disagreement over the cause of the event.

On the one hand, it was suggested that the event must have been a meteorite, as nothing else would explain the lights seen. On the other hand, the view from British Geological Survey (BGS) was that, given the preliminary magnitude value of around 4, it would have needed such a large meteorite that it would have been unmistakable.

The possibility of finding the meteorite drew in geologists to search the area over the weekend, including staff and students from Leeds, Liverpool and Durham universities, and various amateurs. *Nothing* was found. In fact, all searches were without fruit, and the last one was called off at the end of that day. BGS concluded that data was *insufficient* to suggest a causative fault for the earthquake. More imaginative members of the public decided (and still believe) that an UFO had crashed and a government cover-up certainly followed.

However, this UFO crash couldn't have accounted for the *energy* released registered by the seismometers. What the public and geoscientists alike ignore, is that the *real* cause of this "earthquake" came from Russia with premeditation and malice, using torsion waves delivered through a powerful beam weapon.

How did they manage to accomplish that? Firing with the right angle and direction to the *ionosphere* which deflects the beam, reaching target. Probably in that occasion, the Soviets were running a test. This shows that *russkaya nauka* is something that cannot be underestimated, but the West stubbornly, still does.

The *real* cause of Earth's magnetic field

Now we are in conditions to understand, what *really* causes Earth's magnetic field. In fact, the *difficulties* which stand in the way of basing terrestrial magnetism on *electric currents* generated inside the Earth according to the Dynamo Theory are insurmountable. Even scientists admit that it's a very complicated, highly idealized and chaotic system.

This theory is based on the idea of Larmor, in 1919, that magnetic fields of celestial bodies could be due to dynamo effects. However, the giant planets in the outer parts of the Solar System all have *strong* magnetic fields. It was not even expected that they would be sufficiently electrically conducting, to allow dynamo action.

Dynamo Theory presents as many *difficulties* in mathematical terms as well as in predictions. It proves to be extremely difficult to calculate, the dipolar magnetic moment of the extrasolar planets using this model. However, the "current consensus" is that the Earth's magnetic field is the result of a regenerating (self-sustained) dynamo action in the fluid outer core.

The immediate mathematical *problem* with developing Larmor's explanation was one of *symmetry*, or rather the need for symmetry breaking in a fluid dynamo. The geometry of astrophysical objects is very simple, and this means that the fluid flows required to amplify magnetic fields have to be complicated.

Dynamo action, that is the amplification of magnetic fields, *cannot* be obtained if too much symmetry is imposed on the magnetic field or on the fluid flow. The mechanism of generation of the field is closely linked dynamically with the rotation of the Earth. Similar ideas are believed to apply to the solar magnetic field, to other planetary fields, and perhaps to the magnetic field permeating the cosmos.

The alleged dynamo at Earth's liquid outer core is "self-sustained" which alludes to the type of generator that makes its own magnetic field *without* any need for an externally-applied current or *field.* Ever since Larmor suggested this hypothesis, hundreds of geoscientists have worked hard on developing the hypothesis into a theory that works.

They have done theoretical analysis, numerical simulations with large computers, and laboratory experiments with NO avail. In chapter two of this book, we already examined the reasons of the *impossibility* of the existence of this "iron core" and the alternative seismological interpretation of a gaseous layer covering the inner core.

Moreover, in 1933, a young English astronomer with good training in mathematics, T. G. Cowling threw a big monkey wrench into the machinery of dynamos. With straightforward mathematics, applied first to the magnetic fields of sunspots, Cowling showed that simple forms of the dynamo idea couldn't work. He wrote:

"A field which resembles an axially symmetric field in certain respects cannot, in general be maintained by the currents it itself sets up. We are led to conclude that the magnetic field of a sunspot is not self-maintained. For the same reason the general magnetic fields of the Sun and the Earth cannot be self-maintained, as was suggested by Larmor."

The magnetic field outside the Earth is generally symmetric around its north-south magnetic axis. This means that if a dynamo is going to work, the field inside the Earth could not have such symmetry because this symmetry is caused by a field *surrounding* Earth. Instead it would have to *deviate* strongly from symmetry from place to place. That means dynamo theorists have had to abandon grand global explanations and examine specific local phenomena. Their theories have had to become much more complicated.

Responding to the proofs that axisymmetric dynamos can't work, theorists began trying models without such symmetry. One broad class of those were the "kinematic dynamos". Theorists tried to find specific large-scale core fluid flows that would produce the observed field. Bullard and Gellman (1954) used an infinite series of spherical harmonic functions to describe the motions.

The motions, though mathematically regular, were very *complex.* The series appeared to give a solution for the fields when truncated at the 12th harmonic. Because it was the first dynamo model to produce numerical results, it became very influential.

However, later examination showed that the infinite series does not converge, *invalidating* the solution (Gibson and Roberts, 1969). Other variations using different initial flows have *failed* to get around this problem (Merrill and McElhinny, 1983, p. 250). The only kinematic dynamos that appear to work (on paper, at least) are existence theorems that have *unrealistic* geometries, such as an infinitely-large conducting medium.

Apparently, none have been found for spherical geometries like the Earth's core (Merrill and McElhinny, 1983, p. 253). Looking carefully the troublesome and complicated working approach, used by these geoscientists to explain how Earth's magnetic develops, it is clear that they forgot to apply the *Occam's razor.*

In fact, Occam's razor, can be used as a general test of the soundness of a theory, as the general trend of *successful* science is from disorder and complexity toward order and *simplicity*. Occam's razor test indicates that geoscientists took a very wrong turn since the inception of Dynamo Theory.

On top of that, this highly idealized Dynamo Theory has *not* yet predicted *any* successful cosmic field. Its use today rests in the *assumption* that no alternative theory corresponds more closely to observations, which of course is false. Although at the dawn of the 21st century, numerical modelling of the Earth's magnetic field has *not* been successfully demonstrated, but appears to be (hopefully) in reach.

Scientists continue to refine and evolve the numerical model using ever more sophisticated and faster computers. From the scientific method standpoint, this is an idiocy. Doing the same thing, several times when it hasn't worked the first is indeed foolish. Perhaps, these scientists are indeed a bunch of idiots. As explained in chapter one, if a theory *cannot* explain the observed facts, let alone predict new ones, it must be *discarded* and replaced for a *new* and reliable one.

Nonetheless, this is not the case which shows the level of *incompetency* of the mainstream scientific community, when dealing with Earth's magnetic field. To endorse this, mainstream science claims that the origin of the magnetic field might be due to a "seed" magnetic field which will develop to produce the large-scale magnetic field, observed for celestial objects.

Also, there is a possibility that in some stars, magnetic field is "fossil" a relic from a distant past. Both hypotheses *don't explain* how the observed magnetic field originated, in the first place. A huge setback came when new helioseismic observations (Hanasoge et al., 2012) suggest that convection in the Sun is a hundred times *slower* than previously thought.

That makes a dynamo explanation of the observed fields nearly *impossible*. By implication, this reduces the odds that planetary magnetic fields are produced by dynamos. Despite all these *shortcomings*, Dynamo Theory is still the accepted explanation for the origin of magnetic fields observed in planets and stars, although is useless.

In fact, a report published in Nature by early 2018, stated the World Magnetic Model (WMM) based on data provided by magnetic survey satellites, was in *trouble*. In fact, researchers from NOAA and the British Geological Survey in Edinburgh had been doing their annual check of how well the model was capturing all the variations in Earth's magnetic field. They realized that it was so *inaccurate* that it was about to exceed the acceptable limit for navigational errors. It showed that the Earth's magnetic field is shifting so rapidly and unpredictably, that it's rendering the models on which our lives depend, *obsolete* at a faster rate than ever before. Scientists, faced with the new rate of change, are scrambling to keep up.

The reality is, they are indeed *clueless* of what's really causing Earth's magnetic field, let alone to *predict* its behaviour. Their lame excuse is that in regards to Earth's magnetic field and other celestial objects, the mechanism of its origin is poorly understood. Bear in mind that Dinamo Theory is a strong argument used *against* Earth being hollow, because a fluid iron core is "essential" in order to Earth can generate a magnetic field.

The truth is that many scientists *ignore* that space flights and some developments of astrophysics, show a remarkable and earlier unknown fact that *magnetic moments* (μ) of all planets of Solar system, as well as some their satellites, stars and galactic magnetic fields in general, are *proportional* to their *rotation* or angular momentum (L). This indeed, offers a *clue* of the origin of terrestrial magnetism. In fact, this approach was already proposed at the end of nineteenth century, but with some exceptions, it was *ignored* by mainstream science.

Revisiting an idea, he had already aired in 1891 before the Royal Institution, German-British physicist Sir Arthur Schuster (1851-1934) in 1912, discussed different mechanisms by which the *rotation* of any type of body could give rise to a *magnetic field*. He was opposed to the idea that electric currents deep underground, should be the cause of Earth's magnetism, because no suitable electromotive force was known and because for the currents to be decaying remnants of some ancient initial state, unreasonably strong magnetic fields had to be assumed for the remote past. Schuster asserted that, the magnetic fields of planetary and stellar bodies arise solely from their rotation. That is, that electrically *neutral mass* current, at least in the case of rotation, generate magnetic fields.

Schuster held that the *close* alignment between the axis of the dipole and of rotation, could not be a *coincidence.* In fact, recently it was found that the Saturn's magnetic field practically coincide*s* with its rotational axis. As an alternative to the currents, Schuster suggested that any *spinning* body would have a *magnetic* field. The dependence would not be a straightforward relation between angular velocity and field intensity, because the effect of such a mechanism would already have been observed.

On the other hand, the field would be too difficult to detect at the laboratory scale, if it were proportional to the product of angular velocity and the radius squared. Schuster proposed that the rotation would instead give rise to a magnetic force that may or may not induce magnetization, depending on a body's chemical composition.

He also pointed out that magnetic molecules (aether electron-positron dipoles) should *align* with the *axis* of rotation like gyrostatic compasses and in so doing precess about that axis, a very plausible explanation for the *secular variation* of the field, which the Dynamo Theory *fails* to account for. In fact, the correlation of the secular variation of the geomagnetic field and oceans flows is *opposite* with the concept of the core fluids, which was theorized reasonable for self-consistent Dynamo. (Ryskin, 2009).

In 1947, news where made in the scientific community when 78 Virginis, was the first star other than the Sun in which a *magnetic field* was discovered by US astronomer H. W. Babcock the previous year. It was the first reported detection of a magnetic field in an extrasolar object.

Few are aware that Babcock's discovery was indirectly made possible through Schuster's hypothesis, which led US astronomer George Ellery Hale to construct the great 100-inch telescope on Mt. Wilson. Hale in 1915 said: "It is not improbable, as Schuster has suggested, that every star, and perhaps every rotating body of whatever nature, becomes a magnet through the fact of its rotation. It is hoped that the 100-inch reflector will enable this test for magnetism to be applied to certain stars." It did. Babcock's discovery stimulated British experimental physicist Patrick Blackett, to *review* Schuster's work. But Blackett had only three objects on which to confirm the hypothesis.

These were the Earth, the Sun, and the star 78 Virginis. Blackett had noted that the magnetic fields attributed to the Sun and Earth seemed proportional to their angular momentum: "Could it be that every large, rotating gravitational mass generates a magnetic field according to a simple, elegant relation"?

Planets and satellites of the Solar System are neutral *rotating systems* which, according to the Blackett hypothesis, should be magnetized, and indeed they are, as revealed by the data of a number of spacecrafts.

He was the first to claim the existence of a *correlation* between magnetic moment and angular momentum in astrophysical objects. It has been known for a long time, particularly form the work of Schuster, Sutherland and Wilson, though lately little regarded, that the magnetic moment P and the angular momentum U of the Earth and Sun are nearly the square root of the gravitational constant G divided by the velocity of light *c*.

Blackett first presented a sketch of his theory in a seminar in November 1946. The impression is that Blackett originated the idea that the magnetic field of a planetary or stellar body, is related to the body's angular momentum or more generally to its *rotation*. He speculated that there exists an *unknown* universal physical law (in fact it is one of the forgotten original Maxwell's equations). Blackett certainly adopted the idea, and spent considerable effort on testing it.

Since 1947, there has been increasing *evidence* for this effect by the measurements of the magnetic fields of the solar planets, the Sun, other stars and even pulsars; as well as the galactic magnetic field. However, the attempt to *measure* this effect in the *laboratory* depends on the *ability* to measure extremely *weak* magnetic fields and the shielding of extraneous magnetic fields.

Blackett performed his experiments on the premise that his golden rotating cylinder, should show a magnetic moment greater than 10^{-9} Gauss. But the experiments did not get detectable magnetic fields (the reason will be explained shortly). After negative result of the rotational magnetism theories, the self-consistent dynamo was developed by scientists. However, this apparent failure does *not* contradict observations, because the effect of this field can be *observed* mainly for very *heavy* rotating objects, such as stars and planets.

About 32 years later, Sirag tested the Blackett hypothesis using new available data for Mercury, Venus, Jupiter, Saturn, the Moon, and the pulsar Her X-1. Seven magnetic planets are known to exist in the Solar System: Jupiter, Saturn, Uranus, Neptune, Earth, Mercury and satellite Ganymede. From the measurements of the magnetic moment and the mass of these seven planetary objects, a straight line may be fitted to the data. The relationship between the magnetic moment and the mass (m) is a power law:

$$M = 2x10^{-21}m^{-1.7365}$$

hence, increasing planet mass increases magnetic field. This relationship between the data has a correlation coefficient of 0.9804, reinforcing the idea that a *physical relationship* exists between these two variables. However, the generation of the magnetic field does not only depend on the mass.

If we apply the above equation to the case of the planet Venus, according to its mass it should have a magnetic moment similar to that of the Earth ($3.953x10^{22}$ Amp-m^2; Cox, 2000). However, in reality, its magnetic moment is either non-existent or less than $5x10^{17}$ Amp-m^2 (Cox, 2000). This makes us conclude that even though planetary mass may be significant; it is not the *only* factor intervening in the generation of a magnetic field.

The functional relationship between the magnetic moment and the rotation period (p) represents an exponential function, meaning, that when the inverse of the rotation period is increased or the rotation period is decreased; the magnetic moment increases and can be expressed as:

$$M = 4x10 \exp(579696/p)$$

this relationship between the data has a correlation coefficient of 0.9742, *reinforcing* the idea that a physical relationship exists between these two variables. However, the generation of the magnetic field does not only depend on the rotation period. If we were to apply the above equation to the case of the planet Mars, its rotation should induce a magnetic moment similar to that of the Earth: $3.953x10^{22}$ Amp-m^2 (Cox, 2000).

However, in reality, its dipolar moment is non-existent or less than $5x10^{18}$ Amp-m^2 (Cox, 2000) Thus, we can conclude that although the rotation period is important; it is not the *only* factor which intervenes in the generation of the magnetic field. There is something else.

In a paper published in the *Astrophysical Journal* (1995) by US astrophysicists C.N. Arge, D.J. Mullan and A.Z. Dolginov, they demonstrated a *definite* lineal *correlation* between magnetic momentum μ and with angular momentum (L), which cannot be ruled out as mere coincidence. The study was made using published data on magnetic fields, radii, masses, and rotation.

They have compiled a data set of magnetic moments μ and angular momenta L for stars and planets. In the class of hotter stars (classified as A, B, and O), there were 171 objects. In the class of cooler stars (classified as F, G, K, and M), there were 54 objects. For the hot and cool stars, the correlations are *positive*.

However, the hot-star correlation is significantly shallower than for the cool stars. In the Solar System samples (planets), the correlation has essentially the same pattern as for the cool stars, although the magnetic moments are two to three orders of magnitude smaller for the Solar System objects at a given L value.

This is due to the big difference in size (radius) between the stars and planets. Another study (Merrill et al. 1996) arrives at *similar* conclusions, as the following figure shows the lineal proportion between magnetic field strength and angular momentum:

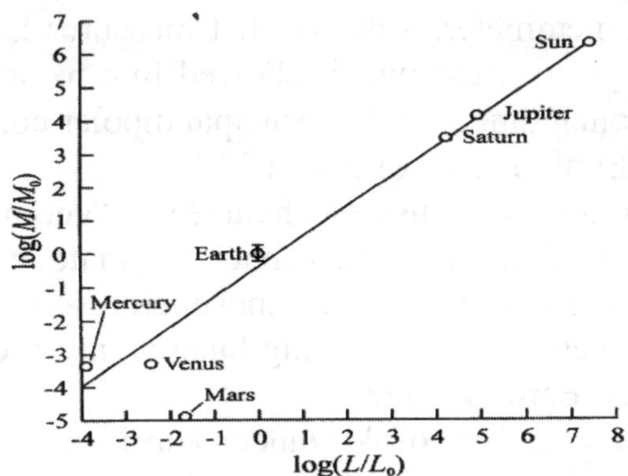

Fig. 92- This graphic shows clearly, that there is a link between the Magnetic Field Strength (M) in terms of Earth's Field vs. Angular Momentum (L) in terms of Earth's Angular Momentum. (Merrill et al. 1996).

At this point, it is important to highlight that Dynamo, Seed and Fossil theories, neither predict nor explain (as any good theory does) the general correlation between magnetic and angular momenta observed in celestial bodies. In fact, the 2016 Juno's mission to Jupiter, the magnitude of the observed magnetic field was 7.766 Gauss, significantly stronger than expected.

In 2017, the Cassini's probe showed that Saturn's magnetic field, appears to be surprisingly well aligned with the planet's rotation axis. Previously, mission scientists thought that 0.06 degrees would be the lower limit of tilt that could generate the observed magnetic field. However, the results show the tilt may be much less than this number.

Scientists currently think that planetary magnetic fields, require some degree of tilt in order to "sustain" currents flowing through the liquid metal deep inside the planets (in Saturn's case, thought to be liquid metallic hydrogen). With no tilt, the currents would eventually subside and the field would disappear.

Given the above, it clearly shows without a shadow of a doubt, that indeed the *problem* of the magnetic field of the planets must be considered as part of a wider subject, the magnetic field of *rotating bodies*. Although the problem has been with us since the publication in 1838, by German polymath Carl Friedrich Gauss (1777-1855) in his book *Allgemeine Theorie des Erdmagnetismus* and despite the subsequent numerous publications, its fully satisfactory solution was rather elusive until recently.

For an extensive survey of terrestrial magnetism, Gauss invented an early type of magnetometer, a device that measures the direction and strength of a magnetic field which allowed him to study the Earth's magnetic field, concluding that the principal dipolar component had its origin inside the Earth instead of outside.

This observation lead Gauss to conclude that "geomagnetic storms" or strong magnetic field changes, cannot originate from the Earth's interior but somewhere in the upper atmosphere. He demonstrated that the dipolar component was a decreasing function inversely proportional to the square of the Earth's radius.

Gauss, with the support of Alexander von Humboldt, founded the *Göttinger Magnetischer Verein* located at what's known the *Gaußhaus,* the centre of which was the Earth Magnetic Observatory from 1836 to 1941. The records kept in this observatory, allowed to acknowledge that Earth's magnetic field is *fading*.

Today it is about 10 percent weaker than it was when Gauss started keeping tabs on it in 1845, scientists say. Currently, there is a computer-assisted magnetometer in the *Gaußhaus*, which measures and plot the temporal changes in the direction of the horizontal Earth's magnetic field (changes in declination).

The definitive answer for the origin of magnetic fields in rotating celestial bodies such as Earth, came at last in 1977 when French physicist Maurice Surdin (1911-2003), while working at the Centre National de la Recherche Scientifique; resumed *successfully* the experiments carried out by Blackett with rotating bodies.

Surdin was a remarkable case of thinking outside the box. He devoted himself to development of various avant-garde physics theories that were close to his heart since long, such as that of the magnetic field of rotating bodies, or that of Stochastic Electrodynamics, based on the concept of a "zero - point field" a novel version of the aether.

On this occasion, Surdin was able to *detect* measurable magnetic fields **(H)**. In his experiments, for evident reasons the configuration of the rotating body should have a symmetry of revolution.

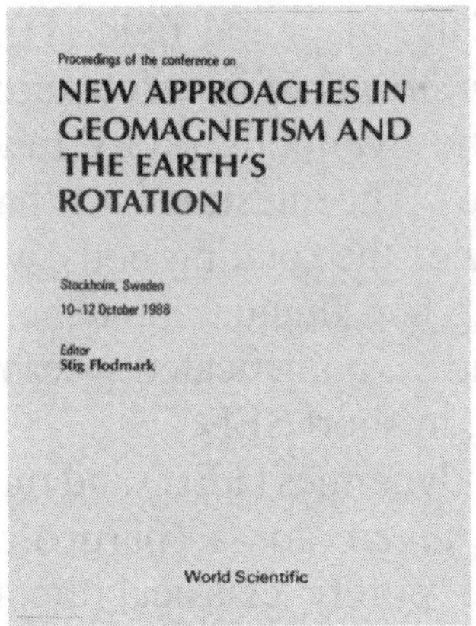

Fig. 93- Frontispiece of the book containing papers presented in a symposium held in Stockholm, in 1988 including novel ideas correlating geomagnetism and the Earth's rotation. This symposium gave clear indications that the accepted theories in geomagnetism suffer from serious fundamental deficiencies. The physics of the Earth's interior would have to be entirely revised with respect to the results presented in this book. This of course was received unsympathetically by most geophysicists.

Rotating a sphere has technical issues with the mechanical set up so instead, a cylinder was chosen. It was desirable to experiment with two different cylinders: Cylinder A was made of tungsten alloy of density ρ = 17.6 and the other cylinder B was made of bronze of density ρ = 8.8 Both cylinders had the same geometry.

Considerations involving mechanical constrains and the requirement that the peripheral velocity, should be lower than the velocity of the sound in air at the target angular velocity of $\Omega = 2\pi \times 1000\ sec^{-1}$, led to the choice of r = 5 cm for the radius of the cylinders.

To avoid flexural oscillations of the body, the height of the of the cylinders was chosen to be h = 15 cm. To rotate such cylinders at the angular velocity mentioned above, air bearings and air turbine were used (electric motors were obviously banned because of possible magnetic perturbations).

Preliminary considerations, supported by order of magnitude calculations, showed that *direct* measurements of the magnetic field presented insuperable *difficulties*, the same ones that Blackett faced three decades earlier and rendered his experiments null. This was due to Blackett's inadequate choice of the experimental dispositions. This has *not* proved that a magnetic field, due to a rotating body, doesn't exist as is it wrongly and currently accepted.

The existence of the magnetic field (**H**), was detected by Surdin *indirectly* from *measurements* of its effect through *induction*; within the atomic structure of the cylinders, pretty much as Schuster predicted (chemical composition). The question is: What caused this induction?

Is the *interaction* of the rotating body with its *environment* or the universal all-pervading and fluctuating aether. However, Surdin uses a more technical and sophisticated term: *Vacuum Stochastic Electrodynamic Field*, in short SED.

Stochastic Electrodynamics (SED) and random electrodynamics are the names usually given to a particular version of *classical electrodynamics*. This purely classical theory is Lorentz's classical electron theory into which one introduces random electromagnetic radiation (classical zero-point radiation or aether) as the boundary condition, giving the homogeneous solution of *Maxwell's equations*.

This mechanism would allow the activation of an *electron-positron pair* (aether dipole) which would behave like oscillators independent harmonics exchanging energy with the zero-point field. Which is indeed the case.

According to Maxwell, Ampère's Circuital law is about how a moving element interacts with the *surrounding luminiferous medium* in such a way as to *induce* vortex rings that amount to *magnetic lines* of force. These lines of force form concentric solenoidal rings around the direction of motion.

The manner of the physical interaction that induces these magnetic lines of force is dictated by *hydrodynamics* in the context of a sea of tiny *aethereal vortices* that are pressing against each other with centrifugal force while striving to dilate. Based on the concept of the random electromagnetic field with dynamical properties (aether) of the Universe, a Stochastic Electrodynamics is developed. Schrödinger's equation is obtained using strictly *classical deterministic* arguments.

The derivation of Schrödinger's equations forms the link between Vacuum or Aether Stochastic Electrodynamics and Quantum Mechanics. SED *induces* electric dipoles of short duration in the constituent atoms of the, otherwise electrically neutral, *rotating* body.

Due to rotation, these electric dipoles are ordered in planes perpendicular to the axis of rotation, pointing towards the axis or in the opposite direction the *rotation* around the axis of these ordered electric dipoles, *creates* in the centre of the rotating body a *dipolar magnetic field* (**H**), parallel or antiparallel to the direction of rotation.

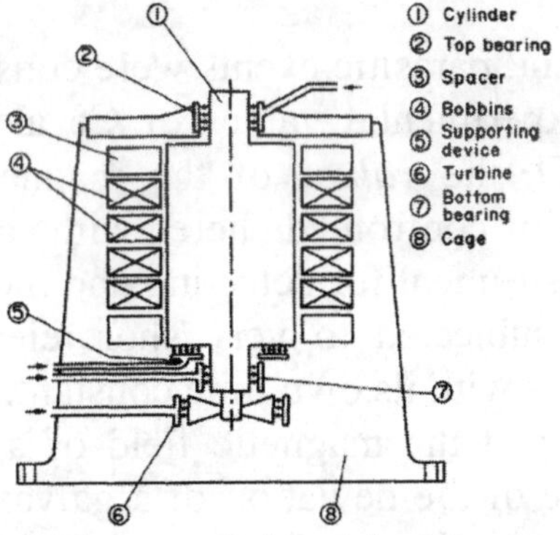

Figure. 94- Experimental setup designed by Maurice Surdin to successfully detect a dipolar magnetic field, through a neutral fast-rotating cylinder applying the principles of Stochastic Electrodynamics (SED). The cage, is made of non-magnetic bronze. Non-conductive bobbins hold the rotating cylinder in place. An electromotive force is induced onto a cylindrical detecting coil surrounding the rotating body, which feeds a very low-noise preamplifier (PA) connected in circuit with an amplifier (A) with noise filters and an oscilloscope. (not shown in the figure).

The reversal of the electric dipoles, from the ordered situation where they point towards the axis of rotation to the inverse direction, is according to SED a random process and corresponds to the reversal of the magnetic field. During the reversal, the field decreases from +H becomes zero and the increases into the opposite direction to become -H then the field goes through reverse succession of events to become +H.

Behaving similarly to a cylindrical coil, where its inductance depends on its geometrical characteristics; the axis of the atomic structure which coincides with the axis of the rotating cylinder, the *reversals* of magnetic field will *induce* an e.m.f. in a form of sequences of short pulses (Et). All with same amplitude alternatively positive and negative, occurring at instants. With these values, one obtains a correlation (Ct) as a function of Et.

When the cylinder is *rotating* at a constant angular velocity Ω, one plots the values Ct as a function of Et. When he cylinder is at *rest* one plots the values of the noise (Cn) or variations in voltage or current that are often random, usually of relatively low amplitude.

The independence of signal and noise leads to calculate the *correlation* function (Ce) necessary to *evaluate* the magnetic field **H**:

$$\mathbf{Ce = Ct - Cn}$$

In the experiment, parasitic events were considered and taken care of. Knowing the experimental values of Ce along the mathematical formulae from SED, the *values* of the magnetic field (**H**) can be determined. We are confronted here with a case similar to a *galvanometer,* an instrument for detecting and measuring small electric currents; which is subjected to very short electric fluctuations but fluctuating as a *whole* with its own time constant.

The *fluctuations* of the magnetic field of a rotating body differ, however, from those of the deviation of a galvanometer from the fact that this magnetic field is due to a "cooperative phenomenon" originated from the *interaction* of the rotating body with the all-pervading aether, so that reversals of the magnetic field may occur in a *shorter* time than the Lamb T_L time constant or the time where motions of electricity, produced in a spherical conductor by any electric or magnetic operations outside it, do occur.

In 1979, Surdin published a study where the *predictions* of a model of the magnetic field of rotating bodies, where compared with the results of observations of the magnetic fields of the planets, showing *positive correlations*. Also, in regards to Earth's magnetic field reversals, the comparison between the results of palaeomagnetic measurements and the theory shows that a substantial *agreement* exists between the two. From this perspective, SED can *predict* when the next Earth magnetic field reversal, will occur.

Surdin's *experiment* applying SED, points out the *true* origin of dipole magnetic fields in *rotating* bodies such as Earth, several planets and stars in general. Also, its internal origin as first noticed by Gauss. It explains the *real* cause of magnetic pole *reversals* observed in stars and the Earth through the paleomagnetic record. However, geoscientists still believe that Earth's pole reversals are caused by "instabilities" occurring in the fluid flow of the core.

On the other hand, the Earth's magnetic field *must* continuously be fed with energy (aether); otherwise its interactions with matter in the planet would cause it to decay and thus, disappear.

Most the researchers working on Stochastic Electrodynamics (SED), hope that it will provide an *accurate* description of atomic physics and *replace* or explain Quantum Theory, because it includes a novel version of the outlawed *aether*.

In fact, despite of the impressive predictive power and strong mathematical structure of quantum mechanics, the theory has always suffered from important conceptual *problems*. Some of these have never been solved. SED is aimed at finding a conceptually satisfactory, realistic explanation of quantum and other phenomena.

Stochastic Electrodynamics reaffirms atomic physical processes could be accurately described within *classical physics* provided one considers the appropriate classical (Maxwellian) electromagnetic stochastic radiation fields acting on classical "charged particles".

The key idea here is that if thermodynamic equilibrium of classical charged particles is at all possible, then the particles must follow fluctuating trajectories in space.

Fluctuating trajectories of charged particles necessarily imply that fluctuating radiation must also be present (aether disturbance), and we again come back to the realization that if equilibrium is possible, then the presence of both particles and radiation, all fluctuating, is a *natural*, *necessary*, and *essential* component of this situation.

Moreover, this property must be true at whatever temperature characterizes the equilibrium situation, including the temperature of absolute zero, T = 0. These properties correspond to what we observe in

Nature, since at T = 0, atomic and molecular motion does not cease, but rather there is a zero-point field motion.

Thus, the "problem" of atomic collapse for a classical hydrogen atom that helped turn physicists, such as Planck, Bohr, Heisenberg and others, from classical physics in the early 1900s and *reject* the aether, might not be the unaddressable "roadblock" for classical physics, but instead might actually be the central element for understanding how quantum mechanical phenomena emerges from fully *classical physics.*

In fact, Lorentz was perhaps the first to propose in writhing that otherwise "empty space" is filled with *Planck harmonic oscillators* (aether electron-positron dipoles). A quantum oscillator can absorb or emit energy only in multiples of this smallest-energy quantum (*h*) and that these oscillators, could be responsible for the action of all forces trough the vacuum of space.

Thus, the duality wave-particle and particle statistics can be replaced by *deterministic* finite precision Wave Mechanics. This was the aim of Austrian physicist Erwin Schrödinger in 1925, before it was twisted and *corrupted* by Max Born with his "probability interpretation" which was accepted by almost everybody, to this day.

SED shows that what is accepted as the "electron spin" can be considered to be generated by the *interaction* of the particle with the zero- point field (aether); in particular, the two-valuedness of the spin projection is associated with the existence of just two independent states of *polarization of the field* (oriented aether dipoles).

Again, the main mechanism for generating the geomagnetic field and the energy source that powers it, which mainstream science cannot figure it out, is satisfactorily explained through SED applied to rotating bodies. Also, it can predict magnetic field reversals of planets.

It must be acknowledged that, if it were not for this *aether* confirmed by experiments but so despised and ignored by mainstream science, there would be *no* magnetic field on Earth protecting us from Extreme Ultraviolet (EUV) and X-ray solar radiation, coming from solar wind and cosmic rays.

The more serious, long term impact would have been the erosion of the atmosphere. Therefore, we along with the rest of life, would never have existed. Quoting Roman emperor Marcus Aurelius (121-180): "It was for the best, so Nature had no choice but to do it".

This denial of aether by mainstream science has led to a chain of misconceptions of the nature of reality that can only be described as incomplete at best and paradoxical at worst. After the initial misinterpretation of 1887 and further endorsement of it in 1905 by Relativity, one misconception led to another, widening the non-reality, deepening the unresolvability.

Physicists know, or suspect, there is something wrong here. The point is they have problems. Serious problems. In the mainstream science magazine, *New Scientist* issue of November 2th, 2019; an article called *Return of the Aether* explains that Relativity has run into *difficulties* of its own due to the wrong assumption that the all-pervasive aether is not needed.

The fact is that incompleteness and the paradoxes that have arisen are too easily demonstrated to be ignored. Unfortunately, the rejection of aether and the consequential incomplete theory of gravity has led theoretical physicists to propose *highly speculative* universes of mathematical genre, abstractions devoid of *reality*.

Without aether, theoretical physicists are led to a totally unrealistic picture of the Universe, thus they are selling a fake product to the public. Without aether, the Universe is contrary to Nature, contrary to reason and common sense. Without aether, the Universe is utterly *absurd*.

Chapter V

How determining properly, a key physical parameter of Earth, will tell us about its interior. A new paradigm of the internal structure of the Earth emerges.

I often say that when you can measure what you are speaking about, and express it in numbers, you know something about it; but when you cannot measure it, when you cannot express it in numbers, your knowledge is of a meagre and unsatisfactory kind.
William Thompson, Lord Kelvin (1824-1907)

Moment of inertia (MoI), mass and gravity

Moment of inertia (MoI) of a rotating body shows us, how much *resistance* to spin it has. This resistance depends on *how* this mass is distributed in the spinning body. It is a scalar value which tells us, how difficult it is to change the rotational velocity of the object around a given rotational axis. A solid sphere has *less* moment of inertia than a spherical shell. This can help us to determine whether the internal structure of Earth is a solid sphere or a *spherical shell*.

In fact, the figure of the Earth's interior is one of the fundamental issues in "solid earth" sciences today. The traditional theories of studying the figure of the Earth's interior usually start from HSE (hydrostatic equilibrium) and one-dimensional Earth model like PREM, obtaining the equipotential surface with symmetrical shape in function of density. Because density is limited by only providing the amount of mass per volume, a planet's moment of inertia can be used to provide more information about the differentiation of a planet's mass.

Particularly, the MoI factor can be used to identify if the planet has the internal shape of a uniform sphere, *spherical shell*, or a sphere and shell. This information can be useful in identifying whether the planet has large *structural variations* that can represent the composition, such as a dense, rocky core, and the radial proportion of the core to the total size.

Much of what scientists "know" about the distribution of mass within planets comes from their (theoretical or modelled) moments of inertia. The moment of inertia *controls* the rate at which a planet *spins* and the location of the spin axis depend on the *distribution* of mass in each body. We illustrate this concept by "racing" a metal ring and disk, which have the same mass but *different* moments of inertia, down an incline (fig. 101).

First, objects with more mass at the centre have lower moments of inertia. Because the relationship of the moment of inertia to rotation is exactly analogous to that for inertia and motion, a body with a *lower* moment of inertia will rotate *faster* than a body with a higher moment of inertia, even if they have the same masses and sizes. Second, if the mass is distributed *differently* with respect to different axes, the body will have a different moment of *inertia*. This is important not just for planetary studies, but for things such as satellites; more than one satellite has been lost because of an incorrectly calculated moment of inertia.

Before analysing the internal distribution of Earth's mass, it is important first to know its value. The standard method is using Newton's Law of Gravitation:

$$\mathbf{F = \frac{GM\,m}{r^2}}$$

F = force of gravity
G = gravitational constant (6.67×10^{-11})
M = mass of one object
m = mass of other object
r = distance between the two objects

Applying the Newton's Second 2nd Law of motion for and object on the Earth's surface:

$$\mathbf{F = ma = mg}$$

Substituting F in the above's Newton's Law of Gravitation and rearranging:

$$\mathbf{mg = \mathit{GMm/r}^{2}}$$

$$\mathbf{M = g\mathit{r}^{2}/G}$$

Knowing that the mean distance (radius) is 6.378×10^{6} m and substituting values, we get the Earth's mass: 5.972×10^{24} kg. Which is roughly the value found in any textbook of Physics/Astronomy.

However, of the many methods adopted up to the present for accomplishing this, none appear to be as free as they might be from many sources of *error*.

This is because the mass of an object *depends* on where it is in relation to the other entities with which it interacts, through a field with dynamic properties (aether). Also, gravity in fact, is *not* a true force as it will be discussed shortly. As in all other scientific systems of taking measurements, especially those in which the object of measurement is *not directly* comparable with our established units, special instruments have to be constructed by which certain measurements are taken, these serving us by the aid of the known laws involved with sufficient data to make calculations from which we derive the answer sought.

This necessitates the selecting of a *method* out of the possible many at our disposal, which, with the same degree of care taken in those measurements, will lead, all things considered, to the most reliable result. This becomes extremely difficult when trying to "measure" the *mass* of celestial bodies such as Earth.

There is evidence that the accepted mass of the Earth has been *overestimated*, which leads to its high average density (5.5g/cm^3) compared to the average density of crustal rocks (2.8g/cm^3). If this is correct, then the postulation of an "iron core" to fill this gap, would have been unnecessary and *misleading*.

In particular, according to the conventional methods using Newton's Universal Law of Gravitation and the Third Kepler's Law, the masses and densities of other planets of our Solar System, including the Sun, were determined. The Sun has a density of 1410 kg / m^3, the force of gravity on the surface F = 273.1 N. Saturn has a density of 687 kg / m^3, gravity force F = 10.44 N.

Planet Earth has gravity on the surface *less* than the indicated celestial bodies, but the average density of the Earth greatly *exceeds* the density of both the Sun and Saturn. Sealing of any bodies, including celestial bodies, occurs under the influence of external forces. These forces can only be attributed to the forces of its own gravity coupled with rotational centrifugal force. Consequently, neither the Sun nor Saturn can have a density several times less than that of the Earth. In addition, the average density of Saturn is less than the density of water.

Then on the surface the density of these celestial bodies must be equal to the density of the gas. This is an *absurd* conclusion. Such *contradictions* appeared within the framework of the theory of universal gravity. More than three hundred years ago, Newton's hypothesis was imposed on the world of scientists that bodies "create" the force of gravity. All scientists, in their calculations, were based on this *erroneous* concept. Therefore, the results of these studies were erroneous.

Newton himself was not sure about the gravitational properties of bodies (matter). He expressed that the cause of attraction may be a change in *density* in the space environment referred as *aether*. Therefore, aether is not just a passive universe prevailing medium. Rather, it is actively *flowing* into matter *constantly*.

This is similar to the views that of Fatio de Duillier's (1690) and Le Sage's (1748) and more importantly that of Newton's (1675), as he proposed a remarkable *mechanism* of kinetic character of aether relating to *gravity*, with some important and crucial differences. Newton supposed that there is in the aether a component of a sticky nature, which is continuously streaming towards the surface of the Earth, where it is *partly* absorbed.

Here it seems that Newton is trying to describe electrostatic *dipolar nature* of aether which is causing attraction as "sticky" and the "streaming towards the surface of the Earth" as in inward flux of aether. Thus, the downward aether *stream* which exerts upon the material bodies a force that Newton identifies with *gravity*.

However, the *definite* and most elaborate aether version of gravity that can explain *observable facts*, is the one proposed by Swiss physicist Georges-Louis Le Sage (1724-1803) in the mid-eighteenth century. The reasons for the present *resurgence* of various Le Sage-type models of gravitation are their *simplicity* and *depth*, features desirable in any physical theory. Whereas Newton's theory and (later) General Relativity were essentially mathematical descriptions of the motions of bodies in gravitation, Le Sage's theory aims to arrive at the very *cause* of gravity.

These suggestions led to the *Push-Gravity Theory*. The basic idea runs like this: Space is *filled* with minute particles or waves of some description (the all-pervading, space filling aether) which *strike* bodies from all sides. These corpuscles have the following properties: minute mass, enormous speed, and complete inelasticity (rigidity). Now, all apparently solid objects, such as books and planets, are mostly *void space*.

Consequently, gross objects absorb but a minuscule fraction of the aetheric corpuscles that are incident upon them. From these premises Le Sage *deduces* an attractive force between any two gross objects.

A single body will not move under this influence, but where *two* bodies are present each will be progressively pushed into the shadow of the other.

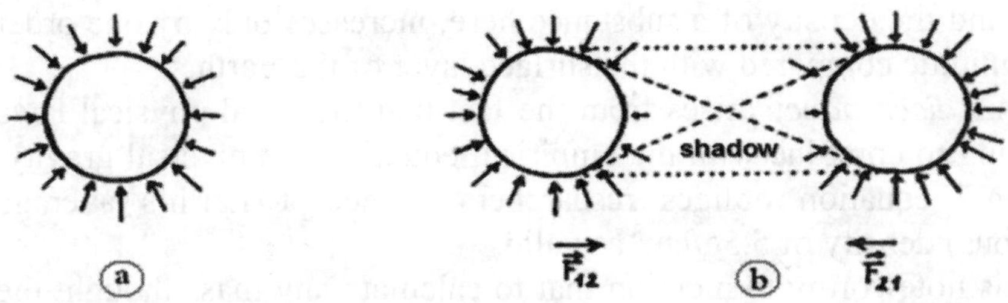

Fig. 95- ***a)*** *Aether particles hit the lone body from all sides with the same average intensity: the resultant of the reaction forces is zero.* ***b)*** *Two bodies located in a space filled with chaotically moving ether particles create a "shadow" relative to each other. Fewer particles fall from the direction from one body to the other than from the other (unshaded) directions, which creates the resulting force pushing the two bodies towards each other.*

The two bodies which appear to pull on one another are actually *pushed together*. Le Sage's theory of gravitation explains why gravity is always an *attractive* phenomenon, also *eliminates* the issue of that celestial bodies must be continually losing energy as a result of the "gravitational emission," causing gravity.

If any theory of gravity can be said to satisfy Occam's Razor, it is surely Le Sage's. Its simplicity and clarity guarantee that it will be conjured up again and again by those who seek to *understand* gravity's mechanism, as opposed to merely its rules.

However, the abolishing of the aether in 1905 led to the current *flawed* theory of gravity, with its unsurmountable difficulties swept under the carpet by alleged "proofs," based on *cherry picked* and *misinterpreted* data. This cosmic, grainy, extremely low dense substance is the *real* cause of gravity.

Today, we know that there exists such a uniform, continuum field: The cosmic vacuum or *zero-point field* (aether). Russian physicist Andrei Sakharov has proposed in 1968 a suggestive model in which gravity is *not* a separately existing fundamental force, but rather an *induced effect* associated with zero-point fluctuations (ZPF's) of the vacuum, in much the same manner as the *van der Waals* and *Casimir* forces are.

On the other hand, the conventional values of the density of the cores of celestial bodies raise great *doubts*. Static (technical) pressure in the center of the Earth in modern science is calculated to be 3.7 x 10^{10} kgs/m^3. That is, the pressure in the center of the Earth increases a million times, and the density of a substance here, increases only by one order of magnitude compared with the surface layer of the Earth.

This *discrepancy* arises from the fact that the usual physical laws are trying to combine with the empirical equation of universal gravity. Newton's equation obliges researchers to accept Earth's average (fictitious) density of 5.5g/cm^3 as valid.

It is noteworthy to mention that to calculate any mass thought the Newton's formula, it is key to experimentally define the value of the *constant of gravity* G. Nevertheless, as a constant using ordinary units (e.g., metric), the gravitational constant G doesn't occur in the well-known Newton's gravitation formula until late in the 19th century. Scientists in Newton's time and well into the 1800's, were quite happy to work in terms of *proportionalities* and *ratios*; the constant of proportionality (G) never needed to be written down.

For example, one does not see a gravitational constant, in any form, in the original Cavendish's description of his experiments to "determine the density of the Earth". In 1798, English chemist and physicist Henry Cavendish, performed an experiment to determine the *density* of the Earth, which would be useful in astronomical measurements. He used a *torsion balance* invented by English geologist and astronomer John Mitchell, to accurately measure the force of attraction between two masses.

From this measurement, he determined the mass of the Earth and then its density. In Cavendish's published paper on the experiment, he gave the apparent value for the density and mass of the Earth but *never* mentioned the value for G. It wasn't until 1873, that other scientists repeated the experiment and documented the value for G. This value for G implied from Cavendish's experiment was very accurate and within 1% of present-day measurements.

Because his experiment ultimately determined the value for G, Cavendish has been often *incorrectly* given credit for determining the gravitational constant. Today, the currently accepted value is 6.67259 x 10^{-11} Nm^2/kg^2. The value of G is an extremely small numerical value. Its smallness accounts for the fact that the force of gravitational attraction, is only appreciable for objects with large mass. Cavendish is credited for being the first man who “weighed” the Earth tough his aim was to determine its density for which he found a value of 5.5 g/cm^3.

Newton's gravitational constant, G, has been measured about a dozen times over the last 40 years, but the results have *varied* by much more than would be expected due to random and systematic errors. Now scientists have found that the measured G values oscillate over time, like a sine wave with a period of 5.9 years. It's not G itself that is varying by this much, they propose, but more likely something else is affecting the measurements. As a clue to what this "something else" is, the scientists note that the 5.9-year oscillatory period of the measured G values correlates almost perfectly with the 5.9-year oscillatory period of Earth's rotation rate, as determined by recent Length of Day (LOD) measurements. Although the scientists do not claim to know what causes the G/LOD correlation, they cautiously suggest that the "least unlikely" explanation may involve circulating currents in the Earth's core.

The changing currents may modify Earth's rotational inertia, affecting LOD, and be accompanied by density variations, affecting G. But this explanation is not valid, because as it was discussed before, the alleged Earth’s molten core *doesn’t* exist. Therefore, the cause of this variation of G remains an open question. The gravitational constant of the Universe, *G*, was the first constant to ever be measured.

Yet more than 150 years after we first determined its value, it is truly embarrassing how poorly known, compared to all the other constants, our knowledge of this one is. We use this constant in a whole slew of measurements and calculations, from gravitational waves to pulsar timing to the expansion of the Universe. Yet our ability to determine it is rooted in small-scale measurements made right here on Earth. The *tiniest* sources of uncertainty, from the density of materials to seismic vibrations across the globe, can weave their way into our attempts to determine it.

Is there any *independent* source that can “confirm” Earth’s mass? Yes, pretty much. The 3rd Kepler’s law: The squares of the periods of the planets (T) are proportional to the cubes of the semi-major axes (a) of their orbits (i.e. their mean distances from the Sun), or:

$$\mathbf{a_1^3/a_2^3 = T_1^2/T_2^3}$$

This equation holds also for the motions of satellites about their planets.

To include mass of planets and satellites, from *Celestial Mechanics*:

$$\mathbf{a_1^3 / a_2^3 = T_1^2 (M + m_1)/T_2^2 (M + m_2)}$$

In the case of Earth, the mass of the planet, Mp, is given by the formula:

$$\mathbf{Mp = \frac{T^2 a_1^3 M}{T_1^2 a^3}}$$

Where T_1 and a_1 are the period and the distance of the satellite from the planet, M is the mass of the Earth, and T and a refer to the Moon. We could also use *M* as the mass of the Sun and let T and a refer to the Earth.

We can determine the mass of Earth Me, in comparison with the mass of the Sun M, comparing the movement of the Moon about the Earth, with the movement of the Earth about the Sun.

The data is:

$T = 365{,}26$ days, $T_1 = 27.3$ days, $a = 149{,}500{,}000$km, $a_1 = 384{,}000$km

$Mp = (365.26\text{days})^2 \times (384{,}000\text{km})^3 \times M/(27.3\text{days})^2 \times (149{,}500{,}00\text{km})^3$

$Mp = 3.02 \times 10^{-6} M$

This is the *proportion* between the Earth mass in relation to Sùn. If we substitute the current Sun's mass of 1.989×10^{30} kg:

$Mp = (3.02 \times 10^{-6}) \times (1.989 \times 10^{30}\text{ kg}) = 5.99 \times 10^{24}\text{ kg}$

Which is practically the same Earth's mass value, using the standard Newton's gravity law.

However, whereas the Newton's Law of Gravitation works well with the planet's orbits around the Sun, is within or nearby a celestial object when *anomalies* had been observed, that don't fit the gravitational theory.

Therefore, the mass of the object or planet is questionable. The *first evidence* comes from a sophisticated experiment which has been suggested and carried out from more than four decades ago, in an Earth's laboratory, using neutron interferometry.

This experiment is known in the scientific literature as the "COW" experiment due to the initials of the researchers. It has been suggested by R. **C**olella, A.W. **O**verhauser, and S.A. **W**erner in 1974 and carried out many times starting from 1975 to 1997. The results of this experiment indicate clearly that there is a real *discrepancy* with existing theories, though some scientists would deny it. Before discussing this discrepancy, we give a simple account on the experiment.

The experiment studies quantum *interference* of thermal neutrons moving in the Earth's *gravitational field*. A neutron interferometer is used for this purpose (see Figure 96). A beam of thermal neutrons is split into two beams A1, A2 at the point *a*. The beam A1 is reflected at *b*, while A2 is reflected at *d*. The two reflected beams A1 and A2 interfere at the point c of the interferometer.

Assuming that the path lengths *ab* = *dc*, *ad* = *bc;* and that the trajectory of the neutrons is affected by the Earth's gravitational potential, a *phase difference* between the two beams A1 and A2 is *expected* which should fit theory. This is due to the *difference* in the Earth's gravitational potential affecting the paths *ab* and *dc*, (since *ab* is closer to the Earth's surface than *dc).*

Using the interference pattern, one can *measure* with precision, the phase difference and consequently the *difference* in the Earth's gravitational potential.

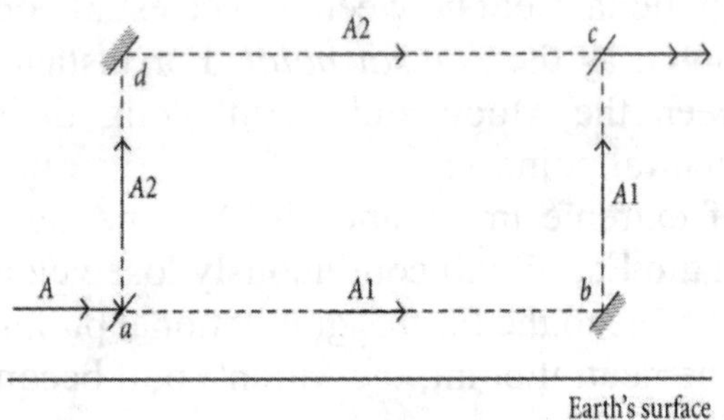

Fig. 96- Diagram of the trajectory of the neutron beams used in the COW Neutron Interferometer, in a non-inertial frame of reference (rotating Earth). The experimental results showed a discrepancy between theoretical and experimental values of the Earth's gravitational potential. In further experiments, a violation of the weak Equivalence Principle was present, even if the ratio of the inertial mass mi and gravitational mass mg is a universal constant.

The theories used for calculating the phase shift have been Quantum Mechanics and Newton's theory of gravitation (The Earth's gravitational field is a weak field). Newton's theory of gravitation is considered a limiting case of General Relativity (GR), in the weak field regime. So, both theories will give about the *same* prediction.

The results of this experiment show clearly that the Earth gravitational potential measured, is *different* from that predicted by Newton theory of gravity (or by GR), even in the weak field regime. The absolute value (measured) of this potential, is less than the corresponding value predicted by known theories of gravity.

It has been found that the experimental results are *lower* than theoretical predictions by eight parts in one thousand, while the sensitivity of the interferometer used is one part in one thousand. Consequently, there is a *real* discrepancy between the results of this experiment and theoretical predictions. A lower gravitational potential might imply a *lower mass*.

Therefore, the calculation of Earth mass is not very accurate to say the least and it might be *less* than the accepted value. To explain this discrepancy either it can be considered as indicator for the existence of a *new interaction* or fifth force, different from four know fundamental forces (interactions) or that one or more of the four known interactions is not well understood. The former is more likely, to be the case.

The *second evidence* is that if we use the Newton's Law of Gravity, we should be able to theoretically calculate the exact point at which the *resultant* gravitational field between two celestial bodies is zero. This is technically known as the *neutral point*. For instance, if we know the distance between the Moon and Earth along their masses, we can calculate the neutral point.

This is of extreme importance in Astronautics, as a lunar probe launched from the Earth; will continuously lose velocity until it reaches the neutral point due to the Earth's gravitational pull. However, after this probe passes the neutral point, the Moon's pull becomes stronger and it begins to accelerate, increasing in velocity.

Thus, it must have the proper trajectory to assume a lunar orbit or to score a direct hit. Fuel requirements to orbit or soft-land on the Moon, will depend on the neutral point' *exact* location. Therefore, is paramount to know exactly where the Moon's pull overcomes Earth's.

If the neutral point, hence the Moon's gravitational pull, deviated considerably from the predicted value derived from Newton's Law of Universal Gravitation, a series of *failures* would be expected in attempts to send successful lunar probes. It is also reasonable to conclude that a discovery of a significant difference in the expected Moon gravity, would require many more years of reprogramming, rocket design, lunar probe design, and so on.

The theoretical value using Newton's Law of Gravity, for the neutral point of the Earth-Moon system is about 23,900 miles from the Moon

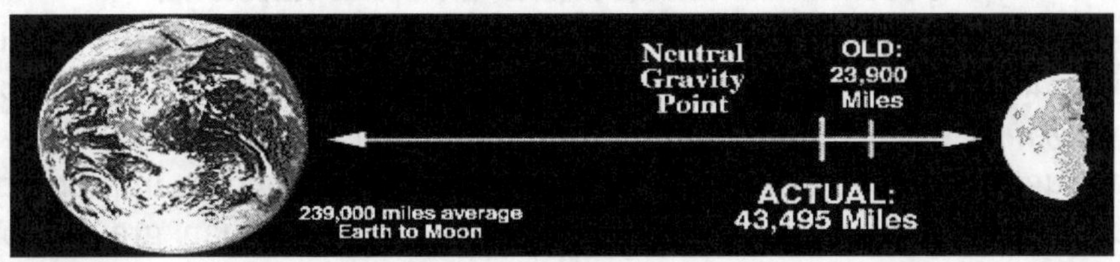

Fig. 97- Theoretically, using Newton's Law of Gravity, the neutral gravity point in between the Earth-Moon system should be at 23,900 miles from the Moon. After several space probes crashed on the Moon, it was found to be at 43,495 miles. This implies a lesser gravitational potential of Earth, and thus it has a lower mass than the conventional value attributed.

However, this value discreps with the *real* value of about 43,495 miles from the Moon, which was found by probes at the beginning of space exploration, by *trial and error*. This real value of the neutral point was reported in *Time* magazine in 1969.

Bear in mind that if the right physical principles are applied and subsequent calculations are correctly done, there shouldn't be any problem at all. Any mission or project would be accomplished smoothly and on scheduled time.

Nevertheless, the first attempts in launching probes to orbiting the Moon, *missed* it before the Luna 1 on January 2, 1959 succeeded. When the first probes where sent to soft-land in the Moon, these *crashed*. It took several years and ten crashes, before Luna 9 landed successfully on the Moon on February 3, 1966.

The above analysis of lunar probes indicates that the U.S. and Russia probably had a clear picture of the *discrepancies* of lunar gravity as early as 1959.

However, it is a certainty that due to their initial mishaps, both countries learned how to work with lunar gravity and make soft-landings by 1966. This suggests that Moon's mass isn't exactly, the conventional calculated value, gravity looks *stronger* than predicted by theory and it may not be the one sixth of Earth as we are led to believe.

It is curious that despite the well-known Newton Law of Gravity and its constant *G* has been established, there is another number that is usable and even *preferred* in space technology. That is called the *Gaussian gravitational constant k*. It was derived by Gauss based on the average orbital period of the Earth. This number *does not* require a specific *mass* of a celestial body, and is in fact the number that space scientists will use in their formulations and space missions.

The Gaussian gravitational constant (symbol *k*) is an astronomical constant, first proposed by Gauss in his 1809 work *Theoria Motus Corporum Coelestium in Sectionibus Conicis Solem Ambientum.* Although he had already used the concept to great success in predicting the orbit of asteroid Ceres in 1801. Also, the Gaussian gravitational constant was used in the calculations made by French astronomer Urbain Leverrier, that led to the *discovery* of planet Neptune in 1846.

So, in essence, space missions *successfully* work because scientists use the right proportions or ratios for the mass of planets and asteroids, even though the specific *mass* is not accurately known.

$$\frac{2}{\sqrt{p(1+m)}}\frac{\Delta A}{\Delta t} = \sqrt{GM} \equiv k$$

Fig. 98- The Gaussian gravitational constant (k) for all bodies orbiting the Sun. $\Delta A/\Delta t$, the areal velocity, is the area ΔA swept out by the body's radius in time Δt, and is a constant for a single body according to Kepler's 2nd law; p is the parameter in the formula for the elliptical orbit of a body and m is the body's mass given in proportion or ratio of the solar mass M. This is based on a two-body solution for an orbit about the Sun and gravitational perturbations by other bodies under permissible circumstances, are ignored.

Ratios are wonderful things: They avoid the need to work out constants and fiddle with many details. For example, the gravity on the surface of a planet of radius r and density ρ is:

$$\mathbf{g = 4G\pi r\rho} \text{ or } \mathbf{\propto r\rho}$$

Now, we know g on Earth is 9.81 m×s^2. What is g on Mars? Well, Mars is rocky, just like the Earth, so its density is going to be about the same. Its radius is 1/3 that of the Earth so on Mars g ~10/3 ~ 3. We never needed to know G or mass, only the density and radius of either planet.

In general, the problem with gravity is that it is poorly understood despite mainstream science saying otherwise. Gravity, in fact, is *not* a true force. Because it needs NO force carriers. A force of gravity, if it is to be consistent with what it means to be a "force," requires a force carrier sort of a messenger particle relaying the force's influence to other bodies (or between bodies) through a *medium*. By standard definition, force is: $\boldsymbol{F} = m\boldsymbol{a}$. It implies *acceleration*.

Something (an external force) must pull/push a body either with a stick or rope. With gravity, the problem was no force carriers have ever been found not then, and not since. For many years, Newton had pondered the question of the *nature* of the agent (modern "force carriers") necessary for the mediation of his gravitation force but could draw no conclusion; unable even to decide whether such agent is material or immaterial, particulate or ethereal. Newton by his own admission gave no cause for his force: “hypotheses non fingo” he said.

As discussed earlier, soon after the appearance of Isaac Newton’s Principia, describing the law of universal gravitation, Newton’s young friend Nicolas Fatio de Duillier (1664-1753) conceived the idea that the apparent force of gravitational attraction between material objects might be due to an *imbalance* of repulsive forces arising from the impacts of tiny rapidly moving corpuscles from the *aether* regions of space.

Aether *flows* in bulk, like a wind. It comes from space and flows into cosmic bodies. As aether approaches cosmic bodies, it *converges* and *accelerates*. The aether collides with cosmic bodies and the transfer of *linear momentum* from the aether to cosmic bodies is the dominant cause-of-gravity. The energy of the impacting bulk aether is dissipated into cosmic bodies.

The energy is converted into heat and the heat causes cosmic bodies to expel individual aether cells into space. The expulsion of aether cells *lowers* the pressure of aether in cosmic bodies below the pressure of aether in space. The result is that higher pressure aether in space *flows* into lower pressure aether in cosmic bodies. pushing them. This flow is like that of a gas that flows from where its pressure is greater to where its pressure is lower.

Newton's gravity became known as "spooky action at a distance." Spooky because it was a complete mystery *what* was conveying the force-effect across vast spans of absolute *space;* and furthermore, the force-effect was conveyed *instantaneously*. In the case of the Solar System, bear in mind that it is *dynamic*. The Sun “emits” a gravitational signal to Earth “pulling” our planet in its direction to where the Sun is now. The Earth also “sends” a gravitational signal to the Sun and “tugs” it (even though much more lightly) to where it is now. According to mainstream science, these signals take 8 minutes, travelling at the speed of light c to reach their targets.

Meanwhile, both the Sun and the Earth *moved* and are somewhere else. The Sun rotated around the Milky Way's centre and the Earth around the Sun. By the time the signal arrives from the Sun to where the Earth was 8 minutes ago, the Earth is already *gone*, travelling at 108,000 km/h ~70,000 mph. This is the "time delay" or aberration problem that gives theoretical physicists, a big *headache*. It cannot be. Gravity must connect Sun and planets *instantaneously* to confer *stability* to the Solar System (and the whole Universe) otherwise it would collapse. As the reader, should recall from chapter one, Nature is characterized for *veranstalten und gestalten* (organizing and shaping) stable and efficient systems, from atoms to galaxies.

If one postulates that gravity travels no faster than the speed of light, we wouldn't never be able to *explain* the orbits of the planets, tough we may delude ourselves in doing so, with outlandish theories. In fact, it is like the Sun and planets are somehow *physically entangled* through space. For finite speed (c) of gravity, according to Special Relativity, there are *serious problems*, increasing total angular momentum, velocity, and energy causing the planets to *escape* from the Sun and planets of other planetary systems. Nothing like that has ever been *observed* for any planet of any planetary system, and systems with satellites.

In 1805 remarkable French astronomer and mathematician Pierre Simon Laplace (1749-1827), considered that if gravity were due a *fluid* whose particles move towards the force center at a finite speed of propagation, this would imply a *retarding force* on the Moon such that the radius of its orbit around Earth would "quickly" drop to zero. The Earth – Moon system would become *unstable* and so the Solar System.

Observational astronomy *does not* support this scenario. Laplace had used observations of the Moon to demonstrate that the speed of Newtonian gravity *must* be 7 million times *greater* than the speed of light. Modern constraints are now 20 billion times the speed of light. Also, the gravitation source emission *differs* fundamentally from light radiation. Gravity sources in opposition to light sources *do not* consume energy, no matter their intensity. Light sources consume energy proportionally to the light emitted. Thus, gravity's propagation speed *cannot* be the same as light.

However, two centuries later after Newton, theoretical physicists *indoctrinated* by General Relativity, developed a Universe that was a *mathematical construction*, not an observable, in which the curvature of space (a mathematical concept) was intimately connected to the presence of matter; that is, the quantity and density of matter determined the degree of curvature of space. Now since this "space" was a *mathematical abstraction* not a physical reality, the curvature distortion of such space must also be a mathematical abstraction as well. But if the gravity effect is purely an abstraction, how is it *connected* to the brick-and-mortar world of r*eality*; in the same way, the well-known equation $F = Gm_1m_2/r^2$ is? In fact, the so-called "space-time" is a mathematical concept-of-convenience and has NO connection to *reality*.

No *causal* mechanism has ever been found to explain, how matter "curves" space and how "curvature" sustains a body's orbital motion. Scientists must believe this gimmick not because of *empirical* evidence but because of *sola fide*, by faith alone which is a total defiance of the *scientific method*. The reader should recall the example of weak analogy, explained in chapter one.

In current theories, gravity is not a force; yet there is the nonsense of researchers who are looking for force-carriers called gravitons. If one asks a physicist what *causes* the gravitational effect: What causes objects to freefall to the ground? The technical answer will be: The curvature of space caused by the presence of the Earth's mass. ... But what is curvature? ... Numbers, a geometric concept, geometrized numbers. ... And what do the numbers mean? ... Well, they symbolize and quantify the curvature (circular reasoning).

If physicians would use such procedure to find the *cause* of a patient's symptoms and try to *diagnose* the illness and further treatment; the medical profession wouldn't never have flourished. Yes, of course, theoretical physicists are aware of the theory's *deficiency*, its missing cause. In fact, it is their painful awareness that is driving the perennial search for gravity waves and gravity particles. Here's where the *fake* gravitational waves "discovery" came in and following the same fashion, the graviton force-carrier particle might be just around the corner.

Nevertheless, in recent heterodox theories, applying the *aether* concept, gravity has now become an inhomogeneous aether *flow* towards mass/matter. A similar theory to EPOLA, the Dynamic Steady State Universe (DSSU) states that gravity is not a force. It is a *side effect* of a heretofore *unrecognized* natural process. Gravity is a side effect of the way in which all matter mass and radiation is conducted by and through a non-material aether; an almost-but-not-quite-material medium. Similarly, Newton came up with the idea that gravity was somehow caused by a *flow of aether* into celestial bodies. From this perspective, gravity is a simple side effect and not a force rather a pseudo-force, in a similar fashion the centrifugal force which appears to act on a body moving in a circular path; it is considered "fictitious" because it is an *inertial force*, meaning that it is caused by the motion of the frame of reference itself and not by any external force **F**.

Given all the details about the *anomalies* associated with gravity, we must assume that the current accepted mass of Earth is an *inaccurate* value of perhaps 70-80% of the real thing. Thus, we still can use this value to calculate its moment of inertia. According to conventional science, the Earth's internal distribution of mass is considered hypothetically; a *solid spheroid* with density ρ increasing towards its centre. But a solid Earth would make difficult to explain the anomalies observed in polar regions discussed so far, in chapter four.

The accepted model of Earth' interior, derived from the *interpretation* of the seismological data available, is based upon assumptions, inconclusive evidence and has an inconsistent hypothesis of the alleged Earth's iron core. Also, this model has NO independent support from other sources. An Earth's interior distribution of mass, based in a *thick spherical shell model* is more *consistent* with recent astronomical-astrophysical research and laboratory experiments, than the unrealistic chaotical accretion-collision hypothesis. A shell model, can account for the polar anomalies as it will be explained later.

The aim of the following dissertation is to throw more light on this subject and propose an alternative distribution of Earth's interior mass. Again, is noteworthy to mention that the Earth as a planet *must* follow the same astronomical rules or principles, that gives the *raison d'être* to other planets and celestial bodies in general. The correct use of the concept *moment of inertia* of a rotating body, will be used to elucidate Earth's internal distribution of mass.

Rotational motion, angular momentum and moment of inertia

In a physics course, we often deal with an object moving from one place to another. One might see problems that involve situations such as a ball being dropped from a roof or a book sliding down a ramp. These are examples of translational motion, meaning an object travels from one point to another in straight line.

However, this is not the only type of motion an object can experience. There is curvilinear motion: This category refers to motion in any curved path, such as elliptical orbits of planets around the Sun. Note the ever-changing velocity for such paths, be it due to speed, direction or both. Another common type of motion we see in physics is *rotational motion*. It is more complicated than linear or curvilinear motion and happens when an object spins around an axis. We can see rotational motion in objects such as a bicycle wheel spinning around its centre point or the Earth's rotation around its poles.

When we compare translational and rotational motion, we find that the two have similar physical quantities. For example, translational motion has linear velocity, force, and linear momentum (mv). Correspondingly, rotational motion has angular velocity, torque, and *angular momentum*. Here, we're going to focus on angular momentum and its relationship with torque and *moment of inertia*.

Our concern here is the rotational motion of a 3D rigid body within its *surroundings*. A rigid body is an object with a mass that holds a rigid shape such as a rocky planet like Earth. The distances between every pair of points in the rigid body are constant.

When a rigid object rotates, as it moves around a fixed axis, every part of it (every atom) moves in a circle; covering the same angle in the same amount of time (angular velocity). No real solid body is perfectly rigid. A rotating nonrigid body will be distorted by *centrifugal force* or by interactions with other bodies. Nevertheless, most people will allow that in practice some solids are fairly rigid, are rotating at only a modest speed, and any distortion is small compared with the overall size of the body.

Objects rotate when a force acts against a point that is not the *center of mass*. When the forces acting on a rigid body combine to maintain the body in a state of rest or of motion with constant angular velocity the body is said to be in a state of equilibrium. The total work done by all the forces acting on an object is equal to the change in the object's *kinetic energy*, according to the work-energy principle.

The *torque* is the basis for the work-energy principle in rotational motion. When a force is applied, the object is said to be in a balanced state if its displacements and rotations are equivalent to zero work. This is important because many studies involving Earth, assume it is in a "hydrostatic equilibrium" and *neglect* its kinetic energy, which is wrong as it will be discussed in this chapter.

If the centre of mass of the body is at the axis of rotation, known as *balanced rotation*, then acceleration at that point will be equal to *zero*. In designing machinery with rotating parts, utmost importance is given to a balanced rotation, and most fixed axis systems will be intentionally built to be balanced. With the acceleration of the centre of mass being zero, the sum of the forces in both the x and y directions must be also be equal to zero. When the centre of mass is *not* located on the axis of rotation, the centre of mass will be *accelerating* and therefore forces will be exerted to cause that acceleration. In perfectly anchored systems these will be forces exerted by the bearings, though these forces can often be felt as *vibrations* in real systems. Therefore, any celestial body including Earth, *must* have a balanced rotation, in order to stay *stable* and avoid destruction trough *vibrations*.

Usually the study of rotational motion of a rigid body, seldom includes the interaction with its *surroundings* such as air for example. This can be understood if the friction is negligible. The problem arises when a celestial rotating rigid body such as Earth *interacts* with the *entrained aether* as Stokes suggested and the Michelson-Morley experiment proved. Aether is dragged by or entrained within moving matter. This is known as the *Lense-Thirring effect* or "frame dragging" wrongly attributed to be a "direct consequence" of General Relativity.

It describes massive rotating bodies distorting and dragging "space-time" similar to spinning an object in water causing the surrounding water to spin around it. General Relativity replaced the *dynamic properties* of space filled with aether and *substituted* it with a mathematical abstraction devoid of physicality called "space-time" (Dingle, 1972). This construct hijacked the physical *causality* due to the aether, acting as an *impostor* and fraudulently makes General Relativity, apparently "successfully tested" in explaining several phenomena.

In fact, "frame dragging" is consistent with the idea of the aether being dragged like a liquid as the alternative to an *inexistent* "space-time," being dragged around by the rotating mass. Under the Lense–Thirring effect, the frame of reference in which a clock ticks the fastest is one which is revolving in the *same* direction as the rotation of the massive object as viewed by a distant observer. This is the case of the notorious Hafele–Keating experiment performed in 1971, in which four cesium atomic clocks were flown twice around the world aboard commercial airliners, first eastward and then westward, and compared with similar ground clocks at the United States Naval Observatory; is often said to be the best confirmation of "time dilation". Earth rotates from west to east. The "official" result was that the eastbound clocks lost time (59 nanoseconds) while westbound clocks gained time (273 nanoseconds) both compared to a reference clock at the Naval Observatory in Washington. In fact, it is NOT the "time" that is affected. It is the *physical process* within devices to measure "time" that are affected.

Flying with the Earth's rotation, eastward, is flying against the 'flow' of aether, relative to the surface of the Earth, causing a greater aether force on the atomic clock, which causes the atomic clock to tick slower. Atomic clocks depend on the emission frequency of cesium atoms.

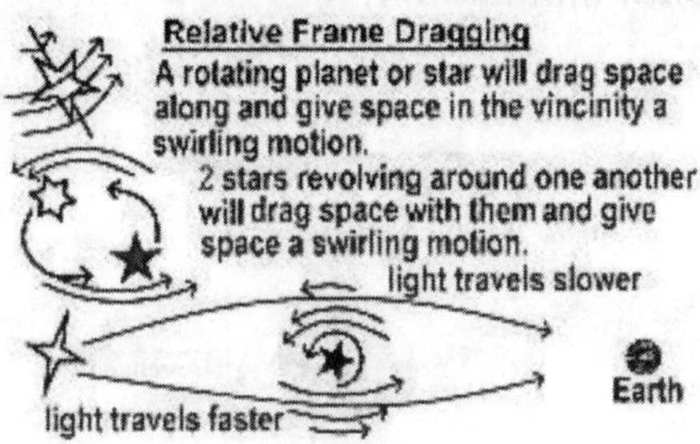

Fig. 98– Typical misconception of relativistic "frame dragging" which is attributed to an imaginary, non-physical "space-time;" when in reality it is caused by entrained aether around Earth and other celestial bodies. This is because the gravitational fields are significant enough to entrain a pocket of the background sea of tiny aetheric electron-positron dipoles with them along their orbital paths.

Consider a photon emitted by the clock. As it flows *against* the aether, its resistance to change velocity causes it to lose energy, reducing its frequency, thus the clock is running *slower*. Flying against the Earth's rotation, westward, is flying with the 'flow' of aether, relative to the surface of the Earth, causing a *lower* aether force on the atomic clock, consequently it ticks *faster*. That's all.

Let's examine now a key concept in assessing the distribution of mass, in a rotating rigid body such as a planet. During childhood, everyone should have played with a spinning top. The faster one spins it, the longer it takes to stop. A spinning top is remarkable because it tends to defy gravity while it is spinning.

If spun fast enough, a spinning top will rise to a vertical position and happily stay there, despite the fact that it will fall over if it stops spinning. This is due to *angular momentum*. When not moving, most of a top's weight rests as low to the ground as possible. If you let go of a top without spinning it, it will topple to the side until its wide part comes to rest on the surface. We're used to this effect of gravity in all aspects of our lives: Objects fall until their centre of mass reaches the Earth, unless there is something in the way.

A top will keep falling and repositioning until Its centre of mass can't get any lower. The goal of spinning a top is to counteract this tendency for as long as possible. The more angular momentum an object has, the more it wants to keep rotating. The reader should be familiar with the formula for linear momentum (p) as mass (m) times velocity (v). Angular momentum works similarly but uses physical quantities specific to *rotational momentum*.

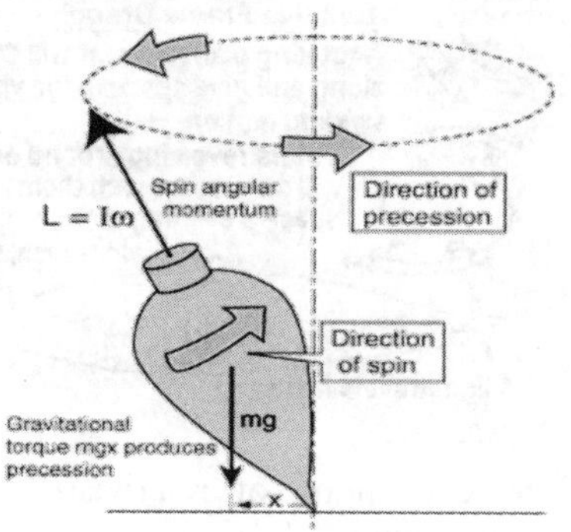

Fig. 99- A spinning top. Usually, the torque (force) acting on a spinning top is just due to the weight of the top. When you spin a top into motion, whether by hand or a string, you are applying a force that causes the top's potential energy (energy at rest) into kinetic energy (energy in motion). You will see spinning device's top end slowly revolve about the vertical direction, a process called precession. As the spin of the top slows, you will see this precession get faster and faster. It then begins to bob up and down as it precesses, and finally falls over.

Angular momentum (L) is defined as moment of inertia (I) times angular velocity (ω). An object's moment of inertia is a measurement of its ability to *resist* angular acceleration. In this way, it works similarly to mass (inertia) for linear momentum.

$$p = mv$$
$$L = I\omega$$

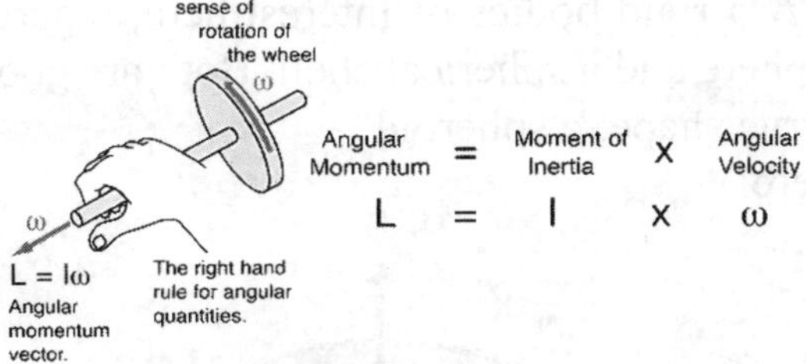

Fig. 100- Definition of moment of inertia (MoI) of a rigid body. It is a quantity expressing a body's tendency to resist angular acceleration.

Another way to define moment of inertia (I) is the tendency of a force to cause a body to rotate, or the tendency of a body to remain rotating once it started to rotate. Often abbreviated as I also, A, B, C for planets. Moment of Inertia (MoI) must always be defined about an axis.

The concept of moment of o inertia is useful because we can measure it *remotely*, and it tells us about distribution of mass (around an axis). The larger the MoI the smaller the angular acceleration about that axis for a given torque. This gives us more information than density alone. A body with the *same* density and/or mass can have *different* moments of inertia.

Having defined the Moment of Inertia of a rigid body, we can now know its unit. The unit of moment of inertia of a rigid body is a composite unit of measurement. In the International System (SI), *m* is measured in kilograms and *r* is measured in meters, with I (moment of inertia) having the standard dimension kilogram-meter square (kg×m^2).

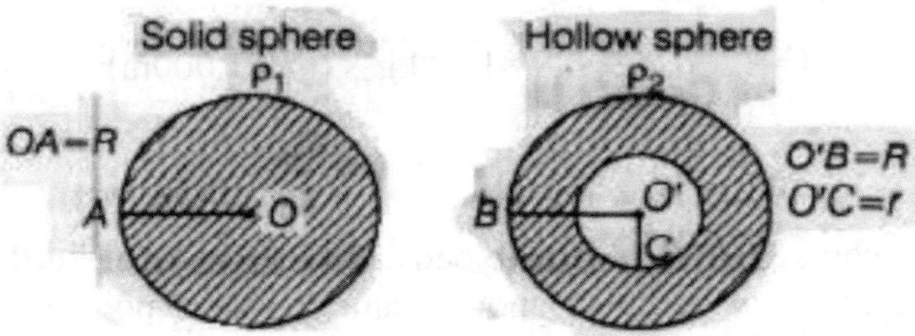

Fig. 101- Both spheres, have the same size, mass and mean density but due to its internal distribution of mass, MoI of solid sphere A is different (lower) from hollow sphere B. If we roll down on an incline both spheres at the same time, sphere A rolls down faster and reach the bottom first. This speed is related to its moment of inertia. The smaller I is, the more concentrated is the mass.

There are *two* rigid bodies of interest here, regarding the value of MoI. A solid sphere and a spherical shell. Both are good *approximation* of the Earth's true shape: a spheroid.

Solid Sphere:

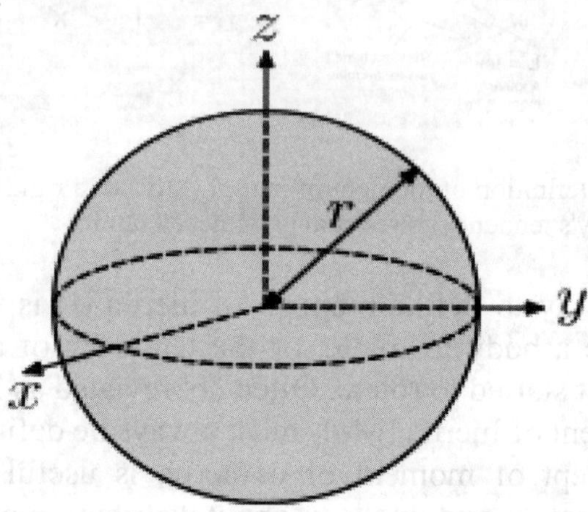

Fig. 102- A solid sphere with constant density (ρ) and radius (r)

A solid sphere rotating on an axis, that goes through the centre of the sphere with mass M, *constant* density ρ and radius r has a moment of inertia determined by the formula:

$$\mathbf{I = (2/5)Mr^2}$$

Assuming Earth as a solid, homogenous sphere MoI is found to be:

$$I = (2/5) \times 5.988 \times 10^{24}\ \text{kg} \times (6{,}371{,}000\text{m})^2$$

$$I = 9.722 \times 10^{37}\ \text{kg} \times \text{m}^2$$

This of course is an ideal value because the density of Earth is not constant along its radius but is enough as a *reference* point for future calculations.

Also, MoI can be expressed in function of density ρ. So, in a solid sphere with uniform density, the moment of inertia is given by:

$$I = \frac{8}{15}\pi\rho R^5,$$

In the case of Earth, making the previous assumptions and using the *mean* known density of Earth:

$$I = 8/15 \times \pi \times 5{,}515 kg/m^3 \times (6.371 \times 10^6 m)^5$$

$$I = 9.7 \times 10^{37} kg \times m^2$$

This value is practically the same as the calculated using Earth's mass.

Hollow Thick-Walled Sphere

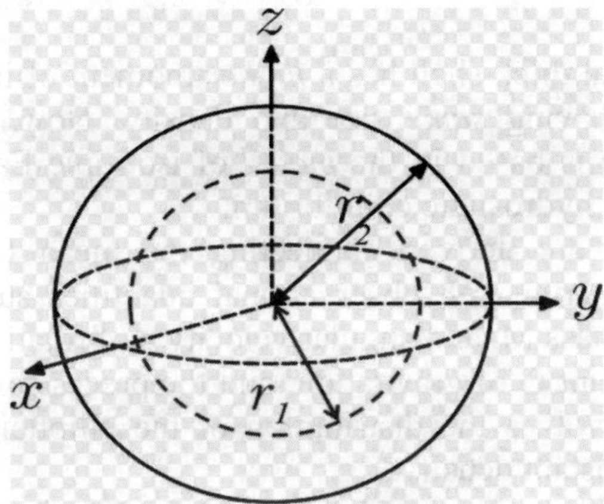

Fig. 103- A spherical thick shell with constant density (ρ), radii (r_1) and (r_2).

A hollow spherical thick shell rotating on an axis that goes through the centre of the sphere, with mass M outer radius r_2 and inner radius r_1, has a moment of inertia determined by the formula:

$$I = (2/5)M \times \frac{(r_2^5 - r_1^5)}{(r_2^3 - r_1^3)}$$

In the case of Earth, let's suppose that the inner radius (r_1) is 3.1×10^8 m which is the one of applied to the "core" and the shell's thickness is the mantle + crust. The outer radius is that of the Earth: 6.371×10^6 m.

Substituting:

$$I = (2/5) \times 5.988 \times 10^{24} kg \times [(6.371 \times 10^6 m)^5 - (3.1 \times 10^6 m)^5 / [(6.371 \times 10^6 m)^3 - (3.1 \times 10^6 m)^3]$$

$$I = (2/5) \times 5.988 \times 10^{24} kg \times 1.021 \times 10^{34} m^{5} / 2.28 \times 10^{20} m^{3}$$

$$I = 1.073 \times 10^{38} kg \times m^{2}$$

This value is assuming the total Earth's mass is in the shell. We can now establish how much the moment of inertia *differ*, between a solid sphere and a spherical thick shell:

$$I \text{ spherical shell} > I \text{ solid sphere}$$

$$1.073 \times 10^{38} kg \times m^{2} > 9.7 \times 10^{37} kg \times m^{2}$$

A spherical thick shell has a *higher* moment of inertia than a solid sphere, both having *same* size and density. Now, the conventional standard value is based in the *model* of isotropic Earth with constant rotation and in hydrostatic equilibrium.

Equations are derived connecting the mean I of the principal moments of inertia with observational quantities such as the Earth's mean radius, mass M, and angular velocity n of rotation about the polar axis, the constant G of gravitation, and the precession constant H. The theory also involves the distribution of the ellipticities € of internal surfaces of constant density.

It is important to remember that Earth, and planets in general, are not spheres, they are ellipsoids. This means that their shape is described by two moments of inertia. The moment of inertia around the axis aligned with the spinning axis is usually given then symbol *C*, while the moment of inertia around an axis through the equator is given the symbol *A*.

Following the standard procedure mentioned above in addition to values given by Earth's PREM model. The Earth's moment of inertia is estimated an average of $8.04 \times 10^{37} kg \times m^{2}$ which is 17.3%, less than the theoretical value calculated using the first equation assuming a homogenous solid sphere. This is because the variation of Earth's density. Also, it is 25% less than the calculated value for the spherical thick shell.

Apparently, Earth is a solid sphere or rather a spheroid. To find out which one is the most *reliable* distribution of the Earth's mass from the two options mentioned above, the key is to compute the Earth's momentum of inertia trough *reliable* astronomical/geophysical means.

The moment of inertia of the Earth depends on its rotation rate or angular velocity. Once a reliable/real value is found, we *compare* this number with the two moments of inertia of interest: the solid sphere and the spherical shell, theoretically calculated and check out which one, closely matches the *real value.*

Flaws in the conventional determination of the value of Earth's moment of inertia (MoI)

Planets are flattened (because of rotation – centripetal force). This means that their moments of inertia (A, B, C) are different.

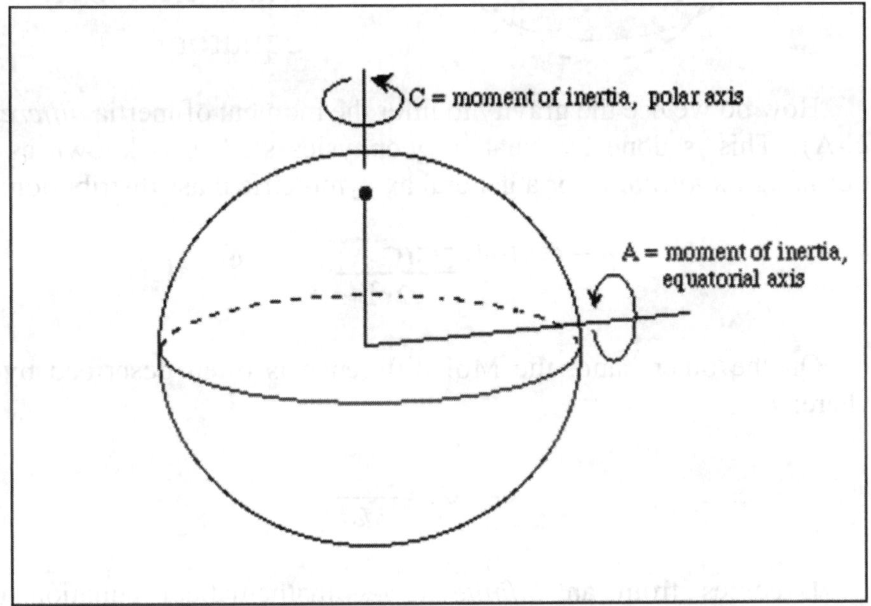

Fig. 104- Earth's two principal moments of inertia (PMOI). When calculating MoI of Earth, C equals A. From the aspect of observation, the PMOI can be calculated from the spherical coefficients of observed gravity field, according to its definition with the figures of the Earth's interior; derived by a generalized theory by assuming a hydrostatic equilibrium figure of the Earth.

By convention C>B>A. The Earth can be considered to have two moments of inertia: one defined about the polar axis (C) the other about an axis drawn through the equatorial plane (A).

The difference in moments of inertia (C-A) is an indication of how much excess mass is concentrated towards the equator. Because a moment of inertia (MoI) difference indicates an excess in mass at the equator, there will also be a corresponding effect on the gravity field.

So, we can use observations of the gravity field to infer the moment of inertia difference. The effect on the gravity field will be a function of position (+ at equator, - at poles).

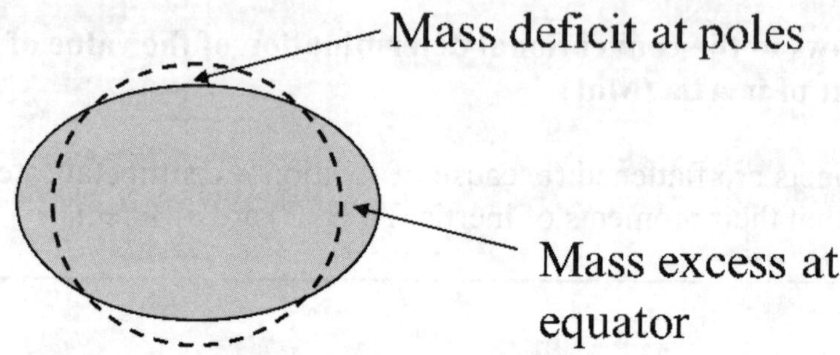

How do we use the gravity to infer the moment of inertia *difference* (C-A)? This is done by what in geophysics studies is known as the *McCullagh's formula,* for a general axisymmetric mass distribution:

$$\mathbf{g = GM/r^2 - \frac{3G(C\text{-}A)[3\sin^2 \Phi - 1]}{2r^2}}$$

On the other hand, the MoI difference is often described by J_2, where:

$$J_2 = \frac{C - A}{Ma^2}$$

J_2 comes from an *infinite series* mathematical equation that describes the perturbational effects of oblation on the gravity of a planet. J_2 refers to the (standardized) zonal quadrupole of the Earth's density and related to C_{20}, the degree-2 order-0 Stokes coefficients in the spherical-harmonic expansion of the Earth's external gravitational field, by $J_2 = -\sqrt{5}\, C_{20}$.

At 1.082636×10^{-3}, J_2 represents by far the largest deviation from a (hypothetical) spherically symmetric Earth among all Stokes coefficients. J_2 is determined for the Earth from observations of satellite orbits and is dimensionless, *a* is the equatorial radius, M is Earth's mass.

It is noteworthy to mention that this J_2 equation, comes from a mathematical derivation *assuming* a uniform density sphere. In other words, this equation presupposes a *solid sphere* or planet. Unless we have other information about the planet, there are *many* different density distributions that can give us the measured value of C.

On the other hand, on a rotating Earth a body on the surface of the planet experiences rotation and thus a centripetal acceleration:

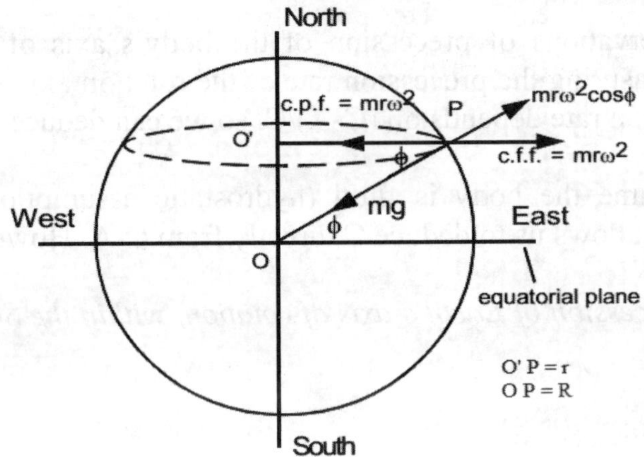

Fig. 105- Centripetal acceleration on point P caused by Earth's rotation.

Radial component of acceleration: - $\omega^2 r \cos^2 \Phi$. So, the complete formula for acceleration g on a planet is given by:

$$\mathbf{g = GM/r^2 - \frac{3GMa^2J_2[3\sin 2\,\Phi - 1]}{2r^4} - \omega^2 r \cos^2 \Phi}$$

Therefore, we can use a satellite to measure the gravity field as a function of distance r and latitude Φ, and obtain C-A. Because for a given body, there are *many* distributions of density that can give us the measured value of C; scientists normally express C as C/MR^2, for comparison with a *uniform solid* sphere:

$$C/MR^2 = 0.4$$

This is also known as moment of inertia *factor* and theoretically has a maximum value of 0.4. Generally, if a celestial sphere has a moment of inertia factor less than 0.4, then its mass must be distributed more towards its "core", and it will be denser at its centre (solid sphere).

If it has a moment of inertia factor greater than 0.4, then its mass must be distributed more towards its *outer layers*, and it will be denser toward its surface (hollow sphere). In the case of Earth, the accepted calculated value is 0,338.

Now, how do we get C from C-A? Traditionally, scientists consider *two* possible approaches:

a) Observations of precession of the body's axis of rotation. It involves measuring the precession rate of the rotation axis of the body. The precession rate depends on (C-A)/C, so we can deduce C.

b) Assume the body is fluid (hydrostatic assumption) and use theory. This allows us to deduce C directly from C-A. However:

The precession of Earth's axis of rotation, within the Solar System is fictitious.

The ratio:

$$\frac{C - A}{C} = H$$

In the case of Earth and other planets, is called the *dynamic ellipticity*, and controls the rate of motion of the rotational axis in space. The ratio of axial lengths describing the Earth's shape is its ellipticity. Considering the problem in an inverse way, if we know the rate of motion H, we should be able to calculate the dynamic ellipticity. By carefully observing the rotation of a planet, one can "detect" the precession (wobbling) of its rotation.

The rate of precession depends on a parameter called the moment of inertia and it is *usually* attributed to the "jointly forces" of the Sun and Moon acting on Earth, thus *tilting* it axis. Over history, different constellations have been visible overhead at different times in the past. At the Spring Equinox, Pisces is directly overhead at the equator, while it has been recorded that in 120 B.C. Aries was overhead at the Spring Equinox.

To understand this motion, imagine a rod through the Earth's axis of rotation. Hold the rod at its middle point (the centre of the Earth) and make one end move around in a circle. That's what the precession describes. The axis of Earth makes a 23.5° angle with a direction perpendicular to the plane of Earth's orbit. This axis precesses, making one complete rotation in 25,780 years.

The period can be estimated that way although more accurate present-day calculations, does in fact give a better figure. To calculate Earth's moment of inertia (MoI) as a function of the precession of the Earth, where a steady change occurs in the orientation of the axis of rotation of the Earth, the standard procedure is comparable to calculate

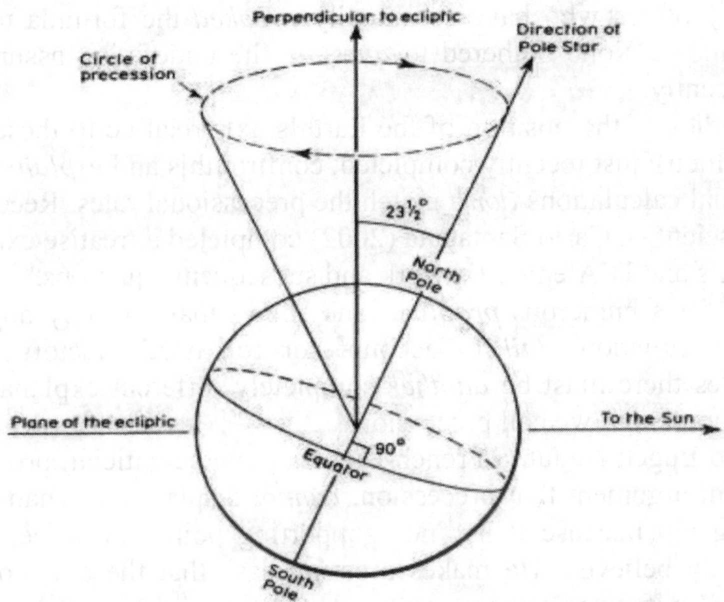

Fig. 106- Precession of the equinox occurs when is viewed against the backdrop of the fixed stars. The Earth's axis complete one revolution o circle in 25,772 years.

the torque of a *spinning top* mentioned before, caused by its weight. In this case the equation will include the *combined* torques made by the Sun and the Moon as it is believed, cause the precession of the Earth's axis and allows to determine H.

The traditional explanation of this phenomenon (lunisolar theory), is that the Moon and the Sun exert a gravitational "tug" on the Earth's equatorial bulge, trying to pull the Earth's equatorial region to be aligned with the Ecliptic plane thus tilting it axis.

Using the combined forces of the Moon, Sun and the precessional speed; enable scientists to devise equations that can estimate the Earth's moment of inertia. The value found is very close to 8.04×10^{37} kg×m^2, mentioned before. However, recent *research* shows that: First, the Sun's tug is *not* enough to cause precession. Second, the Precession of the Equinoxes is *real* when viewed against the backdrop of the *fixed stars.*

But here is the catch: There is NO evidence that this observation is due to any *change* in the Earth's spin axis relative to the Sun, or Moon or Venus or anything "within" the Solar System. In fact, Newton equations *never* did match observed precession rates; so along came French mathematician Jean le Rond D'Alembert (1717-1783), followed by many others who have continually *tweaked* the formula to match observations. None bothered to *question* the underlying assumptions, until recently.

Studies of the position of the Earth's axis relative to these bodies (Sun, Moon), just recently completed, confirm this and *explain* why the traditional calculations *don't* match the precessional rates. Recently, an Italian scientist, Carlo Santagata (2002) completed a treatise examining Newton's and D'Alembert's work and subsequent equations.

He finds numerous *problems* and shows that not only do current lunisolar equations *fail* to account for relativistic factors, but he concludes there must be *another* completely different explanation for the phenomenon, we call precession.

And Eugen Negut, a French-Canadian mathematician provides an insightful argument that precession, *cannot* display the dynamics of a spinning top because it has no "supporting point" in space, as it is commonly believed. He makes a strong case that the axis could not "wobble" without a supporting point and that there must be *another* cause.

Also, two German-Canadian scientists Karl Heinz and Uwe Homann, have produced some compelling time equivalency and related equations to show that the time required to complete lunisolar precession mechanics *does not* fit the observed motions of the Earth.

In fact, is the *movement* of the entire Solar System that *causes* the Precession of the Equinoxes *not* the lunisolar forces. Independent sources rather than confirm the Sun as fixed referential frame *disprove it.* Thus, Newton equations weren't wrong after all but the *frame of reference* was. For this reason, the calculation of Earth's moment of inertia based in the standard lunisolar model, will *not* be considered in this dissertation (see appendix VIII); because it would be based on calculations derived from *inexistent* forces that allegedly, cause precession of the Earth's axis. However, mainstream scientists tend to *ignore* this fact and continue to apply over and over the standard lunisolar model, with consequently *wrong* results. In the beginning, it was reasonable consider the Sun-Moon attraction, to explain the apparent motion of the axis viewed against the backdrop of the fixed stars.

Bear in mind that not only were Copernicus and Newton *unaware* of what a galaxy is, they also assumed a "static Sun" when they first postulated a heliocentric system with a wobbling Earth. They were *unaware* that our Sun is *moving* at great speed through local space or that it could possibly, be gravitationally bound to any other extra-solar system mass. The lunisolar model was developed before knowledge of binary prevalence or any understanding of *binary star* motions.

Indeed, the idea of a single Sun with lunisolar wobble causing precession, was originally developed at a time when the Sun had only recently replaced the Earth as the centre of the Solar System and the Sun was thought to be *fixed* in space. Consequently, any theory to explain the observed phenomenon of precession of the equinox, had to be based *solely* on movement of the Earth.

Fig. 107- This animated model of the helicoidal motion of the Earth around the Sun, shows the *real* trajectory of Earth orbiting the Sun and through our galaxy (Milky Way). Earth's axis changes orientation relative to fixed stars, trough time; completing a cycle every 25,772 years due to this helicoidal motion, when the Solar System orbits the centre of our galaxy.

Basically, the *flaw* of the lunisolar theory consists in taking for granted the frame of reference (Solar System) as *fixed* or static when in reality, it is *moving* around the centre of the Milky Way at an average speed of 828,000 km/h (230 km/s) or 514,000 mph (143 mi/s) along its trajectory around the galactic centre and the Earth's *true* motion around the Sun is spiral shaped.

Therefore, strictly speaking, Precession of the Equinoxes only occurs relative to objects *outside* the Solar System. Earth *does not* precess or change orientation, relative to objects *within* the Solar System. This is why formation mechanism of lunisolar precession and lunar orbit regression, is still a scientific conundrum. This is because Precession of the Equinoxes is true if one measures the position of the equinox relative to the fixed stars "outside" the Solar System, but it is *not* true if you measure the movement of the equinox relative to the Sun or moon or other objects "within" the Solar System; where the lunar data shows us that the Earth goes around the Sun a complete 360 degrees in a tropical year.

Unfortunately, neither NASA nor any other official agency measures the Earth's orientation relative to nearby objects, so the paradox goes *unnoticed.*

The hydrostatic equilibrium assumption is <u>not applicable.</u>

In most cases, the precession rate of the planet is *not* available. How do we then calculate C? In this case scientists as usual, have no option other than recurring to *assumptions*. Before going any further let's examine the *failure* of the hydrostatic equilibrium, assumed for Earth.

Since the time of Newton and Clairaut it was adopted that the Earth is an inert, powerless body. While developing the theory of the Earth's figure and considering the planet as a rotating *liquid body* (which is not) being in the solar uniform force field; it would respond instantaneously to any stresses including those caused by its rotation until it attains a state of *zero stress*. The figure that the Earth would assume if it were in such a state is called the *hydrostatic equilibrium* figure or simply hydrostatic figure or equilibrium figure. In other words: it is state of "balance" between gravity, which wants to pull things together, and pressure, which wants to blow it apart.

Since hydrostatic shape indicates a state of zero stress any *departures* from this state are particularly interesting as they show the extent of available stresses in the interior of the Earth which can be invoked to "explain" any geophysical mechanisms which may be found or assumed to exist in the Earth's interior. On that basis, a *speculative* idea of "hydrostatic equilibrium" of the planet's mass and its inertial rotation has been accepted. However, in 1935 Sir Harold Jeffreys *rejected* it on the basis of existing stress-differences, able to support mountains.

The original purpose of the hydrostatic equilibrium was that, with the then known data, this theory will provide useful "information" about the *distribution* of density within the Earth. However, this approach is *wrong* due the following reasons:

After appearance of the Earth's *artificial satellites* and some special geodetic satellites to observe Earth's shape and gravity field; the situation with the assumed hydrostatic equilibrium has *changed.* The satellites made it possible to determine directly, by measuring of the even zonal moments, the coefficient J_2 in expansion of the Earth gravitational potential by *spherical functions*. In this case at hydrostatic equilibrium the odd and all the tesseral moments should be equal to *zero*.

It was assumed before the satellite era that the correction coefficients of a higher degree of J_2 will decrease and the main expectations to improve the calculation results were focused on the coefficient J_4. But it has appeared, that all the gravitational moments of higher degrees were not zero. The values were proportional to the *square of oblateness*, i.e. $(1/300)^2$ (Zharkov 1978). Further calculations showed an oblateness 1/e = 1/298.25 (Grushinsky 1976; Melchior 1972).

At the same time, if the Earth stays in hydrostatic equilibrium, then applying the solutions of Clairaut and his followers, the planet's geometric oblateness should be equal to e′ = 1/299.25 which is not. On the basis of that contradiction Melchior (1972) concluded, that the Earth does not stay in hydrostatic equilibrium. It represents a simple equilibrium of the *rigid body*. This backs up Earth's mantle solidity.

The majority of researchers dealing with the dynamics of the Earth and its shape come to the unanimous conclusion that the theories based on hydrostatics *do not* give satisfactory results in comparison with the *observations*. This means that the hydrostatic assumption is false.

For instance, Jeffreys straightforwardly says that these theories are *incorrect*. Munk and Macdonald more delicately note that dozen of the observed effects can be called which *do not* satisfy the hydrostatic model. It means that dynamics of the Earth as a theory is *absent.* The above state of art and the conclusion motivated some scientists to search for a *novel* physical basis for dynamics and creation of the Earth, planets and satellites.

Because the Earth does not stay in hydrostatic equilibrium, then the above described *initial* physical fundamentals for interpretation of the satellite observations should be recognized as *incorrect* and the related physical concepts cannot explain the real picture of the planet's dynamics. Theoretical application of the *triaxial Earth* model was not considered because it contradicts the hydrostatic equilibrium hypothesis. But after it was found that the hydrostatic equilibrium *does not* exist, the triaxial Earth alternative is the real one (Ferronsky, 2016).

The *absence* of the hydrostatic equilibrium state of the Earth and the Moon as well, is because Earth and the Moon are *triaxial bodies* and their axial rotation is not an inertial effect.

The meaning of a triaxial body (triaxial ellipsoid) refers to the existence of a *longitude-dependent field*, which implies the existence of *inhomogeneities* and states of *stress* within the Earth which are of *fundamental* importance to all dynamical theories of the Earth's interior.

Observational data from earthquakes demonstrate the Earth's oscillating dynamics with periods from 8.4 to 57 min. Two general modes of the Earth's oscillation are found, namely, spectral with a vector of *radial* direction and *torsion* with a vector perpendicular to the radius. This is another example of the *rigidity* of Earth's mantle.

The main *problem* of a celestial body's equilibrium state, which numerically is a ratio of the kinetic and potential energies, *departs considerable* from conventional theory.

The Earth figure *deviates* from the normal ellipsoid of rotation corresponding to hydrostatic equilibrium of the planet by a value equal to about square of its oblateness by ~ $(1/300)^2$. The other planets, the Sun, and the Moon, the hydrostatic equilibrium for which is also accepted as a fundamental condition, stay in analogous situation.

This is because the hydrostatic approach does not take into account, the *kinetic energy* of the interacted elementary mass particles, which is, in fact, Newton's energy of gravity (and force). As a result, celestial body dynamics have been left *without* kinetic energy and centrifugal force.

A gravitating rotating celestial body is an excellent natural *centrifuge,* that is rotated by the energy of interacting particles. Dynamical effects of such a centrifuge are the centrifugal and centripetal forces which are taken as the *gravity* and *inertia* forces. In every day practice centrifuges are used for separation of the components of matter in gaseous, liquid and solid states with respect to their density (force of weight). In nature, the same forces *separate* the shells and elementary particles of bodies and their systems.

They also provide *expansion* and *creation* of bodies and their systems. Fundamentals of Jacobi Dynamics (derivation of the classical virial theorem, after introducing the volumetric forces and moments) completely *correspond* to the conditions of natural centrifuges such as rotating *celestial bodies* such as Earth and proved by astronomical observations. Notwithstanding the *failure* of the hydrostatic equilibrium, this approach is still used to determine Earth's moment of inertia and prove it "solidity" as follows:

For a fluid, the gravitational. potential is the same everywhere on the surface. Let's equate the polar and equatorial potentials for our rotating shape and let us also define the ellipticity (or flattening) as f:

$$\mathbf{f = (a - c)/a}$$

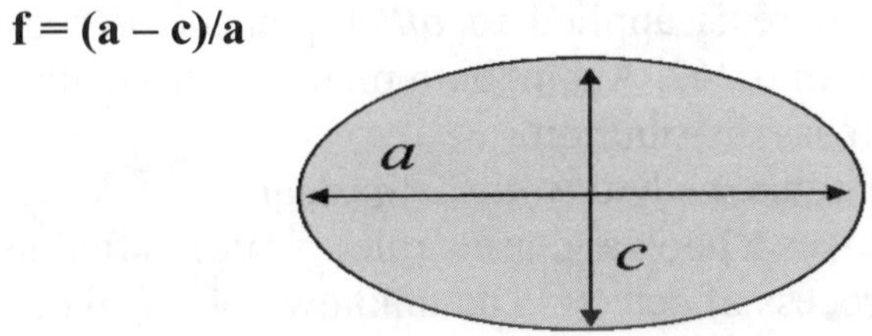

A mathematical derivation, based on the previous equation, gives:

$$\mathbf{F \approx 3/2\ J_2 + ½\ a^3\omega^2/GM}$$

The approximation symbol is to recall that this only works for a fluid body. If we assume that Earth is hydrostatic (i.e. it has no rigidity, which is absurd), then we can derive C directly from C-A using the *Darwin-Radau approximation*:

$$\mathbf{f_{hyd} = \frac{5\ a^3\omega^2/2GM}{1 + (25/4)x(1-3C/2Ma^2)}}$$

Here the flattening depends on C. We also have an equation giving f in terms of C-A We can measure J_2 and thus calculate f (assuming a fluid Earth). All data comes from geophysical tables:

$J_2 = 1.08x10^{-3}$
$w^2a^3/GM = 3.47x10^{-3}$
$f = (3/2) J_2 + w^2a^3/2GM$; therefore: $f = 3.36x10^{-3}$.
The observed flattening: $f = 3.35x10^{-3}$

Traditionally, a fluid Earth is considered a "good assumption", so is common to use *f* and the *Darwin-Radau approximation* to obtain C/Ma^2. Knowing the Earth's values of G, f and ω; we substitute these in the Darwin-Radau approximation. After doing math, we get a result for Earth's moment of inertia *factor* of:

$$C/Ma^2 = 0.331.$$

The current value is 0.3308 obtained using for instance, the planet's gravitational field and the complete *MacCullagh's formula*. C or I can be computed and gives an average value of 8.04×10^{37} kg×m^2, currently accepted.

The *same* procedure is applied to *other* planets such as Jupiter giving a value C/Ma^2 of 0.262. Again, both planets seem to have a solid or massive internal mass distribution.

Bear in mind that this method is *not* a "measurement" of C/MR^2, in the same sense one takes a book and use a ruler (scale) to find its length. Measurement is a process of detecting an unknown physical quantity by using standard quantity. It is a *straightforward* process. And something like this is precisely, what is needed to determine a reliable Earth's moment of inertia (MoI).

Now comes the questioning of the *validity* of this hydrostatic approach given by the *Darwin-Radau approximation*. It depends on the form of the Earth's density variation.

To what extent it is applicable? Indeed, for large objects, the concept of *rigidity* is complex for any use of the moment of inertia. The progress of seismic waves through the solid must be taken into account.

The fact that S as well P waves are transmitted through *all* parts of the Earth's mantle shows that in respect to its response to stresses and time duration, the Earth's mantle is solid, i.e. has a marked *rigidity* (ν) down to a depth of approximately 2900 km.

Astronomical observations yielding data on the movements of Earth's poles together with the observations of Earth's tidal movements, give information of the *rigidity* of Earth as a whole. If Earth 's mantle had considerable "molten lava" then it would actually be subject to *tidal pressures* and the continents, would be broken to pieces as Earth rotated.

This hydrostatic approach, has the same *problem* of assuming hydrostatic conditions to calculate the Earth's inner pressure, discussed in chapter two which didn't comply after the depth of 9 km as the *Kola Deep Borehole* studies showed. Also, orbiting satellites cannot tell researchers, the moment of inertia of a planet until one has already made several *modelling assumptions*; which could of course be wrong.

In fact, models of the interior structure of planetary bodies are generally non-unique, because there are more unknowns than constraints. For example, the *friction* torque results from the relative motion of atmosphere and ocean currents with respect to the Earth surface. Since the friction drag of the Earth's surface is widely *unknown* it is particularly difficult to model.

There are assumptions and there are refinements. This inevitably leads to a *circular reasoning*. Scientists start with an assumption which they consider an incontrovertible fact (the increase of planet's density towards it centre, i.e. it is solid) which is treated as such in their subsequent analysis. By following this trend of thought, they calculate, filter and interpret data arriving at the *same conclusion*; which is the *same assumption* which with they have started.

Thus, planets will be considered *always* solid, regardless of their true internal distribution of mass. So, what the main concern is fundamentally, where the numbers in the books come from. In the case of Earth, it is known that besides the coefficients of *the Darwin-Radau equation*, leading to its approximation; other coefficients known as Love numbers used in different Earth's models, have been *refined* (fudged) to fit the following *assumptions*: the alleged Earth's iron core and the controversial mantle's convection currents.

In fact, geoscientists derived an appropriate numerical value of the pole tide Love number k_2 in the light of mantle anelasticity (delayed elasticity) and the alleged dynamics of the core and oceans.

Therefore, according to conventional science, the calculated and accepted value of Earth's moment of inertia, which tells us Earth is solid, is indeed *a priori* based on unrealistic conditions and thus it is NOT reliable. Indeed, it is artificial.

We have come to a point which is necessary, to have at hand a reliable procedure to find out what is the *real* Earth's MoI. We need to check out the *true* Earth's internal distribution of mass and determine, once and for all, whether our planet it is a solid, massed body or a spherical thick shell. That's why this book is all about. This must be done avoiding the need to work out assumptions and fiddle with many unnecessary details, as scientists usually do.

Successful calculations and predictions, should be sought as evidence for any *physical equation* having been formed to accurately and *directly,* mimic patterns observed and tested in *empirical evidence* measurements.

It is an activity that involves interaction with a *concrete* system with the aim of representing aspects of that system in abstract terms (e.g., in terms of classes, numbers, vectors etc.). For engineers this is paramount but scientists don't care much about this. They are more concerned with theories, models and computer simulations than *empirical* facts and their relationships.

For example, we can calculate (directly) the mas of a specific object using the Newton's Second Law (of motion). It mathematically gives the cause-and-effect relationship between force and changes in motion. Newton's Second Law is quantitative and is used extensively to calculate what happens in situations involving a force. We can write down Newton's Second Law as a simple equation that gives the *real* and exact relationship of force (F), mass (m), and acceleration (a):

$$\mathbf{F} = \mathrm{ma}$$

To find out the mass of the object, we must know the *values* of its acceleration and force or net force applied to it. Then we rearrange this equation, putting force and acceleration as a function of mass:

$$\mathrm{m} = \mathbf{F/a}$$

Substituting the *known* values of F and a, we calculate the mass of the object in question without any fuss and unnecessary assumptions or models. What we do need right now, is a straightforward *equation* where Earth is a part of a physical *system* and a frame of reference as well; where we can determine with empirical evidence, the *true* Earth's MoI.

The proper equations that we use in physics, are relationships between physical quantities observed in a concrete (real) system. Nature has a way of telling us those structured correlations or *Gestaltung*. It is matter to find the right one to get the answers, we are looking for.

A new approach. The Earth-Moon system and MoI

In fact, there is still a *third* way and the only valid to calculate Earth's moment of inertia I, which is ignored by geoscientists for being focused more on Earth *models* than heavenly, *concrete* celestial bodies. It consists in using the orbit's eccentricity of the Moon, around Earth. In addition to its nearness to Earth, the Moon is relatively *massive* compared with the planet, the ratio of their masses is much larger than those of other natural satellites, to the planets that they orbit.

The Moon and Earth consequently exert a strong gravitational influence on each other, forming a system having distinct properties and behaviour of its *own*, despite the pull of the Sun and other planets. In regards to the Moon, the relative acceleration toward the Sun is a small perturbation (less than 1/87th in magnitude), on the Moon's gravitational acceleration toward the Earth.

The Moon has very noticeable effects on Earth such as the tides in Earth's oceans. These tidal effects create friction that slows down the rotation of Earth by less than a second per century. Young Earth's day was only six hours long. Because of this slowing down, the Moon spirals out away from our planet by 4 cm per year. However, some calculations have suggested that tidal forces are not great enough to fully explain this slowing of Earth's rotation. In this system, the Earth and the Moon revolves around their common centre of mass. Since the Earth is much heavier than Moon, this point is inside Earth, but it is not in the middle.

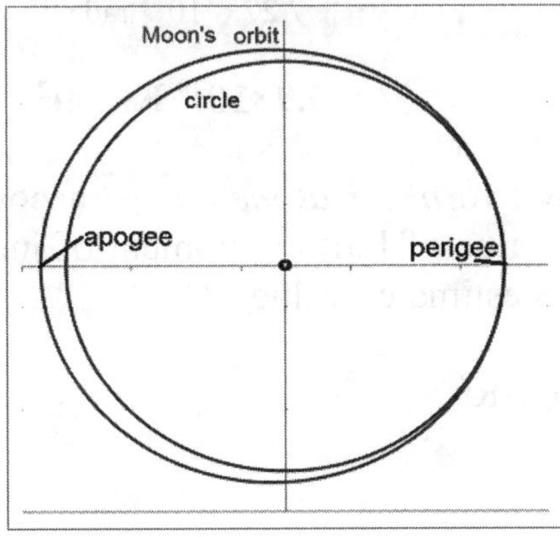

Fig.108- Moon's orbit around Earth. It can be considered as a circle for several practical purposes. The two celestial bodies orbit each other about a common centre of mass. It is thc only moon-planet system with two large, differentiated spherical bodies.

The Earth–Moon system has a large amount of angular momentum (**L**) per unit mass. We can use this fact to find out Earth's MoI considering its orbiting Moon: From *Celestial Mechanics*, the angular momentum and eccentricity are related by:

$$\mathbf{L^2 / GMm^2r = 1 + e\cos\theta}$$

For practical purposes, we can consider the axis of this L to be the same as Earth's and tidal effects are negligible. The Moon's orbit can be estimated circular, meaning e = 0

$$\mathbf{L^2 / GMm^2r = 1}$$

Where G is the gravitational constant, M and m the masses of Earth and Moon, r is the mean distance of the Moon's orbit and L the Earth's angular momentum. Rearranging L:

$$\mathbf{L = m(GMr)^{1/2}}$$

L = I ω, rearranging and substituting:

$$\mathbf{I = \frac{\underline{m(GMr)}^{1/2}}{\omega}}$$

$$I = \frac{7.35 \times 10^{22}\ \text{kg} \times (6.673\times10^{-11}\ \text{N m}^2\ \text{kg}^{-2} \times 5.988 \times 10^{24}\ \text{kg} \times 3.84 \times 10^{8}\ \text{m})^{1/2}}{7.27 \times 10^{-5}\ \text{rad s}^{-1}}$$

$$\mathbf{I = 3.9\times10^{38}\ Kg\times m^2}$$

This value has two *physical meanings*: according to the previous equation, it is the value of Earth's moment of inertia because it is far higher than Moon's estimated value:

$I_{Moon} = 0.3929\ MR^2$

Substituting:

$I_{Moon} = 0.3929\times7.3\times10^{22}\ \text{kg}\times (1.73\times10^{6}\ \text{m})^2$
$I_{Moon} = 8.65\times10^{34}\ \text{kg}\times\text{m}^2$

Also, this calculated vale of Earth's MoI is 20.61% *higher* than the estimated I of the solid Earth through the conventional model:

I spherical shell > I solid sphere

$$3.9\times 10^{38}\ \text{Kg}\times\text{m}^2 > 8.04\times10^{37}\ \text{k}\times\text{gm}^2$$

The *real* moment of the inertia factor C/Ma^2 for Earth is:

$$\frac{3.9\times 10^{38}\ \text{Kg}\times\text{m}^2}{5.988\times 10^{24}\ \text{kg}\times (6{,}371\times 10^6\text{m})^2}$$

$$C/Ma^2 = 1.605$$

Which is 4.75 times *higher* than the conventional accepted value of 0,338 for Earth. This value clearly shows that Earth's MoI indicates a distribution of mass that corresponds to a *spherical thick shell* rather than a solid sphere. This *contradicts* the accepted value of 8.04 $\times10^{37}$ kg$\times$m^2, acknowledged by the mainstream scientific community as the one proving a "solid" Earth.

Indeed, this is an idealistic, *a priori* and artificial value. Therefore, the Earth is *not* a solid sphere rotating about its axis. However, Earth's moment of inertia still is being calculated directly according to the *wrong* procedure, with the figures of the planet's interior derived by a generalized (wrong) theory of the hydrostatic equilibrium figure of the Earth.

Also, it comes from *assumptions* based on the failed and unworkable chaos-accretion collisions hypothesis of Earth and planets formation, *discredited* by recent astrophysical research and laboratory experiments. On the other hand, the spherical thick value agrees with the *Gestaltung* that accurately, determines the Nature patterns that organizes and shapes the current planet formation; trough *dusty plasma vortices* in its early stages, as it is observed in protoplanetary disks.

When the outward motion of particles accumulated by the plasma vortices, is balanced by growing gravitational influence, the matter so collected starts to form a *spherical shell* that later consolidates into a solid matter in the case of a terrestrial planet.

This is a consequence of the dynamical effects related to the Solar System bodies creation, oscillation, rotation (centrifugal force) and *shell separation* by means of the matter differentiation with respect to its *density* according to Archimedes' and Coriolis' laws. Due to centrifugal force being *zero* along the rotational axis coupled with the plasma filament running along it, *no* consolidation of matter, will occur nearby where this axis intersects the shell. This would leave what is known as the planet's *polar openings*.

On the other hand, *recent* findings in our solar System, do *corroborate* a spherical thick shell mass distribution in another planet. The results of the first Juno mission to Jupiter have already challenged long-held scientific theories about gas giants.

In fact, in 2016 during Juno's mission to Jupiter, using microwave equipment, the "core" was measured to extend up to half of Jupiter's 70.000-kilometer radius and has *less* than 20 Earth masses.

It means that Jupiter's "core" is indeed a large and *diffuse* one. Jupiter's core should be 12 to 45 times the Earth's mass, or roughly 4%–14% of the total mass of Jupiter and made of metallic hydrogen, according to mainstream science. That's evidently *not* the case. Jupiter's core is substantially *less* dense than the *outer* layers of Jupiter. This fact threw a pretty hefty spanner, into the current ideas about planetary formation and structure. The measurements made by Juno are 10 times more accurate than all that has been done so far.

They *contradict* the assertion held in planetary sciences, that density (ρ) increases with radius from the planet's surface to its centre. This means that planetary scientists, got it wrong all this long with their assumptions and models. Once more, like in the case of the *Kola Deep Borehole*, experience always beats theory. In fact, it's gotten to the point where planetary scientists wonder if any of their assumptions about gas giants were right.

In fact, Juno's gravity field measurements *differ* significantly from what scientists expected, which *disproves* the distribution of heavy elements in the interior, including the existence and mass of Jupiter's core. To put it simply, Jupiter, along Earth can be considered as practically *hollow*. It is interesting the fact that the ratio between the thick spherical shell and the core, observed in Earth and Jupiter is the same: half of the planet's radius.

The above paragraphs put on display how the current scientific and academic community is doing flawed, substandard and worthless "scientific research" which is being published in prestigious journals with the consequent detriment of society. To endorse it, in 2008, a study using dubious *computer simulations*, “found” that Jupiter has large, rocky core surrounded by layer of ice. The simulation “predicted” the properties of hydrogen-helium mixtures at the extreme pressures and temperatures that occur in Jupiter's interior, which cannot yet be studied with laboratory experiments.

Applying techniques originally developed to study semiconductors, UC Berkeley's Burkhard Militzer, an assistant professor of Earth and Planetary Science and Astronomy, calculated the properties of hydrogen and helium for temperature, density and pressure at the surface all the way to the planet's centre. Co-author William B. Hubbard, professor of Planetary Sciences at the University of Arizona's Lunar and Planetary Laboratory in Tucson, used the theoretical data (not empirical observables or facts) to build a new computer-based *model* for Jupiter's interior. This model could not be validated, in fact it was *disproved*.

This is a fine example of a deceptive GIGO gibberish which is very widespread in the academia community, regularly deceiving both “professors” and students. This includes recent claims to have created a wormhole and a blackhole in a laboratory that are actually based on simulations of each instead of real-world *experiments*. So, where is the *evidence* and why do “scientists” refuse to accept the world as it is?

GIGO stands for *garbage in, garbage out*. It is a computer science and mathematics concept, that the quality of the input determines the quality of the output. In other words, in a processing system, the data’s quality coming out *cannot* be better than what went in.

This principle highlights the critical role of *accurate* data collection and the *risks* associated with neglecting this phase in any *computer simulation*. Therefore, a program will only yield *misleading* results if it is working on *faulty non-real* data, which is precisely the case of this “research”.

A comparison of this model with the planet's known mass, radius, surface temperature, gravity and equatorial bulge implies that Jupiter's core is an Earth-like rock 14 to 18 times the mass of Earth, or about one-twentieth of Jupiter's total mass, Militzer said. Bear in mind that these “scientists” weren’t following the *scientific method* discussed before, which shows their mediocrity in promoting pseudoscience. They are *confounding* models and assumptions with reality.

The *observable* findings of Juno’s mission to Jupiter proved them quite wrong and demolished their “scientific knowledge”. However, this sort of *pseudoscience* is not retracted nor corrected and still is being published in the so-called scientific journals. This disgusting trend still continues to this day.

In fact, despite the accretion hypothesis being proven *untenable* as explained before (chapter three) and *new paradigms* of planet formation have been developed, showing a thick *spherical shell* shape; the majority of the scientific community ignores this fact and continues doing meaningless "research."

This is because of the pressure that so many researchers feel to secure competitive grant funding, which is distracting them from their real job.

For example, recent formation *models* of Jupiter suggest important "modifications" to the old picture of giant planet formation:

i) Cores are formed more "efficiently" via pebble accretion (e.g., (Lambrechts & Johansen 2014; Johansen & Lambrechts 2017)

ii) The *accretion* of solids (heavies) and gas (H–He) leads to a gradual distribution of "heavy elements" within the planetary *deep interior* (e.g., Helled & Stevenson 2017)

iii) Additional "enrichment of heavies" is required to explain the estimated heavy-element masses in Jupiter and Saturn (e.g., Li et al. 2010; Shibata & Ikoma2019; Shibata & Helled 2022).

iv) (Alibert et al. (2018) suggested that Jupiter's formation was characterized by core formation dominated by "pebble accretion" followed by "planetesimal accretion", and that Jupiter reached its final mass 3-Myr after the formation of the Solar System.

v) It was later shown more systematically that Jupiter's formation timescale could have indeed been a few million years if run away gas accretion was delayed by "planetesimal accretion" for a large range of initial formation locations and planetesimal sizes (Venturini & Helled 2020).

The above statements published in peer review journals, are the same *rebranded* nonsense and pure rubbish. The exploration of the giant planets in the Solar System thanks to *accurate gravity* data from the Juno and Cassini missions and *ring seismology* for Saturn by the Juno and Cassini spacecraft provided *new* and exciting data about the interiors of Jupiter and Saturn.

The old view of giant planets having pure heavy-element cores is NO longer valid. The criteria of planet's rotational bulge telling us about the mass distribution inside the planet is *unreliable*.

It is believed that if a planet were a perfectly uniform sphere, that is, the density was the same everywhere, then scientists could predict how bulgy the planet would be given the competing effects of self-gravity and the centrifugal force caused by the rotation. But when they do that for the giant planets, it turns out that they're all bulgier (larger in the equatorial direction) than what would be predicted for a uniform sphere. A thick *spherical shell* allows more freedom for deformation.

An example is Saturn, the bulgiest of the planets, with the equatorial axis about 10% wider than its polar axis. This result according to scientists, tells us that Saturn has denser material concentrated toward the center of the planet and lighter material in its outer layers. This was proved to be utterly *false* and misleading. The new measurements have *challenged* giant planet formation theory and have led to the construction of new formation and internal structure models. It is now more accepted that both planets have fuzzy (hollow) cores, with the core extending out to as much as 60% of the planetary radius.

Finally, for the purpose of calculating the moment of inertia, which shows the distribution of Earth's mass as a *thick shell*; it is noteworthy to mention that in the case of Earth, at its centre, a remaining mass is left.

In chapter two, we learned that Earth has an *independent* inner solid body 2400 km in diameter, less rigid than mantle and spinning faster than it. This inner body *radiates energy* and *particles* that come out through Earth's polar openings. This inner body might be called the inner Sun.

Earth's inner Sun

The inner Sun originates pretty much like its counterpart of the Solar System. Both share the same *Gestaltung,* tough in a much lesser scale. In the spinning vortex of a protoplanet which includes dust, rotation, turbulence, electric currents coupled with magnetic fields, gravity and inertia; denser material coalesces in the outer shell (preferential concentration), whereas lighter "gases" or better said *plasma* material originated form the dissociation of molecules and ionisation of H and He, are left in the *centre* of the protoplanet.

Strong high-frequency electric plasma currents acting through the axis of the polar vortices like jets, would *ignite* this inner Sun once the density, has increased sufficiently for the "gas" to become opaque to infrared photons. Radiation is produced within the central part of the inner Sun cloud, leading to heating and an increase in gas pressure. According to conventional science, the cloud core is nearly in hydrostatic equilibrium and the dynamical compaction is slowed to a quasistatic contraction.

At this stage, we may start to speak of a proto-inner-sun, which starts to radiate heat, light and charged particles. It later fully develops into a body that radiates energy, the inner Sun of our planet. Indeed, Earth is a *system* consisting of its thick spherical shell plus the inner Sun, with a 2000 km of "void" in between. From this perspective and taking in to account the *vortex* formation of planets, it this reasonable that all planets, will be systems made of an outer *thick shell* with polar openings along an inner Sun. However, the structure and composition of these systems, will depend on either these are terrestrial-type or Jovian-type planets.

Back to the inner Sun, according to the Hertzsprung-Russell diagram of stars' classification and taking into account its *high density*, this Earth's inner Sun would be associated or related to the category of white dwarfs; high-density stars: $\sim 10^5$ g/cc. It would have to comply with the equation of state (EOS), that describes the microscopic properties of stellar matter for given density ρ, temperature T and composition Xi.

This is usually expressed as the relation between the pressure and these quantities: $P = P(\rho, T, Xi)$. As a star or its core "contracts" as it is believed, the density may be getting so high that the electrons become degenerate and exert a much higher pressure, than they would if they behaved classically.

The term degenerate in physics means: matter that has been stripped of most of its orbital electrons, so the nuclei are packed close together *increasing* the material's density. Since in the limit of strong degeneracy, the pressure no longer depends on the temperature this degeneracy pressure can hold the star up against gravity, regardless of the temperature. Therefore, a degenerate star does not have to be too hot to be in hydrostatic equilibrium and it can remain in this state forever even when it cools down.

The above exposition, conventionally might approximate how the inner Sun originated. However, many scientists acknowledge that the current status of our knowledge of processes in the life of stars, is far from adequate from their true understanding. Astrophysicists due to their *flawed* training, have shown that although many models have been constructed, their detailed ability to describe observations is limited or non-existent. Furthermore, the general *failure* of all models means that we cannot tell, which are heading in the right direction.

This failure is because as discussed in chapter one and three, astrophysicists due to their specialization or rather *indoctrination*, acquired a tunnel vision leading them to *ignore* studies of other disciplines on the same subject. Also, their main focus is gravity as a driving force, ignoring other key factors. In regards to the inner Sun, a model becomes more complex as the concept of inner suns are unknown or better said, unthinkable in mainstream science.

To understand the origin of this inner Sun, related to the mass/density of withe dwarfs, we must *abandon* the conventional belief of the stellar combustion process, which relies on the fusion alone of hydrogen to helium; along the notion of "empty" space and replace it with *Plasma Physics/Astrophysics*. Indeed, this branch of physics has done a lot of experimental research in this area. Powerful, *high voltage-frequency* electrical currents, passing through the polar vortices in the spinning protoplanetary thick shell, would cause this plasma in the centre of it; to *interact* with the quantum vacuum (aether).

Because of the quantum zero-point motions interacting with this highly energized plasma, it can participate in the density *increase* of the plasma and so the mean separation between its constituent particles *decreases*, thus allowing the formation of *heavy* bodies and not necessarily huge.

Following this pattern, the mean separation becomes of order the de Broglie wavelength of the electrons, and the electron gas becomes degenerate, its properties are determined by quantum mechanics. This matter can behave very *differently* to the matter that we are familiar with; one can call degenerate matter, the counterpart of matter that respects the ideal gas law. When atoms are subjected to extremely high temperature and high electric voltage/frequency, the atoms are stripped of their electrons. In other words, they become *ionized*. The pressure of the gas (plasma) inside of a star is due to the electrons, or the electron pressure.

If the density is high, the particles are forced close together. The law of physics put constraints on the motion of the electrons. Electrons are only allowed to exist at certain energy levels, and no two electrons are allowed to exist at the same level, according to the Pauli's Exclusion Principle (unless they spin in opposite directions).

Therefore, in a dense gas (plasma), all of the lower energy levels become filled with electrons. This "gas" is termed degenerate matter. Electrons in this state exert a degenerate electron pressure which *resists* the force of gravity trying to push them closer together. This is because if they were pushed closer together, their energy would change and they would have to exist in an energy level that is *already filled*, which they can't do.

Also, like any a star or planet, this spinning inner Sun must continually draw energy, interacting with its *environment* or aetheric flow that pervades it, in order to "stay alive." In the case of Earth's formation, the result is a compact, high density star (inner Sun) which can compare to the concept of a white dwarf and it can *explain* the 33% of Earth's total mass and its density of around 8.5, without recurring to the idea of an "iron molten core".

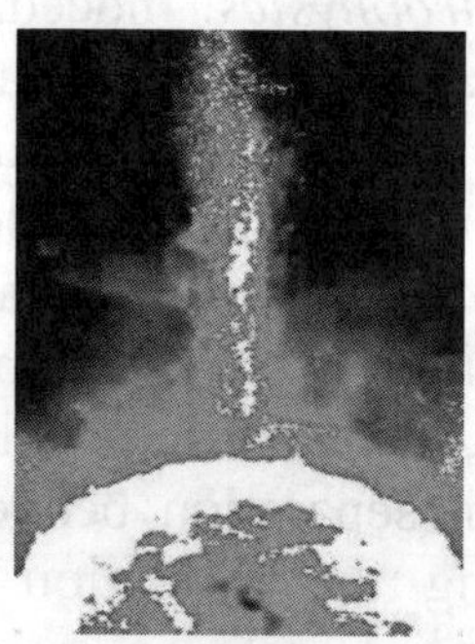

Fig. 109- Earth's inner Sun or plasma core showing a jet of charged particles which is the cause of auroras. This plasma core is a large plasma sphere that has its highest mass density at its surface, rather than at its centre as it would be the case of a sphere of atomic material that its electrically neutral. Highly swarms of protons are shielded by the swarms of electrons and are attracted to them, the result is a dense and tightly locked chain of plasma held together by the electric force that is 39 orders of magnitude stronger than gravity. This mass concentration becomes so dense that atomic elements cannot penetrate it and explains why density at Earth's centre, is higher in comparison to the density of Earth's crust.

Therefore, the rest of Earth's mass would be allocated in its thick 2900 km shell, with a similar density (2.4 g/cm^3) of the analysed rocks coming from the Earth's crust and upper mantle. From this perspective, the Earth's mean density of 5.2g/cm^3, is the *average* sum of the densities of the Earth's thick shell plus the inner Sun.

In regards to the behaviour of seismic waves within Earth, the inner Sun is the real reason of the speeding up of the P waves due to its high density, which currently are *misinterpreted* as proof of Earth's solid inner iron core in the same way earthquake doublets are misinterpreted measuring the rotation speed of it, when indeed these are proving the faster spinning of the inner Sun in comparison to the Earth's shell (mantle + crust).

An inner Sun solves the so-called "inner-core nucleation paradox" which demonstrates an iron core is *impossible* that can exist within Earth. Also, it might explain the "something else" that is affecting the measurements of G which vary over a period of 5.9 years, mentioned before.

This inner Sun possess a magnetic field that *adds up* to the one generated by the spinning of Earth thick shell and it originates *jets* of radiation and plasma that come out from its polar axis, reaching the Earths' polar openings and interact with Earth's lower and upper atmosphere, causing noise and auroras.

These jets of radiation are the *real* cause of surprise, when high energy particles coming off the icy ground, were detected recently in Antarctica and *misinterpreted* as a freak of particle physics as it will be explained shortly.

The inner Sun has a low luminosity which comes from the emission of stored thermal energy. This would explain the faint glow of the inner Sun, observed by several polar explorers, when they reached the outer rim of the north polar opening. Luminosity is the total energy that a star produces in one second. It depends on both the radius of the star and on its surface temperature.

With a radius of 1200 km and a low surface temperature compared with the Sun, the inner Sun faint luminosity however, would be *enough* to heat up and evaporate vast amounts of liquid water inside the Earth's shell, because of its relative proximity (2000 km). Inside this shell, there must be a sort of *atmosphere* to cause the warm and moist winds observed, mostly in the North Pole. In regards of the acceleration of gravity g, on Earth's surface things stay the same way, with g calculated as the total Earth's mass is concentrated at its centre. However, at a point placed 2900 km depth for instance, the gravity will be the *resultant* of the pull upwards of the mantle + crust, as per the Newton's Shell Theorem plus the pull downwards of the inner Sun.

Because the mantle + crust has the 67 % of the total Earth's mass and the point in question is 2000 km from the inner Sun, this resultant would be pointing at the Earth's surface (upwards).

When the system Earth shell + inner Sun formed, it is expected that due to the conservation of angular momentum, the inner Sun would rotate *faster* than Earth's shell, being able to generate jets. Due to its size and isolation, the Earths inner Sun wasn't affected by "magnetic breaking" like the Sun and thus, loose angular momentum.

The jets of radiation ejected by the inner Sun can be compared or related to a fast-rotating body say, a "neutron star" or a galactic core that can generate jets along an axis (of radiation, plasma and even hard matter). These ejections are possible because centrifugal forces allow the nuclear bonding force in the centre of the inner Sun, to be *weakened* allowing a lessening of density, at first in the core but eventually along the axial line; which causes plasma to be "loosened" allowing it, to move more freely along Earth's rotational axis.

We can expect the less dense material, from time to time, to be *ejected* along the polar axis, following plasma physics, because the angular velocity and thus the centrifugal force along the axis is zero. In concert with the increased mass of the equatorial belt, there is an increased angular kinetic energy and the *simultaneous* emission of intense jets of radiation, along the inner Sun axial (polar) line. These jets tough very small compared to stars and galaxies, come out *simultaneously* through Earth's polar openings, causing the observed auroras. This kind of jets are called *polar fountains*.

These emissions do behave as pulsations and support the work of Professor Neil Davis, author of *The Aurora Watcher's Handbook* (1992) mentioned earlier, who found evidence that indicates that auroral energy comes from *inside* Earth. He discovered that pulsations in the auroras occurred *exactly* at the same time in the southern aurora as in the northern aurora, meaning this is a *synchronic* phenomenon.

There are planets in our Solar System where the polar regions appear luminous, auroras are observed and anomalous *pulsating radiations* where recorded. On the other hand, the recent discovery of the high-energy particles, that did burst out from Earth's ground in the South Pole, with 70,000 times higher energy than the energy achieved with the most powerful particle accelerator, seems to be an anomaly.

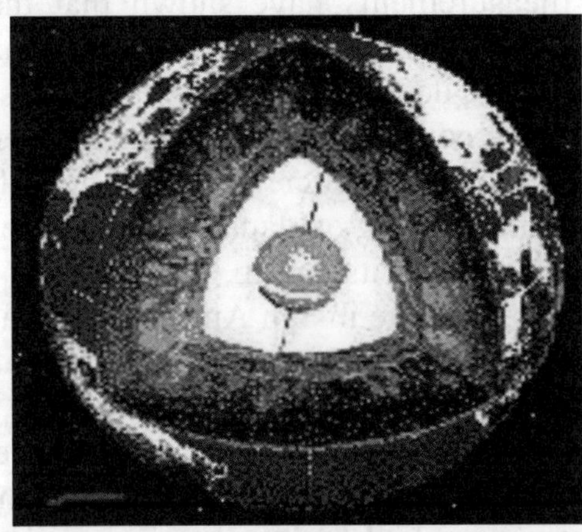

Fig. 110- Diagram showing the faster rotation speed of an interpreted but hardly existent, inner iron core in comparison to the spinning Earth's shell (crust + mantle). Another interpretation of the same seismological data, would be a high-density rotating inner plasma Sun, which explains the weird softness of this alleged iron core, the polar fountains and the warm-moist winds that heat up the polar regions, coming out through polar openings (not shown in this figure).

However, Earth's inner Sun can account for that because as any star, it is an effective *particle accelerator*. It would emit high energy particles as its counterpart the Sun. In fact, Solar Energetic Particles (SEPs) are an important *hazard* to spacecraft systems and constrain human activities in space. Primary radiation sources are *energetic* protons and heavy ions during SEP events, with energies up to few hundreds of mega electron-volts, being protons the most abundant specie.

Such large SEP events, highly random in nature, tend to occur during periods of intense solar activity, although they can also occur around the solar minimum period. These events can lead to high radiation doses in short time intervals. The inner Sun would behave in a similar fashion, with occasionally outburst of high-energy particles.

Furthermore, a *pulsating* hot spot of X-rays has been *discovered* in the polar regions of Jupiter's upper atmosphere by NASA's Chandra X-ray Observatory. Previous conventional theories *cannot* explain either the pulsations or the location of the hot spot, prompting scientists to search for a *new* process to produce Jupiter's X-rays. This result has its own *problems* due to the large distances, required for the source of the ions (originally from Jupiter's moon Io), to travel at least 30 times the radius of Jupiter.

Spacecraft measurements have shown that there are *not* nearly enough energetic oxygen and sulphur ions to account for the observed X-ray emission. Where does the right number of these energetic ions, or better said *jets* really come from; to cause this phenomenon observed only in the polar regions of Jupiter? "The location of the X-ray hot spot effectively retires the existing explanation for Jupiter's X-ray emission, leaving us very unsure of its origin," said Randy Gladstone, of the Southwest Research Institute in San Antonio and lead author of a paper on the results in the Feb 28, 2002 issue of the journal *Nature*.

However, mainstream scientists still believe that the reason is caused by an *external* agent, coming from outer space through the "solar wind" or particles from Jupiter satellites. Instead, they should focus in the Jupiter's *interior*, which originates jets or polar fountains that interact with its atmosphere, causing pulsating X rays and auroras. Chandra observed Jupiter for 10 hours on Dec. 18, 2000, when NASA's Cassini spacecraft was flying by Jupiter on its way to Saturn. The X-ray observations revealed that most of the auroral X-rays come from a pulsating "hot spot" that appears at a fixed location *near* the north magnetic pole of Jupiter.

Bright infrared and ultraviolet *emissions* have also been detected from this region in the past. An aurora of X-ray light near Jupiter's polar regions, had been detected by previous satellites. The X-rays were observed to *pulsate* with a period of 45 minutes, similar to the period of high-latitude radio pulsations detected by NASA's Galileo and Cassini spacecraft.

On the other hand, in 2005 astronomers using the Keck I telescope in Hawaii are learning much more about a strange, thermal "hot spot" on Saturn that is located at the tip of the planet's South Pole. In what the team is calling the sharpest thermal views of Saturn ever taken from the ground, the new set of infrared images suggest a *warm polar vortex* at Saturn's south pole, the first to ever be discovered in the Solar System.

This warm polar cap is home to a distinct compact hot spot, believed to contain the *highest* measured temperatures on Saturn. A paper announcing the results appears in the Feb 4th 2005 issue of *Science*. The puzzle isn't that Saturn's South Pole is warm; after all, it has been exposed to 15 years of continuous sunlight, having just reached its summer solstice in late 2002.

But both the distinct boundary of a warm polar vortex some 30° of latitude from the southern pole and a very hot "tip" right at the pole were completely *unexpected.* This is because if the increased southern temperatures are the result of seasonality alone, then the temperature should increase gradually with increasing latitude, but that is *not* the case. The observations show that the temperature increases *abruptly* by several degrees near 70° degrees south and again at 87° degrees south. What is causing this *sudden* increase of temperature near Saturn's poles?

It is said that the historic Juno mission *rewrote* what was known about Jupiter. Discoveries about its unexpected dilute "fuzzy" core, composition, magnetosphere and polar regions are shattering as explained before. Now it is about the time to rewrite what is known about Earth's *interior.*

Given all the details mentioned before, we can conclude that Earth is indeed a *system* composed of a 2900 km thick *spherical rigid shell* called mantle + crust, with polar openings that do not necessarily coincide with the geographic poles. At its centre is placed what we can call an *inner Sun* 2400 km in diameter softer than the mantle and spinning faster than it. Between the mantle and this inner Sun there is a 2000 km separation which is filled with a sort of *atmosphere*.

There is the fact of the mysterious Earth's *blobs* mentioned in chapter two, where seismic waves generated by earthquakes are slowed down by a tenth to a third of their normal speeds as they sweep across and have scientists puzzled. They know that these two blobs have complicated shapes and structures, but despite their prominent features, little is known about why the blobs exist or why are there just two such huge blobs of anomalous material that lie on opposite sides of the Earth, rather than a continuous anomaly or lots of smaller ones?

Geoscientists have postulated several explanations for what these ultralow-velocity zones are made of and how they're formed. But *none* of those ideas quite fit the data, especially given how differently some of the zones behave from one another. The blobs are made of something different from the rest of Earth's mantle.

Bear in mind that geoscientists try to fit the blob's seismological data, with the *outdated* model of a solid Earth with a molten iron core. Researchers using a type of standing wave called *Stoneley modes* that vibrate *depending* on the density of the blobs, analysed records of ground movement in the days following large-magnitude earthquakes, looking for the low-frequency vibrations of standing waves. Comparing their results with models, they found that the blobs must be *less dense* than the surrounding mantle, to explain several constraints like the slight wobbles on the core's surface.

Despite critical advances in seismology, the quest to understand the blobs is an inherently interdisciplinary problem. Geoscientists are painfully aware they literally *don't know* what they are, where they came from, how long they've been around, or what they do. This is due to their *ignorance* of the behaviour of *dusty plasmas* in planet formation.

As explained in chapter three, studies carried out by plasma astrophysicist Vadim N. Tsytovich demonstrated that often the observed property of a dusty plasma is *not* a homogeneous distribution of dust particles but rather a formation of *structured* dust regions featuring also *stable voids*, i.e., dust-free regions inside the dust cloud.

When a planet is forming immersed in this dusty *plasma* cloud, it acquires an internal structure. There can be no internal architecture if the dust structures are not *separated* from one another by voids. In Earth's case these voids would be mostly, the puzzling blobs. In fact, a *structure* must be separated from the rest of the mass by a space containing far more *less density*, for otherwise it is no longer a distinct structure.

Also, as mentioned before, these blobs *must* be located only antipodally, in the opposite sides of the globe, because this configuration of voids embedded in the rigid mantle; would confer *stability* to the Earth as a rotating body.

When considering the new paradigm of the system Earth (crust and mantle + inner Sun), these blobs would be gigantic hundred-kilometres sized *caverns*, partly filled with water and lighted up by the inner Sun. These conditions coupled with a sort of atmosphere, might allow the existence of *ecosystems*. Through fractures in the rigid mantle such as passages and tunnels, it is likely that these caverns would be connected to Earth's surface.

From this perspective, the ancient legends of many cultures, which claim that there are races of people, entire civilizations that thrive in subterranean cities, may not be legends at all. For in instance, the Macuxi Indians are indigenous people who live in the Amazon, in countries such as Brazil, Guyana, and Venezuela.

According to their legends, they are the descendants of the Sun's children, the creator of Fire and disease and the protectors of the "inner Earth." Their oral legends speak of an entrance into Earth. Until the year 1907, the Macuxies would enter some sort of cavern, and travel from 13 to 15 days until they reached the interior. It is there, “at the other side of the world, in the inner Earth” is where the Giants live, creatures that have around 3-4 meters in height. This wouldn’t come as surprise because deep underground, the resultant of the acceleration of gravity *g* is minor, allowing living organisms to grow without the gravity constraints of Earth’s surface.

So far, there is *evidence* that planets are spherical thick shells that possesses polar openings and a luminous point at the centre. There are strong indications that this is indeed the case, but it seems that data is *not* freely released. Where some evidence does become available, it is explained away or simply not discussed because it does not fit the prevailing view (paradigm).

We should see polar openings leading to the hollow interior of our own and other planets, only we don't at least normally. Very few published images actually show the poles directly. Many of those present signs of having been *manipulated.* The reader should try and find images that show either the Earth's North pole or South pole.

One will see a blob of "white-out" or the electronic equivalent of it, a white spot that has been airbrushed into the picture. In fact, nowadays, digital images can be easily *modified* by using software, which has seriously *debased* the credibility of photographic images as definite records of events.

A common manipulation in tampering with an image, known as *region duplication*, is to copy and paste a continuous portion of pixels to a different location in the same image. However, digital forensic techniques can *detect* a copy-paste trail in a doctored JPEG image, i.e., to check whether a copied area came from the same image or not.

When a copy-paste procedure is done on an image, especially adding or *hiding* an object, the block artefact grid contained in the copy-pasted slice is moved together. Since a slice must be placed properly in the target image to avoid obvious vision flaw, the grid in the slice *mismatches* to the original grid in the target image normally. Hence the photography was tampered.

On the other hand, in polar regions, especially Antarctica and as a rule; scientists are told their scientific presentations must be *cleared* by government agencies in advance, prior of being publicly released. Antarctica has become a central focal point of intense controversy and growing secrecy, that confirms the *withholding* a wealth of information and knowledge, from the ever-curious public masses.

The Expanding Earth

Whereas this topic is not part of the main theme of this book, it would be incomplete without briefly, discuss this concept and its implications. Also, the planet's *spherical thick shell* can be helpful as a starting point in *explaining* the mechanism of the expanding Earth, which can be observed on the crust or lithosphere (rifts).

In the context of global and planetary systems, the lithosphere has a major role as the interface between the inside and the outside. Both are connected and dependant. Mainstream science holds the *assumption* that the dimensions of the lithosphere and thus Earth, have been constant throughout geological time.

Nonetheless, there is *evidence* of planetary expansion in our Solar System. Voyager images of one of Jupiter's moons Ganymede, show clear evidence of global tectonic activity. A variety of mechanisms for creating stresses in planetary lithospheres have been recognized.

However, since Ganymede may have undergone relatively large volume changes, progressive fragmentation of the lithosphere by membrane stresses due to *planetary expansion* would be an attractively simple mechanism to explain *global-scale* rifting. Expansion is consistent with the *observed* distribution of terrain types and may therefore explain at least the latest stage of tectonic activity on Ganymede (Parmentier and Zuber, 1982).

In the past, geologists where pretty convinced that Earth was contracting and cooling down, since its formation according to the now obsolete and useless, chaotic-accretion hypothesis. The Contracting Earth Hypothesis was once a dominant *paradigm* in geology. Over time, from being in a molten state, the Earth cooled, causing the formation of a solid crust, and allowing liquid water on the surface. Features such as mountains formed as Earth cooled and shrank.

Contractionism supposed the Earth's crust, mainly represented by continents; was fixed, static. Oceans were pretty much, the same shape and size since eons. According to mainstream science, in modern times, contraction may have stopped but the cooling still continues in the Earth's interior which is molten, meaning the core.

Contractionism overlooks the fact that it is completely *impossible* to assume that the outermost shell of the rotating Earth, has been able to contract so much as to cover a surface area less than a quarter of the total surface of our planet. Also, it is neglected that a massive planetary body with an incandescent and progressively cooling core, would *crack* into pieces while cooling and shrinking. Besides, the model based on Earth contraction had increasing *problems* accommodating new geological evidence.

During the 1960's, through the study and analysis of principally, palaeomagnetical and oceanographical data, scientists put forward the Theory of Plate Tectonics, mentioned in chapter two. This theory *only* modified the fixed status of oceans and continents, substituting it for a dynamic interplay between the two. The rest was left untouched. Especially, that a solid Earth contracted to its current size and this one is therefore "constant".

Officially, the idea of the mobility (drift) of continents was first proposed by German meteorologist Alfred Wegener in 1912. One of the reasons was to explain the good fitting observed between the eastern margin of South America, with more than half of the western margin of Africa.

He suggested that the continents were once a single landmass and gradually, drifted apart around Earth's surface (assuming constant size). The term drift is misleading because the German word originally used by Wegener was *Streuung,* which means dispersion and is more accurate as we will see later. When the work of Wegener was translated into English, the term used for this word was drift.

However, Wegener's hypothesis was rejected by scientists because he couldn't provide a *mechanism* that explained this continent's mobility. Geologists argued that the oceanic crust was just too tough for continents to "simply plough through." Decades later, Plate Tectonics Theory *seemingly* proved to be convincing in "explaining" a wide range of apparently unrelated geological processes and features observed on Earth.

Nonetheless, there were some *loose ends* that this theory, couldn't account for and were conveniently swept under the carpet. The analysis of those loose ends led to several scientists, to consider the fact that Earth is *expanding* since its origin. Of course, mainstream scientists would dismiss this expansion and rather maintain a fixed Earth's radius.

It's important to notice that the *origina*l idea of a continental drift came from the Italian musician and geologist Roberto Mantovani (1854-1933) and *not* from Alfred Wegener.

Mantovani was a defender of the continental drift and *planetary expansion*. He formulated a mobilistic theory, attributing the moving apart of the continents to the expansion of the entire planet. This theory is more general than that of Wegener.

Mantovani's theory, was published in 1889 and 1909, together with a 'Pangea reconstruction'. In the paper of 1909, he drew a map of the opening Pacific with dotted lines to show the points on the opposite side of the oceans that were once in contact. He concluded that the size the Earth *must* have been smaller in the geologic past than today. Mantovani's theory was officially recognised by the French Geological Society in 1924, which incorporated it in its body of legitimate ideas.

Encouraged by French geologist-oceanographer Jacques Bourcart in 1924, Wegener quoted the Italian in his famous book *Die Entstehung der Kontinente und Ozeane* as one who offered ideas extraordinarily close to his own. He expressed his doubts about the possibility of Earth's expansion because he thought the planet was rigid and "solid".

But Wegener himself said that the idea of an expansion of the Earth is certainly very convincing as far the *huge fractures* that occurred in the outer crust of the globe, are concerned. Mantovani was not only a precursor of the continental drift idea but was also far ahead of Wegener, who was even not considering the possibility of *variation* of the Earth's radius.

Textbooks frequently extol Plate Tectonics Theory without *questioning* what might be wrong with the theory or without discussing a competitive theory. How can students be taught to challenge popular ideas when they are only presented a one-sided view?

The main reason besides *defying* the current paradigm is that this hypothesis of expanding Earth, apparently lacks a mechanism that explains its expansion. It is for the same reason Wegener's hypothesis of continent's mobility (Continental Drift), was rejected more than a century ago.

On the other hand, today it is often forgotten that the epic debate about Earth in the years from about 1955 to 1970 involved *three* and not merely two rival theories. In addition to Continental Drift and the traditional picture of an essentially static Earth, a minority of scientists discussed and sometimes advocated the idea of an *expanding* Earth.

Expansionism was evidently on the side of mobilism and against fixism, but at the same time it was *opposed* to Continental Drift in the sense of Alfred Wegener's arguments. Although many modern Earth scientists may be unaware of the expansion theory of the past, its role in the Plate Tectonics revolution is *documented* in the historical literature.

Basically, in the accepted theory of Plate Tectonics, the Earth's crust or lithosphere; is divided into several large, rigid plates that "move" over a soft layer of the upper mantle known as the (inexistent) asthenosphere. These plates interact at their boundaries, where they converge, diverge or slide past one another.

Such interactions are *believed* to be responsible for most of the seismic and volcanic activity of the Earth. Plates "cause" mountains to rise where they push together and continents to fracture and oceans to form where they rift apart. The continents, sitting passively on the backs of the plates, drift with them at the rate of a few centimetres a year.

According to Plate Tectonics Theory, originally, Earth had a supercontinent called Pangea and a unique huge ocean called Panthalassa. This supercontinent started to split some 250 million years ago, and its parts were drifting until reaching the configuration of the present day. This is an ongoing process at an estimate rate of 5cm/year.

However, one of the main proponents of Earth Expansion Theory, Australian geologist Samuel Warren Carey (1911-2002) during the 1950's; after repeated attempts to fit together accurate tracings of the main continental masses on a large plastic globe, commented finally: "The assembly of Pangaea is not possible on the Earth of the present radius", thus he supported the hypothesis that the Earth's diameter has gradually *increased*.

A similar conclusion was reached *independently* in 1955 by US seismologist Hugo Benioff, who studied earthquakes' strain-rebound characteristics during 1907 to 1951. He concluded from the regular alternation of high activity with periods of three relatively low activity, that great shallow earthquakes are *not* independent events.

Benioff pointed out that a *world mechanism* for locking and releasing the faults in accordance to rebound characteristics, that are related in some form of world-wide stress system. It could involve changes or *increases* in the radius of the Earth. In fact, present-day stress fields, deduced from seismic records and also directly measured, favour e*xpansion* rather that Earth having a constant radius (Holmes, 1965).

Also, during the 1950s, surprising *links* were proposed between cosmological theory and the geological and paleontological sciences. These links were mainly provided by Paul Dirac's hypothesis of 1937 that the gravitational constant G *decreases* with cosmic time. Pascual Jordan in Germany and Robert Dicke in the United States, formulated theories of gravitation that went beyond General Relativity by incorporating a *varying* gravitational constant. Jordan developed its geophysical consequences, concluding that the Earth is expanding.

The Dirac-Jordan dynamic cosmology with creation of matter, requires that the gravitational constant presently decrease at the rate of 0.1 to 0.2 parts per billion (1 to $2x10^{-10}$) per year. As a result of a *stronge*r gravitation in the past, the ancient Earth would have been *smaller* than today. The terrestrial force of gravity, which can be represented by *g*, has systematically *decreased* as Earth has grown older.

Another scientist researching into varying gravitational constant was Hungarian geophysicist László Egyed (1914-1970) in the 1950's. He was not only a leading figure in the expansionist alternative but also an advocate of varying gravity, as the cause of the growing Earth. It was only with his work that the expansion hypothesis *entered* the normal literature of science. Egyed stated that Continental Drift can be explained by an expanding Earth *only*.

Among the *evidence* which inspired Egyed to his conclusion were estimates of the *change* of water-covered continents through the history of Earth. According to Henri and Geneviève Termier at the University of Paris, the area had *decreased*. The Termiers did not intervene in the controversy over the expanding Earth except that they *concluded* that "global palaeogeography does not display any argument against Earth expansion" (Termier and Termier, 1969, p. 101).

During the next two decades Earth expansion was discussed by several physicists, astronomers and Earth scientists. Among those who for a period felt attracted by gravitational expansionism were Arthur Holmes, John Tuzo Wilson and Fred Hoyle.

This idea of a varying gravitational constant had an *impact* on geophysics in the period from about 1955 to the mid-1970s; when the ludicrous Plate Tectonics Theory finally took over as a reigning paradigm. A *conspiracy* of crooked geologists, geophysicists and publishers, was responsible of this.

The expanding Earth and other aspects of the Earth Sciences relating to the varying-G hypothesis had cosmological consequences and also *implications* for the Earth Sciences, this chapter in the history of interdisciplinary science deserves attention.

For one thing, it *changed* the landscape of both the cosmological and geological sciences. For another thing, the question of varying natural constants is still *unsettled* and the subject of scientific investigation.

Despite the Earth Expansion theory was overran by Plate Tectonics in the 1970's, a minority of geoscientists continued doing research on expanding Earth. In February 1981 the symposium titled *The Expanding Earth* was held at the University of Sydney convened by Carey as Professor of Geology at the University of Tasmania.

Expansionist papers have been *published* in scientific journals and *lectures* were dedicated to the Expanding Earth Theory: at the 30th International Geological Congress, Beijing (Dickins, 1996; Hongzhen et al., 1997), 32nd International Geological Congress, Florence (Anonymous, 2004), and in the 34th International Geological Congress held in Brisbane (Anonymous, 2012; Choi and Storetvedt, 2012 a, b).

Recently, the group of researchers include the names of Cliff Ollier, Karl-Heinz Jacob, Jan Koziar and James Maxlow among others. Dr. James Maxlow may be the most famous geologist you have never met. Inside this quiet, unassuming man lies a passion for something he knows is true, that will change all of science: The Earth is expanding.

Maxlow and other researchers show us that the expansion of the Earth is not constant, but is gradually *increasing* over time. Both surface and radius are in fact gradually increasing. This means the expansion of the Earth is accelerating; which can *explain* the recent increase of earthquakes and volcanic activity.

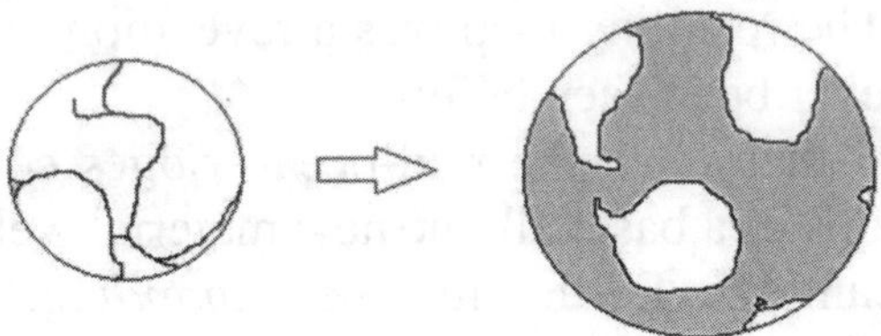

Fig. 111- South pole view of the expanding Earth. Continental Drift assumes Earth having constant size and was based on the striking similarities between types of rock and fossils on opposite sides of the Atlantic Ocean. Also, the jigsaw puzzle-like shapes of the continents, and how well they seemed to fit together. It was suggested that the continents were once a single landmass along a single ocean and gradually drifted apart. Expanding Earth, doesn't require the continents to start on one side of the globe or for subduction to occur. Continents don't "drift." Earth's expansion brought about the splitting and gradual dispersal of continents as they moved radially outwards not sideways, during geological time. Oceans are young but not the oceanic floor, except in the mid-ocean ridges. These young oceans now cover approximately 71% of Earth's surface and have added about 40% to its size. Antarctica remained at a fairly stable position because it is surrounded by a ring of expansion.

The grounds of Plate Tectonic Theory lay upon two remarkable features on Earth: First, *expansion cracks* such as the rift valleys and mid-ocean ridges. 'Rift' is a split or break and a 'valley' is a low point. The rift forms when the Earth's crust divides or diverges. The valley that forms in the divide is usually flat, narrow, and has steep sides, such as the ones observed in East Africa (Great Rift) and Rhine Valley, in the border area of west Germany and northeast France. In the case of East Africa, as both sides of the Great Rift are *pulled apart*, blocks slide down along vertical faults, forming the steep cliffs delimitating the rift valley. The tectonic movements and faults are also associated with volcanic activity. The African continent slowly *splits* into two, paving the way for forming a new ocean.

Seafloor spreading according to conventional science, is a geologic process in which the supposed "tectonic plates", large slabs of Earth's lithosphere, split apart from each other. Seafloor spreading occurs along mid-ocean ridges, continuous submarine mountain chain extending approximately 80,000 km (50,000 miles) through all the world's oceans and act as plate boundaries. The Mid-Atlantic Ridge, for instance, separates the North American plate from the Eurasian plate and the South American plate from the African plate. It is noteworthy to mention that US geologist Bruce Heezen, was one of the pioneers in discovering and mapping the *mid ocean ridges* in the 1950's along Marie Tharp and Maurice Ewing. The resulting map was a revelation because it showed details that had never been seen before.

Heezen noted that since the mid-ocean ridges covered the *whole* Earth, like the seams of a baseball, the new material welling up at all the ocean ridges meant the Earth must be *expanding*. If new oceanic lithosphere is continually being created at the oceanic ridges, the oceans should be expanding indefinitely, increasing Earth's radius. The continents do not really move away from each other, that's an *illusion.* It's the *space* between them that is increasing because of that stretching of the oceanic lithosphere.

Later on, he was *forced* to retract this idea by his peers who were of the opinion Earth always had a fixed diameter. Heezen was an *expansionist*. He was sure the Earth is getting larger day-by-day. In 1959, at the International Oceanographic Congress, Heezen presented several papers on Earth expansion, getting a wide attention.

Second, the alleged *subduction zones* also known as oceanic or marine trenches often associated with island arcs. Where two tectonic plates meet at a subduction zone, one bends and "slides" underneath the other, curving down into the upper mantle or the hotter layer under the crust known as asthenosphere. The older and denser seafloor under thrusts the continental mass, dragging downward into the Earth's upper mantle the accumulated trench sediments. Subduction zones are practically located in the Pacific Ocean along the so-called Ring of Fire. Only a few are in the North-east Indic Ocean and Caribbean Sea.

Whereas rift valleys were well known features of Earth's surface since the 19th century, it is until the middle of the 20th century that marine trenches and mid-ocean ridges; were fully acknowledged trough geophysical and oceanographic studies. After all, oceans cover 71% of Earth's surface. These submarine features demanded a *global* rather than local tectonic explanation. In the 1960s the once discarded theory of Continental Drift proposed by Alfred Wegener was substantially revised and transformed into the modern standard theory of Global Plate Tectonics. For a decade or so the new theory of the Earth and the traditional contraction theory faced competition from a *third alternative*, the hypothesis of the Expanding Earth.

In 1960, US geologist Harry H. Hess proposed an alternative solution that fitted with the Constant Diameter Earth theory. The new ocean floor could spread out to form the central ridge but then as it reached the continents it would "sink down" under the continents and then be consumed deep within the mantle, only to be "recycled" millions of years later as new ocean floor. It formed a giant Convection Cell constantly recycling material to maintain a Constant Diameter Earth. This was in keeping with traditional Earth concepts and is the theory generally accepted and widely taught today. Continental Drift was re-christened as Plate Tectonics and finally became an established "scientific theory" (dogma).

Using the sonar, Hess was able to map a portion of the ocean floor and gave his own interpretation of the mid-Atlantic ridge profiles. He also found out that the temperature near to the mid-Atlantic ridge was warmer than the surface away from it. Hess believed that the high temperature was due to the magma that leaked out from the ridge. To explain this phenomenon. Hess applied the model of "convection currents" in the mantle proposed by British geologist Arthur Holmes in the early 20th century.

At this point is necessary to notice that Hess didn't bother to consider, the *physics* involved in his machinations of recycled sinking slabs within the rigid mantle. His proposal *did not* conform to known physical laws regarding buoyancy, hydraulics, strength materials, rock mechanics and so on; which could allow a solid slab "sink" inside Earth's mantle and be able to melt and "circulate" if so. Hess like many geologists, blatantly *disregarded* physics, indulging in poetry and mythology. Then came the so called "geophysicists" who instead of doing their job properly, *twisted* physics to accommodate the *delusional* geological ideas of Hess and provide "solid ground" for Plate Tectonics theory. This is a disgrace.

Canadian geophysicist John Tuzo Wilson confirmed the production of oceanic crust at ridge crests and its subsequent lateral transfer, when he analysed a special system of faults associated to these ridge crests technically known as transform faults. He concluded that when plates split apart, they do not follow a continuous line. In fact, it is fractured following a zig-zags or staircases pattern.

This is because of differing speeds, creating space anywhere from a few to several hundred miles between spreading margins. Where the difference of speed is significant, adjacent portions of crust move in parallel creating fractures in the plate, which Wilson called as transform faults and run perpendicularly to the spreading margins.

The final supporting evidence considered for seafloor spreading came through palaeomagnetism in 1963. Palaeomagnetism is the study of recordings of the Earth's past magnetic fields. Volcanic rocks and some sedimentary rocks containing magnetic minerals, are able to fossilize the intensity and orientation of the magnetic field at a given epoch. Magnetite is one of these minerals.

Using palaeomagnetism, the shift of the magnetic poles and above all the periodic inversion of the Earth's geomagnetic field were discovered. British geophysicists Fred Vine and Drummond Matthews along Canadian geophysicist Lawrence Morley, surmised in 1963 that new ocean crust would *inherit* the ambient magnetic field at the *time* of its formation; the magnetic field would show either normal (north) or reversed (south) polarity.

If the spreading of oceanic crust moves like a conveyor belt away from the mid-ocean volcanic rift, then the rocks should record successive magnetic reversals. Furthermore, the magnetic anomalies should be symmetric either side of the mid-ocean ridge showing magnetic anomalies that were both parallel to and symmetrical about the ridge.

Later geomagnetic surveys found in general, that the patterns are in fact present, providing strong confirmation to support the hypothesis proposed by Hess. Not necessarily. There is NO scientific reason to connect the fact of spreading of the ocean floor and its magnetic anomalies, with the *unproven* hypothesis of convection currents and of course, the *myth* of "subduction" of the ocean floor. As Isaac Newton rightly said: "I shall not mingle conjectures with certainties."

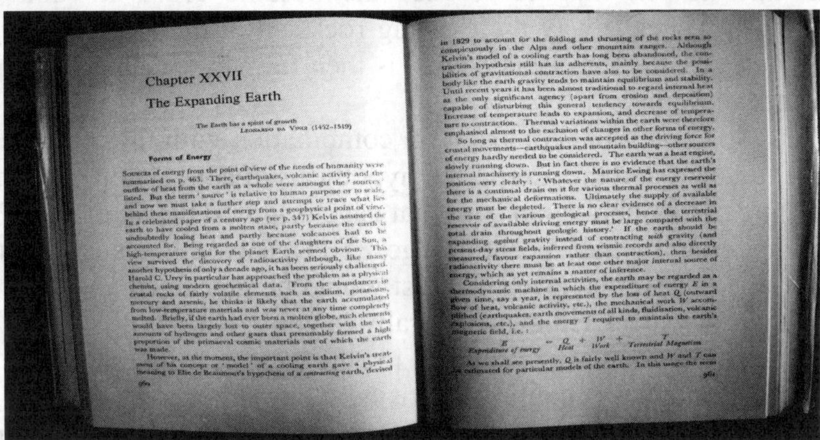

Chapter XXVII

The Expanding Earth

The Earth has a spirit of growth
LEONARDO DA VINCI (1452–1519)

Forms of Energy

Sources of energy from the point of view of the needs of humanity were summarised on p. 463. There, earthquakes, volcanic activity and the outflow of heat from the earth as a whole were amongst the 'sources' listed. But the term 'source' is relative to human purpose or to scale, and now we must take a further step and attempt to trace what lies behind these manifestations of energy from a geophysical point of view. In a celebrated paper of a century ago (see p. 347) Kelvin assumed the earth to have cooled from a molten state, partly because the earth is undoubtedly losing heat and partly because volcanoes had to be accounted for. Being regarded as one of the daughters of the Sun, a high-temperature origin for the planet Earth seemed obvious. This view survived the discovery of radioactivity although, like many another hypothesis of only a decade ago, it has been seriously challenged. Harold C. Urey in particular has approached the problem as a physical chemist, using modern geochemical data. From the abundances in crustal rocks of fairly volatile elements such as sodium, potassium, mercury and arsenic, he thinks it likely that the earth accumulated from low-temperature materials and was never at any time completely melted. Briefly, if the earth had ever been a molten globe, such elements would have been largely lost to outer space, together with the vast amounts of hydrogen and other gases that presumably formed a high proportion of the primaeval cosmic materials out of which the earth was made.

However, at the moment, the important point is that Kelvin's treatment of his concept or 'model' of a cooling earth gave a physical meaning to Elie de Beaumont's hypothesis of a *contracting* earth, devised

960

in 1829 to account for the folding and thrusting of the rocks seen so conspicuously in the Alps and other mountain ranges. Although Kelvin's model of a cooling earth has long been abandoned, the contraction hypothesis still has its adherents, mainly because the possibilities of gravitational contraction have also to be considered. In a body like the earth gravity tends to maintain equilibrium and stability. Until recent years it has been almost traditional to regard internal heat as the only significant agency (apart from erosion and deposition) capable of disturbing this general tendency towards equilibrium. Increase of temperature leads to expansion, and decrease of temperature to contraction. Thermal variations within the earth were therefore emphasised almost to the exclusion of changes in other forms of energy.

So long as thermal contraction was accepted as the driving force for crustal movements—earthquakes and mountain building—other sources of energy hardly needed to be considered. The earth was a heat engine, slowly running down. But in fact there is no evidence that the earth's internal machinery is running down. Maurice Ewing has expressed the position very clearly: 'Whatever the nature of the energy reservoir there is a continual drain on it for various thermal processes as well as for the mechanical deformations. Ultimately the supply of available energy must be depleted. There is no clear evidence of a decrease in the rate of the various geological processes, hence the terrestrial reservoir of available driving energy must be large compared with the total drain throughout geologic history.' If the earth should be expanding *against* gravity instead of contracting *with* gravity (and present-day stress fields, inferred from seismic records and also directly measured, favour expansion rather than contraction), then besides radioactivity there must be at least one other major internal source of energy, which as yet remains a matter of inference.

Considering only internal activities, the earth may be regarded as a thermodynamic machine in which the expenditure of energy E in a given time, say a year, is represented by the loss of heat Q (outward flow of heat, volcanic activity, etc.), the mechanical work W accomplished (earthquakes, earth movements of all kinds, fluidisation, volcanic explosions, etc.), and the energy T required to maintain the earth's magnetic field, i.e.:

$$\underset{\textit{Expenditure of energy}}{E} = \underset{\textit{Heat}}{Q} + \underset{\textit{Work}}{W} + \underset{\textit{Terrestrial Magnetism}}{T}$$

As we shall see presently, Q is fairly well known and W and T can be estimated for particular models of the earth. In this usage the term

961

Fig. 112- When the second edition of Arthur Holmes popular geological text (Principles of Physical Geology) was published in 1965, it devoted an entire chapter on the Expanding Earth. He wrote: "*we must seriously consider the possibility that the Earth's interior is expanding.*" It is perhaps the only official geology textbook to openly discuss this matter; ever published. It is unique. After Holmes passed away, some years later, by the mid -1970's the publishers had decided some ideas presented in this book were unacceptable, particularly the chapter of Earth's expansion. In 1978, the third edition of Holmes book was published and this inconvenient and original chapter, was deleted.

Unlike on the continents, where regional magnetic anomaly patterns (that is, magnetic patterns that deviate from the common rule) tend to be confused and seemingly random, the seafloor possesses a consistently regular set of magnetic bands of alternately higher and lower values than the average values of Earth's magnetic field. These positive and negative anomalies are more less linear and parallel with the oceanic ridge axis, show distinct offsets along fracture zones, and generally resemble the pattern of a zebra skin.

In most cases the sequence on one side is the approximate mirror image of that on the other. In this way, *seafloor spreading*, at least closer to the mid-oceanic ridges; has been confirmed as a scientific fact, supported by experimental evidence as per the scientific method procedure. Right now, along the 80,000 km of mid-ocean ridges, *new* oceanic crust is created. The oceanic crust is composed of rocks that apparently, move away from the ridge as new crust is being formed.

The formation of the new crust is due to the rising of the molten material (magma) from the upper mantle. Plate Tectonics supporters claim this molten magma reaches the oceanic crust, it cools and somehow "pushes" away the existing rocks from the ridge equally in both sides. A younger oceanic crust is then formed, causing the spread of the ocean floor.

This spreading phenomenon, in combination with the existence of marine trenches and the non-tested hypothesis of convection currents in the mantle plus "subduction", is what gave origin to the Theory of Plate Tectonics, roughly half a century ago. The basic outline suggests that, at the ridge crests, new oceanic crust would be generated and then like a conveyor belts, carried away laterally to cool then it will subside and finally be destroyed or melted in the nearest marine trenches. Meanwhile, magma is continually rising along the mid-oceanic ridges, where the "recycling" process is completed by the creation of new oceanic crust.

Consequently, the age of the oceanic crust should increase with distance away from the ridge crests and because recycling was its ultimate fate, very old oceanic crust would not be preserved anywhere. Oceanic crust is made mostly of silicon and magnesium (Sima) whereas continental crust is made mainly of silicon and aluminium (Sial).

Plate Tectonic Theory started with a premature *extrapolation* of *insufficient* data. In fact, the mechanism of convection currents in the upper mantle is poorly understood and it has several loopholes in physics, it's still not clear exactly *how* the asthenosphere, if such thing ever exists, works and in which directions its materials are "flowing" if so. Marine trenches do exist and are used as tools for the concept of subduction zones, but their nature is based upon *interpretations* of seismological data.

Subduction zones are generally depicted as dozens of kilometres' thick slabs descending into the Earth. This "descent" is *controversial* from the perspective of physical laws. Subduction zones have the issue of overlooking the fact that *frictional forces*, required for subduction to occur are *enormous* and rarely do Earth scientists address this problem.

Not only is energy not present to initiate subduction, but the required energy to sustain subduction is *not available*. How could a thick oceanic lithosphere with an average density of about 3.0-3.4 g/cm^3 begin to subduct into a mantle with an average density of between 4-5.7 gm/cm^3 and with greatly increasing geostatic pressures with depth?

Friction causes heath and plate tectonicists admit that it is hard to see, *how* the subduction of a cold slab could result in the high heat flow or island arc volcanism. The very low level of seismicity, the lack of a megathrust and the existence of flat-lying sediments; at the base of oceanic trenches *contradict* the alleged presence of a down going slab.

Finally, the Permian equator now lies 37° north of the equator in North America, 40° north in Europe, and 17° north in Siberia, which is *impossible* on an Earth of constant radius without at least 6,000 km of post-Palaeozoic "subduction" within the Arctic, which has *no* marine trenches. However, scientists still believe in subduction like a religion.

The oldest known rocks from the continents are just under 4 billion years old, whereas according to Plate Tectonics, none of the ocean crust is older than 200 million years (Mesozoic). This is cited as conclusive evidence that oceanic lithosphere is constantly being created at midocean ridges and consumed in subduction zones. There is in fact, abundant evidence *against* the alleged youth of the ocean floor, though geological textbooks tend to *pass over* it in silence.

The numerous findings in the Atlantic, Pacific, and Indian Oceans of rocks far older than 200 million years, many of them *continental* in nature, provide strong evidence against the alleged youth of the underlying crust.

During legs 37 and 43 of the *Deep Search Drilling Project*, Palaeozoic and Proterozoic igneous rocks were recovered in cores on the Mid-Atlantic Ridge and the Bermuda Rise, yet not one of these occurrences of ancient rocks was mentioned in the Cruise Site Reports or Cruise Synthesis Reports (Meyerhoff et al, 1996a).

There are several *inconsistencies* in defining plates by their boundaries. Early articles noted as few as six, but as fieldwork began to challenge proposed boundaries and mechanisms, the number increased rapidly, primarily due to the addition of small 'microplates' in ambiguous areas.

For example, the apparent lack of clear traditional plate boundaries (i.e. rifts, subduction zones, and transform faults) between Africa and Europe, resulted in the addition of *numerous* small plates in that region.

There is fairly good consensus as to the primary and secondary plates using data from geophysics to define their boundaries. However, many secondary and tertiary plates are added mainly on the continental crust as different criteria are used. So how many plates are there? The question is still *unresolved.*

Perhaps the biggest problem with Plate Tectonics Theory is the idea of *mantle convection*. The main issue is the sideway force that pushes plates away, where they rift apart. One law of *physics* states that when a pressure difference exists between two areas, they will equalize if there are no *constraining* factors. For example, volcanoes erupt when the pressure below grows strong enough to breach the constraining surface and then stop erupting when the pressure is equalized or constrained.

Plate Tectonic Theory states that convection currents within the mantle, force liquid magma up through the centre of the ocean basin expansion ridges. This liquid magma is then somehow forced sideways pushing the opposite sides of the ridge apart, moving whole continents in the process.

If pressure was being exerted from below the liquid magma, it would take the path of *least resistance*, which would be upwards building volcanic structures. This is in fact what does happen along the oceanic ridges. The origin of the proposed sideways force is undisclosed and *conflicts* with the laws of physics. There is simply *no* constrains in the *upward* direction, to force the flowing magma to move sideways.

Sir Harold Jeffreys got it right when he made a *quantitative* estimate of the frictional resistance opposing the motion of continental crust crustal blocks (Sial), through the Sima or upper mantle (asthenosphere). This led him to the conclusion that it would require *enormously* large horizontal driving forces to make such a motion possible, and the origin of these could NOT be explained in any physical manner.

This issue made Jeffreys's attempts at computer modelling fed with *real* physical parameters, to consistently run in to *impossible* logic barriers such as the *inability* to keep mantle convection patterns stable, the *inability* to generate mantle flow in any targeted direction for any sustained length of time, the *inability* to obtain seafloor elevations that conform to known maps even after manually forcing plate movements, the *inability* to create subduction zones or seafloor spreading even with manually forced and targeted mantle convection flow patterns, the *inability* to generate enough friction between mantle flow and lithospheric plates to drive one plate under another even when exaggerating any hypothetical driving forces, and the *inability* to align known fossil records and sedimentation data with any conceivable plate relocation scenarios, just to name a few.

Closely associated to mantle convection currents are *mantle plumes,* use in several cases to "explain" volcanic activity. Mantle plumes are believed to be areas where heat and rocks in the mantle, are rising towards the surface. They are supposed to originate near where the Earth's core meets the mantle, almost 3000 km underground. A *hot spot* is the surface expression of the mantle plume.

However *new* seismology data are now confirming that such narrow jets *don't* actually exist, says US researcher Don Anderson, the Eleanor and John R. McMillian Professor of Geophysics Emeritus at Caltech. The study was published in the *Proceedings of the National Academy of Sciences* (2014). In fact, he adds, *basic physics* doesn't support the presence of these jets, called mantle plumes, and the new results corroborate those fundamental ideas.

Thanks, in part to more seismic stations spaced closer together and improved theory; the analysis of the planet's seismology is good enough to confirm that no narrow mantle plumes, have ever been *detected*. Nevertheless, mantle plumes advocates respond to such refutations by making their conjectures ever more complex, unique to each example, and these are *untestable*.

On the other hand, there is the problem of *semantics* as discussed in chapter one. It is an unfortunate fact in geology, that several technical terms for the various structures of Earth are *not* clearly defined. The crust of Earth is often called lithosphere, but its nature and dimensions are not agreed upon. It is far from being homogeneous.

There is a *confusion* involving identification of the boundary between the lithosphere and asthenosphere. For geologists, convection is *assumed* to take place in the mantle as solid-state flow (lasting millions of years), instead of liquid state flow. Then, how something with long-term strength like the mantle can show liquid features? These two conditions are rheologically *incompatible*. It cannot be both.

There is a semantic confusion. To explain a uniform natural phenomenon partly by cause A, but in other parts by a cause B is a dubious procedure, this kind of sloppy "scientific" thinking must not be tolerated.

For these reasons, in the past, physicists considered the work of geologists to be like collecting stamps. Also, most geologists along geophysicists, when inventing and clinging to their fancy and mythical ideas; they are in fact ignorant of Sir Harold Jeffreys's prestigious book, *The Earth, Its Origin, History and Physical Constitution* (6th edition. 1970) quite the most authoritative treatise ever on the physics of the Earth, following the tradition of Osmond Fisher and Lord Kelvin. This book is a broad-ranging account based on *observations* analysed with care, using physics and mathematics as tools.

Jeffreys's scientific method (now known as Inverse Theory) was a logical process. It was his studies of the *strength* of Earth materials, in particular the ocean floor, which led Jeffreys to the conclusion that Plate Tectonics Theory, with the continents ploughing through the oceanic crust, was completely *untenable* on mechanical grounds. Jeffreys's scientific work presents a model of how to do science, but most geoscientists either by ignorance or stupidity; have taken exactly the opposite way, promoting *pseudoscience*.

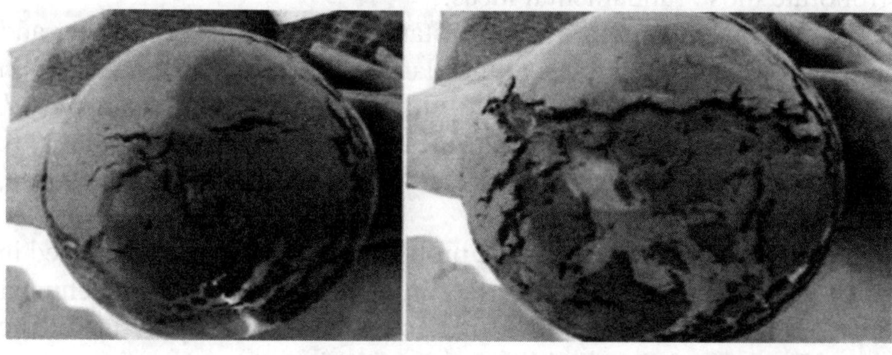

Fig. 113- The dominating phenomenon of the Earth's surface is rifting and extension. The figure shows a scale model of Earth's expansion. Fracture patterns appear here, caused by the increase of internal pressure in layers of plastic solidified on an inflated football bladder. This mimics the expansion of the interior of the Earth, so that solid and differentiated crust breaks, and individual blocks become separated by insertion between them of newly derived mantle differentiate (mid-ocean ridges) or rift valleys and oceanic trenches often associated with earthquakes, due to the appearance of tensile and compressive forces with multidirectional vectors of forces. In fact, there is extension in all oceanic zones. Marine trenches are extensional zones (not subduction zones) analogous to the rifts. $eismic refraction measurements in the upper part of the continental ramp along the east coast of North America revealed sediment-filled trough of analogous dimensions to the present bordering marine trenches, filled with thick, old and undisturbed sediments of Puerto Rico and the circum-Pacific belt. In other words, those sediments were never "subducted" and "recycled" into the mantle, at all.

The *exact meaning* of convection is the transfer of heat through a fluid substance due to density changes. It requires a smooth absorption of heat. Substances that don't transfer heat easily are called insulators and that's the problem. The mantle is pretty much an *insulator*, composed of silicate material which has much lower heat capacity and lower thermal conductivity than water, upon all this fuss of convection currents is based. Therefore, convection currents in the mantle would be an extremely *inefficient* thermal mechanism for any purpose. Also, convection is only observed in *real* fluids (water, air, etc.).

Besides, geologists and geophysicists' conclusions in regards to the alleged convection currents, depend crucially on the *assumptions* of the values of the viscosity (μ) and of the strength (σ) of the mantle. Convection is assumed to be an effective homogenizer. Recent amendments to this idea assert that there must be a radioactive-rich layer deep in the mantle, but this is based on unlikely *assumptions* about upper mantle radioactivity.

Most convection calculations *ignore* accretional differentiation and the effects of pressure and temperature on thermal properties such as thermal expansion and thermal conductivity. Current models *assume* that heat from the core and heat from the mantle are decoupled in the sense that imaginary "mantle plumes" remove core heat and plate tectonics removes mantle's heat.

Although there is no missing energy when direct heat flow data are used, there is a *mismatch* between oceanic heat flow and theoretical predictions from the cooling plate model. In some compilations, about 12 Terawatts is added (fudged) to the measured global heat flow, to match the square-root-of-age predictions.

On the other hand, in regards to scientific knowledge, universities usually lag *behind* industries. The later do have far more financial resources and experienced qualified personnel aimed to a specific task, than many universities. Industries are much more in practical contact with *reality* and have first-hand knowledge of it than their academic counterparts, who rely more on theory.

Oil and mining industries, have combed most of the Earth's surface in search for natural resources. Their surface seismic-velocity, heat-flow and gravimetric studies do show that; the asthenosphere upon the plates apparently "slide", is far from being a continuous layer. In fact, there are *disconnected lenses* (asthenolenses), which are observed only in regions of tectonic activation and high heat flow.

Seismic tomography has merely *reinforced* the message that ancient continental shields, have very *deep roots* and that the low-velocity asthenosphere is very thin or *absent* beneath them. Therefore, the concept of thin (less than 250-km-thick) lithospheric plates "moving" thousands of kilometres over a global asthenosphere is *false*.

The asthenosphere is *not* a continuous layer at global scale. In fact, it is more associated to *local* magma chambers with interconnected channels. Also, there is another problem: the geoscientists' ignorance of the *Gestaltung* which organizes and shapes convection cells.

For example, a convection cell large enough to drive the Pacific plate, would have to have a cell width-to-depth ratio that would be nearly 10:1, which we don't observe in Nature. In fact, in the atmosphere, the convection cell has a width-to-depth ratio of about 1:1. In Nature, from the biggest to the smallest, convection cells like whirls, do follow the *same* pattern of proportions or ratios; that occur everywhere naturally, expressing minimum energy and stable configurations. From this perspective, the proposed convection mechanism for the Pacific plate is *impossible,* which makes the alleged tectonics plate "drift" unfeasible. This makes Plate Tectonics Theory, arguably the most convincing *pseudoscience* of all time, which continues to thrive despite overwhelming contradictory evidence. Plate Tectonics has survived for decades because its proponents have increasingly *detached* their theory from reality by systematically rejecting or overlooking any *contrary* evidence, and *selectively* picking only the data that support Plate Tectonics.

This fact should generate *concern* among all scientists, for historically the rigorous testing of ideas effectively ceases in intellectual environments dominated by a single concept (paradigm). Furthermore, as intensive geotectonic *research* has vastly increased the *database* for Earth-dynamic studies, plate tectonics has NOT adequately and completely *explained* the geology of *many* regions of the world.

Finally, for Earth to remain the same size over time, as mainstream science asserts, the amount of crust being created at ridges would have to be *equal* to the amount of plate being subducted along trenches. If more mass is being created than consumed, then the Earth's volume would be increasing.

There is no evidence to support Earth's growing volume in a such way. However, because there are 46,000 miles of ridges and only 15,000 miles of trenches, this would suggest that the Earth would have to be expanding, otherwise there would be a huge mass-balance problem. Furthermore, why are trenches located primarily in the western Pacific Ocean, whereas ridges are more equally developed over the entire Earth?

Several *mechanisms* have been proposed for Earth's expansion. A change in universal gravity was generally considered the least likely mechanism, and the most popular mechanisms to explain the Expanding Earth were grouped into explanations that allowed the Earth to expand with a constant mass, or were it expanded with an increasing mass. These two major variations of the Expanding Earth theory became known as the Constant Mass Expanding Earth or the Increasing Mass Expanding Earth.

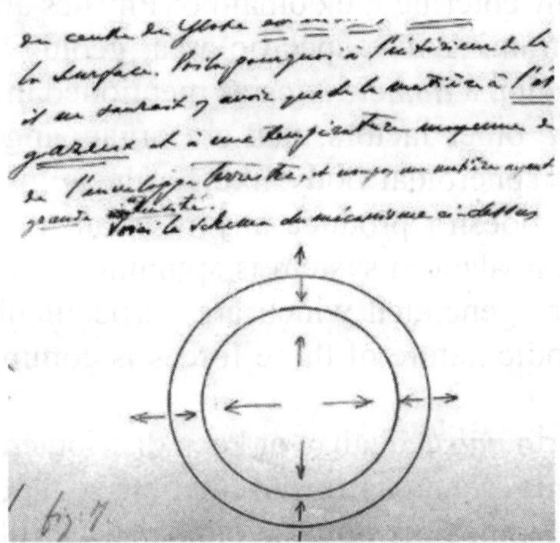

Fig.114- Mantovani's sketch of the mechanism of the planetary expansion by the pressure of the inner Earth's material in a high temperature gasiform-state (manuscript *Dilatation planétaire* of 1932). A hollow Earth along centrifugal force is essential for the phenomenon of planetary expansion to occur. A cavity must be present at Earth's centre to allow a degree of freedom for the thick shell to expand, at an estimate rate of 22mm per year. Since there is evidence Earth's volume has increased 356% its original size, we can be 100% sure Earth is filled with caves, holes or one big void. Think of the blobs found within Earth's mantle (if conservation of mass and conservation of atom volume, are valid).

A striking feature of the oceans and continents today, is that they are arranged *antipodally*: The Arctic Ocean is precisely antipodal to Antarctica; North America is exactly antipodal to the Indian Ocean; Europe and Africa are antipodal to the central area of the Pacific Ocean; Australia is antipodal to the small basin of the North Atlantic and the

South Atlantic corresponds though less exactly, to the eastern half of Asia. Only 7% of the Earth's surface does not obey the antipodal rule.

If the continents had slowly drifted thousands of kilometres to their present positions, the antipodal arrangement of land and water would have to be regarded as purely coincidental. Researchers calculated that there is 1 chance in 7 that this arrangement is the result of a random process. Indeed, it is not. Why? Because it is a matter of *stability*.

Plate Tectonics Theory argues that in a remote past, all the known continents were united in a "supercontinent" called Pangea, which started to split around 200 million years ago. From a geological perspective, it might sound reasonable as a scientific hypothesis. However, besides its geological features, Earth is a dynamic *rotating system.* We are now entering the domain of Physics and its strict laws.

Due to their training in a specific area, geologists like any other scientists, will develop a *tunnel vision* as mentioned in chapter one. This will lead to *ignore* other factors, that are organizing and shaping our planet. A rotating spheroidal body like Earth, is in *dynamic balance* when the rotation doesn't produce any resultant *centrifugal force* or couple. When an unbalanced system is spinning, periodic linear and/or torsional forces are generated which are perpendicular to the axis of rotation. The periodic nature of these forces is commonly experienced as *vibration.*

Therefore, Earth *must* equalize or keep distributed the weight of the combined continents around its axis of rotation, so that it spins smoothly at its roughly 1000 miles per hour, avoiding instability (vibration) and further destruction. It is already known that the *redistribution* of ice-water-vapor masses, currents, winds, hydrology, magnetism, post-glacial rebound, earthquakes, etc. do *affect* the tilt of the axis of rotation of the Earth.

To rotate in a stable way, the Earth's centre of gravity *must* always coincide with its centre of rotation. That's the reason why oceans and continents today are arranged antipodally, from the beginning when Earth originated.

This *Gestaltung* must be always kept. From this perspective, a supercontinent lying in one side of Earth, a very long time ago, would have been *impossible.* It would have displaced the Earth's centre of rotation from its centre of gravity, rendering our planet *unstable* due to vibrations and it wouldn't have lasted much. This is the same physical principle governing the car's wheel balance.

Every time a wheel is first mounted onto a vehicle with a new tire, it must be *balanced*. The goal is to make sure the weight is evenly distributed throughout each of the wheels and tires on a vehicle. This process evens out heavy and light spots in a wheel, so that it rotates smoothly. If there is even a *slight* difference in weight in the wheels, it will cause enough momentum to create a *vibration* in the car.

This is another reason why the Expanding Earth Theory is more acceptable, because it *keeps* the antipodal arrangement of continents and oceans that confers *stability* to the rotating Earth. Even so, Earth axis slightly teeters as observed in the *Chandler Wobble*, a movement in Earth's axis of rotation that causes latitude to vary with a period of about 433 days.

In regards to rotation, if Earth's radius had increased since its inception, to conserve its angular momentum, the Earth's spin must *slow* down through time. This indeed is a fact, tough it is believed is the result of Moon's tidal effects that creates friction, which slows down the rotation of Earth.

On the other hand, the *constancy* of the total area of continental regions (only slightly affected by mountain folding) emphasizes a question for which only a few authors have recognized an importance for an answer. We now ask:

Why does the continental crust only cover part of the Earth's surface and why is it uniformly thick?

Conventional science *cannot* offer a satisfactory explanation of why Earth's crust is shaped in *two* distinct regions: Continents and ocean basins. A sharp boundary between the high-level region, the continental blocks, and the deep sea does exist. There must be a reason for this in the laws governing the structure of the crust, which must be consequence of the history of the Earth.

In fact, Earth Expansion Theory offers a simple answer. Once Earth originated as explained in chapter three, the *thick spherical shell* has consolidated to some extent. On its surface, the Sial layer or continental crust had *already* formed and uniformly covered the whole Earth, though not continuously. This suggests the distinctive *chemical signature* of our continents, was established at the very *beginning* of Earth's history. It probably had chemical features remarkably like today's continental crust.

There were huge grooves covered with shallow seas, that separated the continents we know today. This vast continental crust that cloaked a smaller Earth, was first broken up by *rift valleys* across the globe, where the huge grooves delimited continents or cratons. The first crust likely broke into pieces that became thicker in some areas, forming the beginnings of continents.

The sea between the continents *increased* in width by the continuous addition of new Sima material containing water along the rift, emerging from the depths. Newly discovered *ringwoodite* in the Earth's mantle contains more locked-in water than all the oceans combined together. There is literally enough water trapped in the crystalline structure of ringwoodite emerging from the mantle and then released; to refill all the Earth's oceans a few times over. This makes it much more plausible that the oceans leaked out from inside the cracked expansion zones over millions of years.

This process is *currently* under way. In the case of the continued separation of South America from Africa, is not due to one or both continents "floating" on the underlying Sima material and moving relative to it.

Rather, it is due to the *widening* of the original rift between the land masses and the gradual building up of the ocean bed to the Atlantic, by a swelling up of the underlying magma.

From this perspective, oceans are the *final result* of the expansion of the Earth, covering 71% of its surface, whereas the total area of the continental crustal blocks remain *constant* throughout Earth's history, with the exception of a certain shrinkage produced by mountain folding.

Therefore, the ocean floor must have geological traces left during this expansion. In fact, during the Ocean Survey Program (OSP; 1966-1993) and the World Bathymetry Division (1993-1998) of the US Naval Oceanographic Office, when scientists were mapping the ocean floors, they gradually came to realize that NONE of the predicted physical features of divergent margins and subduction zones existed.

According to the fundamentals of Plate Tectonics Theory, these zones should span 40,000 miles in length, yet not a single mile could be identified. Instead they found *thousands* of under-sea mountains, guyots (drowned atolls), mountain ranges, canyons, and valleys, as well as tens of millions of square miles of relatively flat sedimentary layers that seemed to have existed *undisturbed* for billions of years judging by dated rock samples and detailed maps. It turns out the ocean floors around spreading zones are *ancient*, predating every theorized tectonic scenario of relatively young oceans, less than 200 million years old. (Smoot, 2004). Plate tectonicists swept this issue under the carpet. The fact that the oceanic floor's features don't fit Plate Tectonics Theory is not recent. As part of a 1965 symposium on Continental Drift, Bruce Heezen and Marie Tharp were able to report on their investigations of the Atlantic and Indian ocean floor, revealing new features discovered.

Their paper was titled: *Tectonic fabric of the Atlantic and Indian oceans and Continental drift.* After presenting the evidence, they rightfully concluded that Earth must be expanding:

"A simple convection current patterns is not favoured for the explanation of these features. The tectonic pattern of the Indian ocean floor seems particularly difficult to explain in terms of oceanic spreading. Although there are many unsolved problems and some evidence which apparently is adverse to this concept, the writers believe that a general expansion of the Earth better explains the sea floor tectonic fabric than the recently popular convection current hypothesis"

Apart from Earth's expansion, there is no a single idea which does not presuppose physical impossibilities discussed before, to explain why the Earth is covered today by an incomplete blanket of continental crust (Sial) whose residue has a constant thickness (apart from secondary effects).

Hence the expansion theory, through an unknown mechanism, points the way to the solution of a problem which might be considered one of the great problems of natural science, although for a long time its outstanding significance, has only be realised by a very few research workers. Also, the advantage of the expanding Earth idea is a common *explanation* of several lasting and out-standing *problems* coming from palaeontology, palaeomagnetism, geology and climatology.

Moreover, *gigantic cracks*, due to Earth's expansion, are ripping apart the Pacific Ocean floor, challenging the understanding of Plate Tectonics. The Pacific Plate is being pulled apart by internal faults, not just "subduction", a study finds. For decades, the theory of Plate Tectonics has painted a picture of "rigid plates" gliding across the Earth's mantle.

But a new study by University of Toronto researchers throws a wrench in that image, revealing a Pacific plate riddled with *massive* internal faults that are actively pulling it apart. This finding published in *Geophysical Research Letters* (Gün et al, 2024) suggests that the Pacific plate, which covers most of the Pacific Ocean floor, isn't the monolithic and stable entity we once thought it was.

On the other hand, researchers have uncovered seismic data, suggesting a *split* in the Indian tectonic plate beneath the Tibetan plateau (Liu et al, 2023) The latest analysis, using seismic waves as well as helium isotopes, indicates that the top layer of the Indian plate is not just going under. Instead, it's peeling away like layers of an onion from the denser lower part.

This peeled-off layer then moves horizontally beneath Tibet, possibly causing the region to *sink* and *crack*. In some regions, the bottom of the Indian tectonic plate plunges as deep as 200 kilometres. Conversely, in other areas, it only reaches a depth of 100 kilometres. Seismic data also reveals *widespread fractures* within the Indian plate, possibly foreshadowing future earthquakes. These are typical tectonic features showing Earth's *expansion,* though conventional, brainwashed geoscientists attribute it to the fictitious slab "subduction," which requires the Indian tectonic plate to be warped or torn.

The task remains for future maverick researchers, to find out the *mechanism* of this expansion, based of course on the novel ideas discussed in this book, which will give them a guaranteed success.

Epilogue

The question of what lies beneath our feet has been the subject of much speculation over the course of human history. Virtually every culture and religion worldwide has, at some point, entertained the idea of a subterranean world. The idea of a hollow Earth dates back to at least the 17th century.

The English astronomer and mathematician Edmond Halley suggested in 1692 that the planet was made up of two inner concentric *shells* and an innermost core, with atmospheres separating these shells. Halley envisaged the *atmosphere* inside as luminous (and possibly inhabited) and used the theory to explain earthly phenomena, like the Aurora Borealis, one in which he was right.

However, conventional science claims Halley's theory was disproven in the following century: first, tentatively, by Pierre Bouguer in 1740 and then definitively by Nevil Maskelyne and Charles Hutton around 1774 with the *Schiehallion experiment*. This study attempted to determine the *mean density* of the Earth, from which it was believed its mass and answers about its mysterious interior could be extrapolated.

By measuring the gravitational attraction of Schiehallion, a mountain in the Scottish Highlands, the researchers determined that the Earth is nearly twice as dense as the mountain on its surface, essentially "debunking" the Hollow Earth Theory.

They jumped to the conclusion Earth's interior is denser than its surface. If our planet were hollow, its density would be lower, not greater, than the density of its crust, they reasoned. To top it off, later on fantastic high pressures and temperatures inside Earth, worthy of a science fiction novel, were taken for granted.

This in reality proves *nothing* because as it was shown, the same mean density can be attributed to both: a solid and a *hollow* sphere (different mass distribution). Both spheres can account for the gravitational pull experienced on the Earth's surface.

Don't be fooled by the claim trumpeted by scientists that satellite technology and space missions, which have provided detailed maps and gravitational data of the Earth; do "align" with the model of a solid planet with distinct layers. They don't. The constant use of different methodologies has been shown to result in large *variations* in crustal thickness even when using the *same* data as source. The non-uniqueness of satellite gravity *interpretations* has been usually reduced by using *a priori* information from seismic tomographic *models*. Can you spot the fallacy of this argument, purporting to elucidate Earth's interior?

Mainstream science believes the inaccessibility of Earth's interior makes the depths of our planet so fascinating to people. Then claims that people conjure fantasies about what lies there. But hey, scientists have fantasies too! *disguised* as assumptions and unproven hypotheses leading to false theories. Even so, they *pretend* to have a pretty good idea of what is going on inside our planet. What a joke!

The first scientific hypotheses of Earth's interior were probably based on the fact that molten lava erupts from volcanoes, which suggested (erroneously) the idea that the interior of our planet is red-hot and molten.

Modern science holds for sure that the Earth is an unbroken series of layers, crusts, and liquid magma surrounding a dense, molten core made primarily of iron and nickel. But, haven't we seen that the concept of Earth's molten iron core is a *myth*? That it is physically *impossible* such thing composed of iron and nickel, can ever exist? Check it out.

It is claimed that one of the most compelling pieces of evidence that contradicts the Hollow Earth Theory comes from the study of *seismic waves* generated by earthquakes (Seismology). These waves travel through the Earth and are recorded by seismometers worldwide.

Seismic waves move through them at different speeds, allowing geologists to figure out what type of material the waves are going through. This is *false*. Because seismic velocity depends on the density and elastic properties of the material, *regardless* of its composition.

It is well known that *different* materials can have the *same* seismic velocity. So how the heck you can tell which one is? How you can *distinguish* between fluids such as liquid or gas if seismic waves S cannot propagate trough both? Why are you so picky in choosing a *liquid* phase instead of the gaseous phase for the "outer core"? Because it is more convenient?

By themselves, seismic velocities alone are not particularly *diagnostic* regarding rock type. Didn't the Kola Deep Borehole proved geologists and geophysicists wrong, in regards of the basaltic layer "expected" to be found after certain depth?

According to the overhyped seismic velocities marvel, that should have been the case but instead, *metamorphized granite* was found. What a blunder! The "study" of Earth's mantle is by tradition the province of seismology while mantle composition has historically been a subject for computer models or *simulation*s of fanciful high-pressure and high-temperature mineralogy, petrology and geochemistry; with claims that important advances (educated guesses) have been made in these areas. But wait. Computer models can be designed to show anything. The so-called "scientists" can control the inputs in order to achieve the desired output, a sort of circle reasoning. No real, *observable* and *measurable* data is involved. Also, there is *little* to be known from seismographs and regarding mantle's composition: has anyone taken a *sample* of it?

Because volcanic rocks are inconclusive. These are used to *extrapolate* their mineralogical composition into Earth's mantle. Any honest scientist knows that extrapolation is *unreliable.* Extrapolation of trends can be *very risky*, since you may have no very good way to judge the *suitability* of the model outside what data you have.

The closest opportunity to get a sample of Earth's mantle was attempted with the Mohole Project, before its cancellation in 1966. Without sampling is *impossible* to know for sure, the exact composition of the mantle, let alone the "core." Only speculations can be made, write countless *pseudoscientific* papers, to justify money grants and sell them as "science."

All the raw seismological data with its *theoretical* computed (not measured) seismic velocities, must be *interpreted* to make sense of it. This is what scientists do to "figure out" the hidden properties of Earth's deep interior from surface measurements. Scientists ignore or dismiss the fact that they are always faced with the problem of *non-uniqueness* of their conclusions.

However, for them there is just *one* and only one conclusion: to fit the raw data in the framework of the Earth having layers (discontinuities), *assuming* a liquid magma surrounding a dense, molten core made primarily of iron and nickel. Once this assumption is accepted, a *circular reasoning* arises to "prove" this assumption of Earth being solid. Thus, there is no hollow Earth or any planet for that matter.

Geoscientists while struggling to know what's inside Earth, often neglect what is known as the *inverse problem,* which means: We begin with the results and then calculate the cause. This is the opposite of the forward (or direct) problem, where we begin with the causes and calculate the results.

For example: Given the known (observed) gravitational field on the Earth's surface, what is the density distribution in the Earth's interior? The point is that the inverse gravimetric problem is non-unique. In other words, given a measured gravitational field on the Earth's surface, there are *many* possible density distributions that could exist in the Earth's interior which would generate that *particular* gravitational field.

Methods exist to "solve" the inverse problem. However, when we find a solution, we can only say that the solution we have found is just one of many possible solutions. But we *canno*t be certain what the real density and mass distribution inside the Earth is. It could be solid, hollow or a combination of both. However, we are being told it "increases" with depth so, Earth is "solid". There are ways to help reduce the ambiguity of such inversion solutions, including the use of "constraints". In Seismology, constraints are nothing more and nothing less that *assumptions*. Pure guesswork.

At this point is necessary to outline the state of *bankruptcy* in which most of current science is, because of being immersed in the realm of *fantasy* and *incompetence*. This is an obstacle to really know how Earth originated and got its inner structure.

It is claimed that planets formed by "accretion" from a nebular disc (proto Solar System), in which dust and gas *gravitated* together and "coalesced" to form ever larger bodies. The Solar System's planets grew as small grains *colliding chaotically*, coalescing into bigger ones, colliding yet more until they formed *planetesimals*. The planetesimals then collided until they formed "solid planets" as varied as the Earth and Jupiter. Has anyone ever seen this messy collision theatre, happen in today's protoplanetary discs?

This is just another pathetic academic idea (assumption or unproven hypothesis) and there is NO real evidence found in outer space, that this is how planets are formed, in other words it doesn't comply with the observable *Gestaltung*. a shaping and organizing mechanism in which planets are really forming as per recent research, acquiring a *spherical shell,* a stable and efficient structure; as the Juno mission to Jupiter did prove. Nonetheless, we still have professors parroting this *delusional* accretion and collision nonsense, to their servile students without realizing they have been deceived.

Scientists are among the few professionals who can keep their jobs when proved *wrong*. If engineers build a bridge that collapses, they can be held accountable or sued. Some scientists, though, just rearrange their *ignorance* to fit new data. This is being mediocre.

The motive for this dishonest conduct is rather obvious, it's *money*. Society expends a large sum supporting these "experts" through tuitions paid by parents and *grants* by governments and industry. The donors naïvely believe they are paying for the "teaching" of the young when in fact, they are being *indoctrinated* with pseudoscience.

Innate talent, outstanding mental capabilities and a strict adherence to the *scientific method;* were necessary for leadership in the natural sciences but, for several decades ago (since 1905 to be exact), the rules of a *dogma, ideology* and *mediocrity* have been imposed on the natural sciences and these have been degraded. Thus, anyone can become an "expert." Key scientific facts are either dismissed or supressed by "science commissars", because they don't fit the Party line. Therefore, it became possible for the untalented (incompetent) majority to gain ascendancy.

This becomes a *problem* when the expectations of science students and employers of science graduates seem to be reshaping science education and redefining science work, to make them more relevant to industry's needs. But the *skills*, *attitudes*, and *values* required for science work in industry are *not* been taught. Only *ideological* nonsense. As a result, science teaching is *not* addressing the challenges of preparing science students for a *practically* and socially significant role in industry.

The author faced this situation while working on the Aberdeen Harbour Extension Project (AHEP) in Scotland. Local "geologists" the ones responsible of providing the geological and geotechnical report, to be used for piling foundations, made an *inadequate* and *substandard* ground investigation and geology of the area. The result was that ten constructions of piles *collapsed* on site, each one with an average cost of £ 200,000 (US$ 258,000) totalizing £ 2,000,000 (US$ 2,580,000) in losses. The reputable foreign construction company DRAGADOS in charge of the project, was ordered to leave, without being *directly* responsible of these failures.

However, in planetary science if we can call it that, the situation is different. One might think "experts" would be at risk when spacecraft explore worlds that look very different than expected. But no; sometimes they just let their silly-putty theories morph to fit the new findings, and carry on with their jobs as "experts".

How many decades have planetary scientists taught in textbooks and college courses that the Nebular Hypothesis explains the arrangement of planets in our Solar System? The leading theory explained why we have rocky planets close to the Sun, and gas giants farther out. Indeed, our Solar System *had* to evolve this way because of the "frost line" that burned away volatiles close to the Sun (leaving the rocks) and put them out into the outer orbits. This arrangement was dramatically *falsified* when the Kepler spacecraft began tallying up exoplanets with very different arrangements. So many stars were found with "hot Jupiters" (gas giants close to their star), our system began to look unusual.

Thinking people want *accountability*. Why should we *trust* their expertise when they were so *wrong* for decades? These "experts" punt any theoretical explanations for these arrangements into futureware, and *boast* about how the new scheme opens up opportunities for more storytelling. "Now, for the first time, we have a tool to study planetary systems as a whole and compare them with other systems." They get to keep their jobs.

Over the past decade, the gas giants Jupiter and Saturn have been closely studied by the Juno and Cassini spacecraft, respectively. In their Commentary, Helled & Stevenson [2024] summarize the main findings of these campaigns, which result primarily from measurements of gravity and, for Saturn, ring seismology (Mankovich, 2020).

The authors focus on the fact that Jupiter and Saturn *do not* simply consist of a compact, rocky core beneath a hydrogen-rich envelope as early models *assumed*. Instead, their cores appear to be "fuzzy" (hollow) and large, extending out to about half the planetary radius. How to explain these fuzzy cores is a puzzle. They continued saying: "We therefore conclude that it is very challenging to explain Jupiter and Saturn dilutes core from standard formation models".

Notice the sleight of hand in the way the challenge is worded. It is a challenge to the models, not to the *experts*! This straightforwardly *eliminates* accountability to the human beings who made up the models. Don't they deserve to be called to account for being wrong? Doesn't the *public* have a right to know whose models have been *challenged*? If the models were wrong, shouldn't *new* experts be hired to come up with new models, even if *non-standard*? Would you give the same job in industry to someone whose bridge collapsed, and invite him to make another bridge?

And what about the purported existence of Earth's "iron core" that allegedly, is the source of geomagnetism. According to the so-called experts: "Using constraints from experiments, simulations, and theory, we show that spontaneous crystallization in a homogeneous liquid iron alloy at Earth's core pressures requires a critical supercooling of order 1,000 K, which is too large to be a plausible mechanism for the origin of Earth's inner core made of iron. … yet a sort of inner core structure exists: this is the nucleation paradox." Scientists regularly remain *quiet* about a paradox in their perspective until hopefully, some researcher or group of researchers think they have a potential "solution" to the problem and keep the Earth's iron core *myth* going. The paradox of solidification of the inner core is a case in point. The researchers consider several mechanisms, all with caveats.

A further problem is that the growth of the inner core is considered to be the primary driver or energy source for Earth's "geodynamo" by causing convection in the liquid outer core. But this could not be possible because the alleged iron core, *cannot exist*. However, an iron core is one of the silly excuses used by scientists, to justify a "solid planet" and dismiss an alternative distribution of mass such as a *thick spherical shell*. However, Juno's mission to Jupiter proved them quite wrong. Because of being hollow (fuzzy core) Jupiter cannot have any dynamo, which *demolishes* the conventional "explanation" of the origin of its magnetic field.

The existence of Earth's *magnetic field* during eons is a long-standing problem for the "experts" If the inner, solid core cannot form, the dynamo hypothesis becomes *implausible*. On the other hand, Jupiter and Saturn are characterized by their *intense* magnetic fields, which create a magnetic cavity, the magnetosphere.

Due to the fact that these planets are *hollow*, NO dynamo hypothesis is needed at all, to account for the presence of magnetic fields. Stochastic Electrodynamics (SED) can explain them nicely, upon well-established scientific grounds.

Make no mistake, Academia is bankrupt because of publishing for the sake of doing so (paper is their currency); the low, substandard quality of research being carried out and the lack of knowledge of the *real world*. Earth is hollow and expanding. Get used to it.

Case closed.

THE END

Appendix I

A nineteen-century description of polar fountains

Few are aware that in 1898 a 48-page booklet was produced entitled *The Secret of the Poles*. The author was a certain Henry Campion and his treatise was published in Birmingham, England.

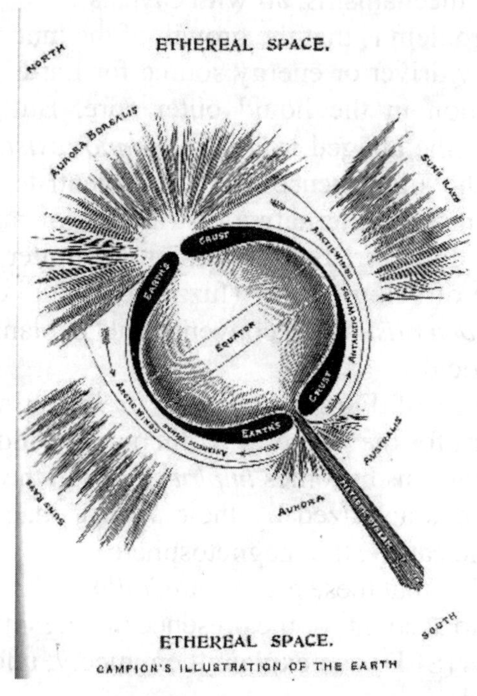

CAMPION'S ILLUSTRATION OF THE EARTH

Campion wrote that the Earth was hollow with entrances at both the north and south poles. To quote Campion: "The Earth's axis is completely hollow, and entirely devoid throughout – from south to north – of *terra firma*; it has two complete openings, one at the South Pole, the other at its exact opposite, the North Pole, it is continuously hollow, and completely open throughout its supposed dense structure from pole to pole, quite literally and completely."

Campion's illustration accurately describes the polar fountains that causes auroras and were discovered by NASA satellites, a century later in the North Pole.

Also, this illustration shows the cause which keeps the scientific community saying that something mysterious is coming up from the frozen ground in Antarctica, claiming it could break physics as we know it. The mysterious signal was detected by the Antarctic Impulsive Transient Antenna (ANITA). The *problem* is that if the signal was astrophysical, whatever it was (cosmic rays. "neutrinos," etc.) it had to pass through a huge chunk of the Earth.

Physicists are still wrestling with strange signals coming in from the ground that, to date, have *defied* easy explanation. Some have proposed that the anomalies arose from radio waves bouncing through caverns or buried lakes in the ice. Other theorists proposed more exotic ideas, such as that the heavy, high-energy particles in line with ANITA's data may describe one candidate for dark matter, the mysterious stuff that's believed to make up 85 percent of the matter in the Universe but has never been detected. Still others hypothesize that the exotic particles fit an existing theoretical model of a "parallel Universe."

The fact is that, a beam of high-energy particles ejected by the *inner Sun* or Earth's plasma core, is the real cause of this anomaly detected in Antarctica. In other words, a polar fountain is the origin of this something mysterious, that is coming up from the frozen ground in the vicinity of the South pole.

Appendix II

The impossibility of the hydrostatic assumption in Earth's structure

Elevations and depressions, mountains and valleys, continents and ocean basins, are rendered possible by the property of *rigidity* of Earth's crust and mantle, which together make a spherical thick *shell*. The impo-

Box 3.1 Collapse of topography on a strengthless planet

Consider a long mountain ridge of height *h*, width *w* and effectively infinite length *L* standing on a wide, level plain. For simplicity suppose that the profile of the mountain is rectangular, with vertical cliffs of height *h* bounding both sides (Figure B3.1.1). The surface gravitational acceleration of the planet on which this mountain lies is *g*, and ρ is the density of the material from which both the mountain and planetary surface are composed.

The weight of the mountain is $\rho ghwL$. If there is no strength, this weight (force) can only be balanced by the inertial resistance of material accelerating beneath the surface, according to Newton's law $F = ma$. The driving force *F* equals the weight of the mountain, $F = \rho ghwL$. The acceleration *a* is equal to the second time derivative of the mountain height, $a = \frac{d^2h}{dt^2}$. The mass being accelerated is less easy to compute exactly, but it is approximately the mass enclosed in a half cylinder of radius *w/2* beneath the mountain (this neglects the mass of the mountain itself, which is not strictly correct, but if *h* is small compared to *w*, the mountain mass is only a small correction). The mass is then $m \approx \frac{\pi}{8} w^2 L \rho$. This yields a simple, second-order differential equation for the mountain height *h* as a function of time, *t*:

$$\frac{d^2h(t)}{dt^2} = \frac{8}{\pi}\frac{g}{w}h(t). \qquad \text{(B3.1.1)}$$

This equation has the solution

$$h(t) = h_o e^{-t/t_{\text{collapse}}} \qquad \text{(B3.1.2)}$$

where h_o is the initial height of the mountain and the timescale for collapse is given by:

$$t_{\text{collapse}} = \sqrt{\frac{\pi}{8}\frac{w}{g}}. \qquad \text{(B3.1.3)}$$

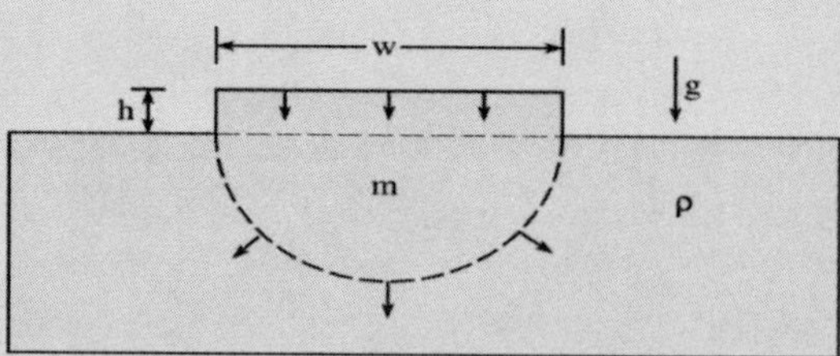

Figure B3.1.1 The dimensions and velocity of a linear collapsing mountain of height *h* and width *w* on a strengthless half space of density ρ that is compressed by the surface gravity *g* on a fluid planet. As the mountain collapses vertically it drives a plug of material of mass *m* underneath it that flows out through the dashed cylindrical surface.

rtance of *strength* is highlighted by a simple computation that Sir Harold Jeffreys included in his masterpiece, The Earth (1970). This computation is summarized in Box 3.1, where it is shown that, without *strength*, a topographic feature of breadth *w* would disappear from the surface of a planet in a time $t_{collapse.}$

Without strength, a mountain 10 km wide on the Earth would *collapse* (literally) in about 20 seconds, and a 100 km wide crater on the Moon would disappear in about 3 minutes. Clearly, such features can and do persist for much longer periods of time. Planetary topography, and the material strength that makes it *possible*, lend interest and variety to planetary surfaces.

However, when seen from a distance, it is clear that the shapes of planets are, nevertheless, very close to spheroids. Only very small asteroids and moons (Phobos and Deimos are examples) depart greatly from a spheroidal shape in equilibrium with their rotation or tidal distortion. Thus, although the strength of planetary materials (rock or ice) is adequate to support a certain amount of topography, it is evidently *limited.*

Such things as 100 km high mountains do not exist on Earth because strength has limits. the ultimate extremes of altitude on a planet's surface are regulated by the antagonism between the *strength* of its surface materials and its gravitational field.

The fact that planets have substantial *non-hydrostatic* contributions to their figures plays an important role in studies of *rotational dynamics* and the tidal evolution of bodies in the Solar System. For example, the present orientation of the Moon with respect to the Earth is due to the distribution of *dee* lavas in the low-lying basins on the nearside. The orientation of Mercury can be controlled by mass anomalies associated with the Caloris Basin and if Europa has too large a non-hydrostatic figure, then its putative slow non-synchronous rotation *cannot* occur.

Appendix III

Jupiter has a fuzzy, dilute core. In other words, this planet is hollow

For decades, planetary science was firm on the fact that Jupiter has a rocky core. New data extracted directly from the king of planets by the Juno spacecraft in 2017, *challenged* that long-held standard. New research shows that Jupiter's core is not very compact at all.

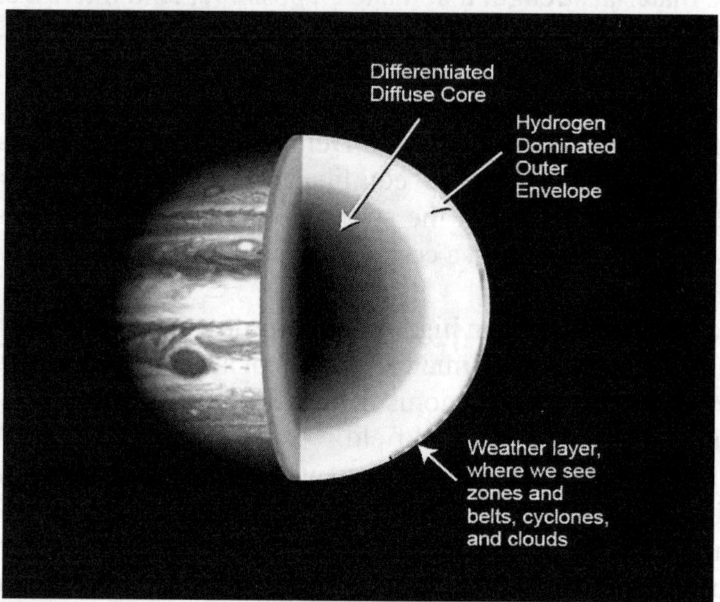

These new results imply that current models of Jupiter's formation and evolution require a *revision*. Jupiter internal structure models that *fit* Juno's gravity *data* indicate that the planet is not homogeneously mixed and that its core is *diluted* from a pure composition of heavy elements. It is very challenging to explain Jupiter's dilute core from standard formation models, based on the accretion-collision hypothesis.

Juno's mission data suggests that Jupiter's dilute core extending to several tens of percent of its total radius, perhaps even beyond half the radius. The core region is extended with a composition that may contain a similar number of heavy elements by mass but is so diluted that hydrogen remains the most abundant material by number and might even dominate by mass.

For ordinary astronomers and astrophysicists Jupiter harbors a deep mystery: Rather than the distinct dense compact core of ice, rock and metal they expected; it has a fuzzy (hollow) center and it lacks a well-defined boundary. This fact *demolishes* the arguments trying to debunk the hollow Earth and hollow planets in general. It is said that any hollow space could not exist without "collapsing" due to intense gravitational pressure in Jupiter's interior. The fact is that Jupiter has a very *stable* spherical shell nearly half of its radius, *contradicts* this ridiculous claim.

The current theory, of gradual accumulation of mass from particles, dust grains and "planetesimals"; does not really explain how this fuzzy Jupiter's "core" came into being. The belief is that gas giants like Jupiter initially form in the same way as terrestrial planets like Earth until their cores become sufficiently massive to begin accreting hydrogen and helium, the two lightest and most abundant elements. Hydrogen and helium will accumulate slowly as the planet heats up, but eventually the planet reaches a critical mass at which hydrogen and helium accrete. The mechanism of how two gases can "accumulate" under *hot* conditions is undisclosed. Kinetic Theory of Gases is clear: Heat makes gases *expand* and *dissipate*. But the so called "experts" just don't care about physics.

This "accretion" would imply the formation of a dense core in Jupiter which obviously it *didn't* show up. Finally, scientists aren't entirely sure why the Jovian magnetic field is so *powerfu*l. One theory claims that, deep within Jupiter (core), a layer of metallic hydrogen is crunched to such a high pressure that it serves as a sort of electrical conductor. However, Jupiter's magnetic field *cannot* be generated by dynamo action at depth (beneath 0.81 Rj) in convective metallic hydrogen; because at this depth, Jupiter's core is fuzzy or *hollow*. No dense metallic hydrogen can be present, to sustain "convection" and induce any magnetic phenomena. This *disproves* dynamo theory as the origin of Jupiter's magnetic field.

Again, Juno's mission to Jupiter and its finding of a dilute (hollow) "core" poses a serious trouble for astronomers and astrophysicists. It *shatters* their cherished credo for planets and stars in general, in which the force of gravity is constantly pulling every object inward, to the center of mass, if any. This is their basis of planetary differentiation, to explain the origin of (solid) planets and refute the Hollow Earth Theory.

Differentiation is the phenomenon of substances settling according to their densities (ρ). According to the obsolete and inaccurate paradigm of planetary formation (collision-accretion hypothesis); during the planetary formation stage of the early Solar System, the process of accretion produced homogeneous objects (planetesimals) which had the same composition throughout. Materials of different densities were able to co-exist throughout these bodies until a certain size (a diameter of approximately 200 km) was reached. At this point, the internal heat (generated through "gravitational compression", energy from impacts, radioactive decay, and perhaps tidal forces) was sufficient to "melt" the interior of the object (protoplanet) or so the "experts" say.

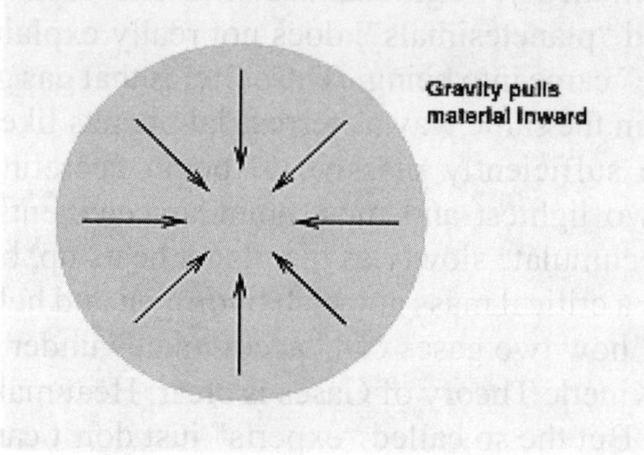

When this happened, the densest material "sunk" towards the center to form a core, while the lighter material floated to the surface to form a mantle and crust in a process called 'gravitational differentiation'. This planetary differentiation is believed to be a crucial process that shapes the internal structure and composition of planets, including the Earth, and other celestial bodies. This process of course, is a *myth*. It is not *observed* happening, anywhere in Nature.

It ignores the *rotation* of planets. Since its inception in a *vortex* of dusty plasma, a planet spins around its axis and all objects within the radius of the planet, experience an *outwards* centrifugal force. Therefore, nothing can "sink" inwards to the planet's center. Also, the Newton's *Shell Theorem* is neglected, there is no such thing as something in the center of a planet, pulling things radially downwards by gravity. Everything above this center will pull it (upwards) too, resulting zero gravity in the center of a planet.

Jupiter's inner "core" is exactly the *opposite* of what scientist thought it might be. It doesn't comply with the current paradigm of planetary formation at all, emphasizing solid planets. Instead they are with some variations (terrestrial or Jovian), practically *hollow*.

Appendix IV

THE AUDIBILITY AND LOWERMOST ALTITUDE OF THE AURORA POLARIS
Chapman, S.; *Nature*, 127:342, 1931.

Some of the accounts of low aurorae seen between the observer and terrestrial objects are very striking and circumstantial. One writer states that he and his party (members of a government radio station, in the winter of 1917-18) were enveloped in "a light mist or fog-like substance in the aurora"; a hand extended could be seen as if in a coloured fog and a kaleidoscope of colours was visible between the hand and the body. It was impossible to feel this visible fog or mist, and there was no dampness. By stooping close to the ground it was possible to see under this light, which did not go below four feet from the ground. The low-hung aurora lasted fifteen minutes, while great streaks and shafts of light came and went in the heavens. The occasion was unique in the writer's experience.

Aurora at earth's surface

Another writer, during a brilliant auroral display saw the light "play" down between himself and a steep glacial deposit bank about 125 feet away; he "stepped right into the aurora". Another saw an aurora between himself and a ten-foot bank no more than one half-mile away; there was "immense light", "on the very surface", and a shaft of light shot up to an immense height. Another in 1915 saw the familiar landscape "beautified by changing coloured lights which came and went rapidly, now bathing us, now withdrawing with a swish to a height of a mile or so"; "it came down in streamers, or again as a fog (only much faster than fogs come, only a matter of seconds)"; "we saw the tree trunks through the aurora", the colours being mostly, though not wholly, greens.

These letters make it difficult to deny that aurorae occur, very rarely, quite near the earth and are sometimes accompanied by noises. Almost all the reports come from a belt of country about three hundred miles wide, lying roughly along the auroral zone; the belt includes the Klondike region, where is situated the township of Dawson, with a population of several thousand; several reports come from this neighbourhood. The only other places along the auroral zone, likely to afford such favourable fields of inquiry, appear to be near Churchill, on Hudson Bay; the extreme south of Greenland; Iceland and the most northerly part of Norway.

Since very low aurorae seem to be very rare and to be confined to localities near the auroral zone, it is perhaps not surprising that Stormer, Vegard, and Krogness should have observed no such case, or that the Polar Year (1882-83) failed to provide evidence establishing their existence; with the better organisation of auroral observation which it is hoped to achieve during the proposed new Polar Year (1932-33), there is more chance that opportunities of critical examination of these appearances will occur.

These low aurorae must obviously be very different in character from those observed in the upper atmosphere, though connected with them. Inability to understand their physical nature is not a sufficient ground, in the present state of knowledge, for rejecting the possibility of such occurrences; such an attitude would, for example, forbid acceptance of the reality of globular lightning. The observations cited by Mr. Johnson constitutes at least a case for active further inquiry, and renders it highly desirable that auroral investigation near the auroral zone should include not only visual and photographic observations, but also atmospheric electric registration at a well-equipped observatory. (Nature, 127:341-342, 1931)

Appendix V

Archimedean Proof of the Physical Impossibility of Earth Mantle Convection

by

J. Marvin Herndon
Transdyne Corporation
San Diego, CA 92131 USA
mherndon@san.rr.com

Abstract: Eight decades ago, Arthur Holmes introduced the idea of mantle convection as a mechanism for continental drift. Five decades ago, continental drift was modified to become plate tectonics theory, which included mantle convection as an absolutely critical component. Using the submarine design and operation concept of "neutral buoyancy", which follows from Archimedes' discoveries, the concept of mantle convection is proven to be incorrect, concomitantly refuting plate tectonics, refuting all mantle convection models, and refuting all models that depend upon mantle convection.

1 Introduction

Discovering the true nature of continental displacement, its underlying mechanism, and its energy source are among the most fundamental geo-science challenges. The seeming continuity of geological structures and fossil life-forms on either side of the Atlantic Ocean and the apparent "fit' of their opposing coastlines led Antonio Snider-Pellegrini to propose in 1858, as shown in Fig. 1, that the Americas were at one time connected to Europe and Africa and subsequently separated, opening the Atlantic Ocean (Snider-Pellegrini, 1858).

Fig. 1 The opening of the Atlantic Ocean, reproduced from (Snider-Pellegrini, 1858).

Half a century later, Alfred Wegener promulgated a similar concept, with more detailed justification, that became known as "continental drift" (Wegener, 1912). Another half century later, continental drift theory was modified to become plate tectonics theory (Dietz, 1961;Hess, 1962;Le Pichon, 1968;Vine and Matthews, 1963).

Any theory of continental displacement requires a physically realistic mechanism and an adequate energy source. In 1921, Bull suggested the idea of mantle convection being involved in mountain formation (Bull, 1921). In 1931, Holmes elaborated upon the concept of mantle convection and suggested it as a mechanism for continental drift, publishing the illustration reproduced as Fig. 2 (Holmes, 1931). Mantle convection was later adopted as an inextricable part of plate tectonics theory, as illustrated by the U. S. Geological Survey diagram reproduced as Fig. 3.

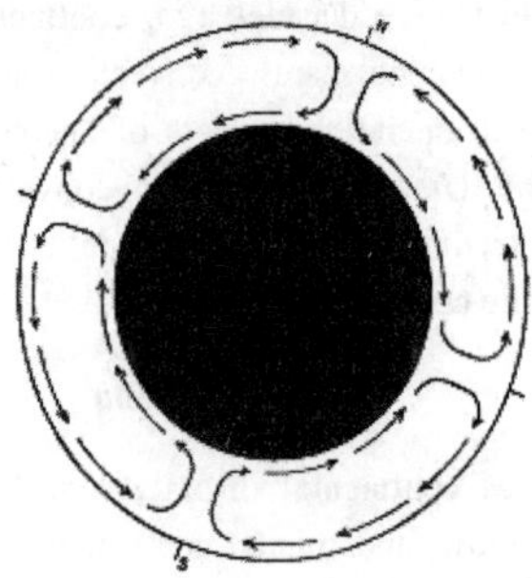

Fig. 2 Schematic representation of mantle convection, from (Holmes, 1931). Reproduced with permission of the Geological Society of Glasgow.

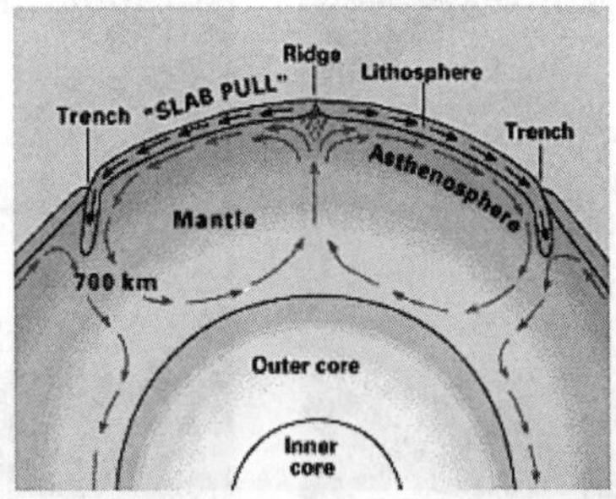

Fig. 3 U. S. Geological Survey schematic representation of mantle convection associated with plate tectonics theory.

2

For eight decades, mantle convection has been taken for granted, its existence assumed without proof and justified by (flawed) numerical calculations. Herndon (2009, 2010a, b) first discovered the physical impossibility of mantle convection, revealed the widespread mis-application of the Rayleigh Number to the Earth's mantle, and pointed to flaws underlying mantle convection models.

Modeling mantle convection typically involves complex computer-mathematical calculations based upon approximations applied to the Eulerian equations of fluid dynamics and generally begins with the (false) assumption that mantle convection occurs in nature. Whereas sophisticated calculations may appear elegant and impressive, their use can be misleading, especially when parameterization techniques are applied. Rather than moving toward increased complexity and abstraction, the author attempts to move toward increased simplification and ease of understanding, reducing a problem to its fundamental elements, intimately connected to the properties and behavior of matter. In the following section, the author proves in a new, readily understandable, yet mathematically precise way, that convection in the Earth's mantle is physically impossible. The author then points to popular research topics that are no longer valid as a consequence of their being critically dependent upon physically-impossible mantle convection, and refers to a different development of geodynamics which is independent of mantle convection.

2 Physical Impossibility of Mantle Convection

The lava lamp, invented by Smith (1968), affords an easy-to-understand demonstration of convection at the Earth's surface. Heat warms a blob of wax at the bottom, making it less dense than the surrounding fluid, so that the blob floats to the surface, where it loses heat, becomes denser than the surrounding fluid and sinks to the bottom. The lava lamp model is not applicable to the Earth's mantle due to compression caused by over-burden weight (Herndon, 2009, 2010a, b).

Earth-mantle density as a function of radius is shown in Fig. 4. Because of the weight of the rock above, the mantle is about 62% more dense at the bottom than the top. The Rayleigh Number, as discussed by Herndon (2009), was derived on the basis of constant density and is thus inappropriate to apply to the Earth's compressed mantle.

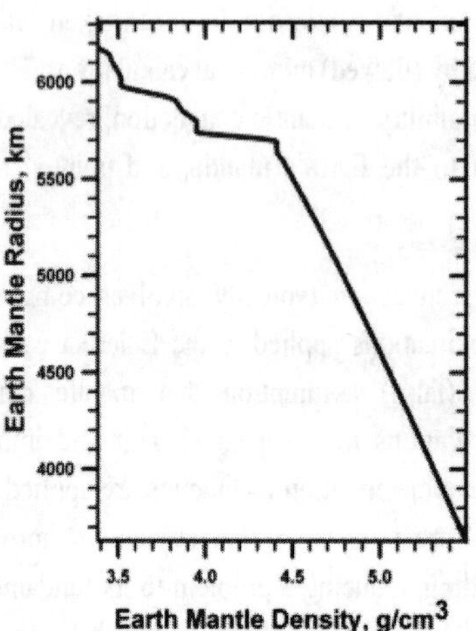

Fig. 4 Mantle density as a function of Earth radius. Data from Dziewonski and Anderson (1981).

Consider a "parcel" of matter at the base of the Earth's lower mantle existing at the prevailing temperature, T_0, and having density, ρ_0, indicated by the data upon which Fig. 4 is based (Dziewonski and Anderson, 1981). Now, suppose that the "parcel" of bottom-mantle matter is selectively heated to temperature ΔT degrees above T_0. The "parcel" will expand to a new density, ρ_z, given by

$$\rho_z = \rho_0 (1-\alpha\Delta T)$$

where α is the volume coefficient of thermal expansion at the prevailing temperature and pressure.

Now, consider the resulting dynamics of the newly expanded "parcel". Under the assumption of ideal, optimum conditions, the "parcel" will suffer no heat loss and will encounter no resistance as it floats upward to come to rest at its "neutral buoyancy", the point at which its own density is the same as the prevailing mantle density. The Earth-radius of the "neutral buoyancy" point thus determined can be obtained from the data upon which Fig. 4 is based; the "maximum float distance" simply is the difference between that value and the Earth-radius at the bottom of the lower mantle.

The relationship between "maximum float distance" and ΔT thus calculated for the lower mantle, shown in Fig. 5, proves conclusively the physical impossibility of lower mantle convection. At the highest ΔT shown, the "maximum float distance" to the point of "neutral buoyancy" is <25 km, just a tiny portion of the 2900 km distance required for lower mantle convection, and the 3750 km required for whole-mantle convection. Even with the assumed "ideal, optimum conditions" and an unrealistically great $\Delta T = 600°K$, an error in the value of α by two orders of magnitude would still not cause the "maximum float distance" to reach 2900 km, the top of the lower mantle.

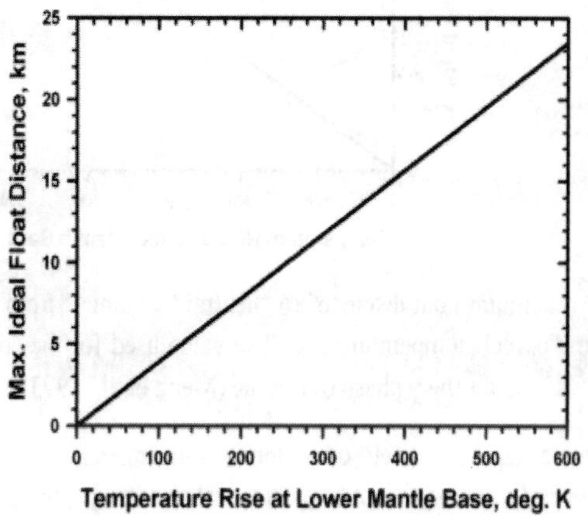

Fig. 5 The "maximum float distance" to "neutral buoyancy" from the base of the lower mantle as a function of "parcel" temperature rise. The value used for the coefficient of thermal expansion, $\alpha=0.37\text{x}10^{-5}\ K^{-1}$, is from the standard reference state value of $MgSiO_3$ perovskite (Oganov et al., 2001), reduced by 80% to take into account lower mantle base temperature and pressure, according to (Birch, 1952).

The same reasoning can be applied to prove the physical impossibility of upper mantle convection. Here the upper mantle is taken to begin at the seismic discontinuity at radius 5700 km, which separates the endo-Earth (lower mantle plus core) from the matter above.

The relationship between "maximum float distance" and ΔT thus calculated for the upper mantle, shown in Fig. 6, proves conclusively that convection is physical impossibility in upper mantle. Even with the assumed "ideal, optimum conditions" and an unrealistically great $\Delta T = 600°K$, an error in the value of α by an order of magnitude would still not cause the "maximum float distance" to reach the top of the upper mantle. Upper mantle convection, like lower mantle convection, is physically impossible.

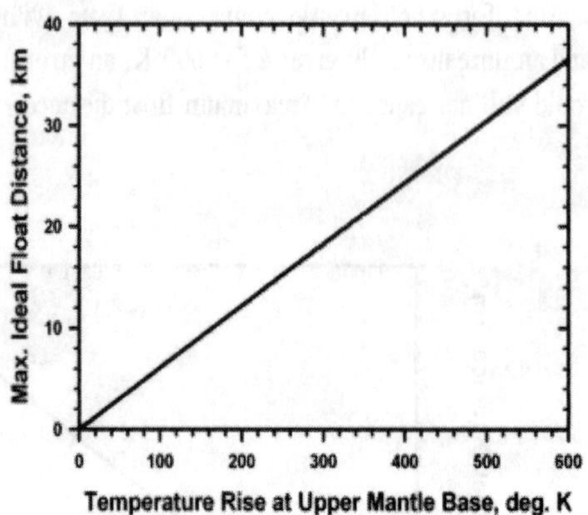

Fig. 6 The "maximum float distance" to "neutral buoyancy" from the base of the upper mantle as a function of "parcel" temperature rise. The value used for the coefficient of thermal expansion, α=1.71x10^{-5} K^{-1}, is for the γ phase of olivine (Meng et al., 1993).

The idea that a heated "parcel" of bottom mantle matter, under ideal, optimum circumstances will float upward to come to rest at its "neutral buoyancy", the point at which its own density is the same as the prevailing mantle density, is based upon the well-founded concept of submarine operation and design (Kormilitsin and Khalizev, 2001), which follows from Archimedes' discoveries. The abstraction is in assuming ideal, optimum conditions, i.e., the heated "parcel" will suffer no heat loss and will encounter no resistance as it floats upward. Because the mantle is solid, not liquid, the heated "parcel" would in fact suffer heat loss and encounter resistance; thus, the "maximum float distance" shown in Fig. 5 and in Fig. 6 would be severely reduced.

As proven above, mantle convection is physically impossible; thus, all investigations that depend upon or assume mantle convection are concomitantly refuted, which includes plate tectonics theory (Dietz, 1961;Hess, 1962;Le Pichon, 1968;Vine and Matthews, 1963), mantle convection models (Schubert et al., 2001), and models which assume mantle convection, such as heat emplacement at the base of the crust involving convection (Phillips and Coltice, 2010). Herndon has presented a fundamentally different development of geodynamics which does not require mantle convection (Herndon, 2010a, b).

References

Birch, F.: Elasticity and constitution of the Earth's interior, J. Geophys. Res., 57, 227-286, 1952.

Bull, A. J.: A hypothesis of mountain building, Geol. Mag., 58, 364-397, 1921.

Dietz, R. S.: Continent and ocean basin evolution by spreading of the sea floor, Nature, 190, 854-857, 1961.

Dziewonski, A. M., and Anderson, D. A.: Preliminary reference Earth model, Phys. Earth Planet. Inter., 25, 297-356, 1981.

Herndon, J. M.: Uniqueness of Herndon's georeactor: Energy source and production mechanism for Earth's magnetic field, arXiv.org/abs/0901.4509, 2009.

Herndon, J. M.: Impact of recent discoveries on petroleum and natural gas exploration: Emphasis on India, Current Science, 98, 772-779, 2010a.

Herndon, J. M.: Inseparability of science history and discovery, Hist.Geo. Space Sci., 1, 25-41, 2010b.

Hess, H. H.: History of Ocean Basins, in: Petrologic Studies: A Volume in Honor of A. F. Buddington, Geological Society of America, Boulder, 599-620, 1962.

Holmes, A.: Radioactivity and Earth movements, Trans. geol. Soc. Glasgow 1928-1929, 18, 559-606, 1931.

Kormilitsin, Y. N., and Khalizev, O. A.: Theory of Submarine Design, Riviera Maritime Media, Enfield, UK, 339 pp., 2001.

Le Pichon, X.: Sea-floor spreading and continental drift, J. Geophys. Res., 73, 3661-3697, 1968.

Meng, Y., Weidner, D. J., Gwanmesia, D. G., Liebermann, R. C., Vaughan, T., Wang, Y., Leinenweber, K., Pacalo, R. E., Yeganeh-Haeri, A., and Zhao, Y.: In situ high P-T X ray diffraction studies of three polymorphs of Mg_2SiO_4, J. Geophys. Res., 98, 22199-22207, 1993.

Oganov, A. R., Brodholt, J. P., and Price, G. D.: Ab initio elasticity and thermal equation of state of $MgSiO_3$ perovskite Earth Planet. Sci. Lett., 184, 555-560, 2001.

Phillips, B. R., and Coltice, N.: Temperature beneath continents as a function of continental cover and convective wavelength, J. Geophys. Res., 115, B04408-B04421, 2010.

Schubert, G., Turcotte, D. L., and Olsen, P.: Mantle Convection in the Earth and Planets, Cambridge University Press, Cambridge, 940 pp., 2001.

Smith, D. G.: Display Devices, U. S. Patent 3,387,396, 3, 1968.

Snider-Pellegrini, A.: La Création et ses mystères dévoilés (Creation and its Mysteries Unveiled) Paris, 1858.

Vine, F. J., and Matthews, D. H.: Magnetic anomalies over oceanic ridges, Nature, 199, 947-949, 1963.

Wegener, A. L.: Die Entstehung der Kontinente, Geol. Rundschau, 3, 276-292, 1912.

Appendix VI

Earth weaves its own invisible cloak

Polar fountains fill magnetosphere with ions

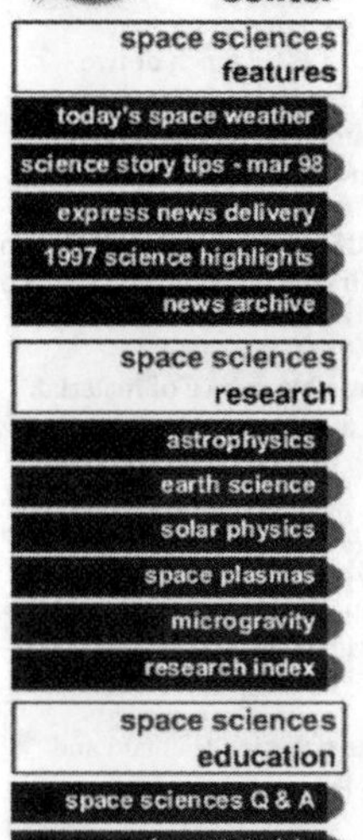

December 9, 1997

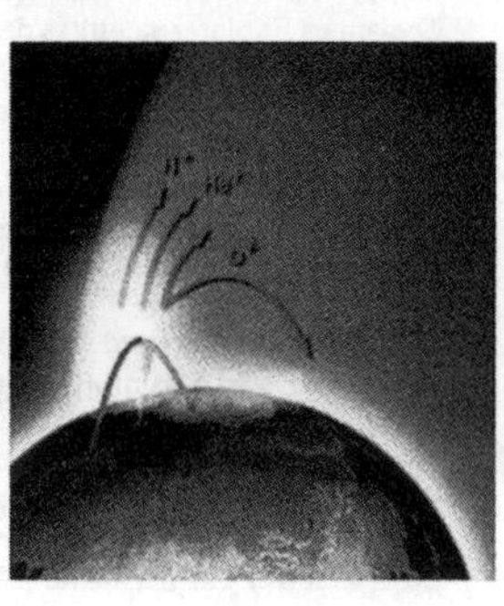

Since the late 1950s and '60s, scientists have believed that the Earth's magnetic field has captured a lot of the solar wind - charged particles flowing out from the sun - and formed an immense, comet-like cloud of electrified gas that surrounds our planet. An accidental discovery in the 1980s and new data collected since 1996 indicate that this magnetosphere may well be filled by a fountain of energized gas (right) blowing from the north and south poles.

You don't need to stock up on bottled air - the leakage is tiny compared to Earth's atmosphere - but the studies are making us take a new look at 40-year-old assumptions about how the Earth interacts with the space environment.

The new look will be discussed in two papers by Drs. Rick Chappell and Barbara Giles at the fall meeting of the American Geophysical Union in San Francisco. Chappell and Giles are plasma physicists at NASA's Marshall Space Flight Center.

Starting with the discovery of the Van Allen radiation belts by the first U.S. satellite in 1958, scientists have known that the Earth is surrounded by a cloud of electrified gas. This cloud, called the magnetosphere, is constrained by Earth's magnetic field, compressed by the solar wind to within about 100,000 km of the Earth on the sunward side, and drawn out by the solar wind to more than a million km on the night side.

"The conventional wisdom from the time I was in school 30 years ago," said Chappell, NASA/Marshall's associate director for science, "was that because these ions have so much energy (about 1,000 electron-volts [1 keV], they must come from the sun."

"Why should it be from the ionosphere?" Giles asked, repeating the thinking of the 1960s and '70s. The ionosphere, the top layer of Earth's atmosphere from about 60 km and up, is composed of ionized atoms and molecules. "It's so cold. What they were measuring they assumed came from the solar wind because the energies were right."

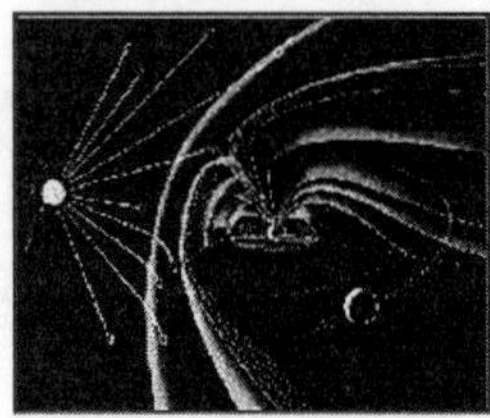

The magnetosphere (left) is a complex structure that changes with the solar wind. Major features include the polar cusp where the magnetic field lines are almost vertical and leave the Earth's poles exposed to space, and a neutral plasma sheet which extends outward from around the magnetic equator. This complex region is being explored by the International Solar-Terrestrial Physics (ISTP) campaign comprising satellites such as Polar (right) which carries the TIDE instrument.

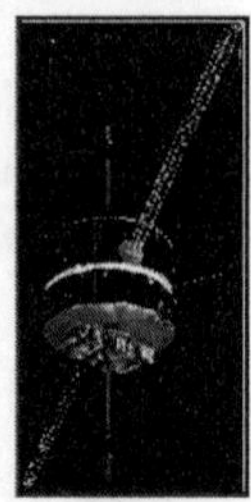

The perception started to change in the mid-1980s following the Aug. 3, 1981, launch of two Dynamics Explorer satellites designed to study the magnetosphere near the Earth. DE-1 carried Chappell's Retarding Ion Mass Spectrometer (RIMS), designed to measure the population of the plasmasphere, a torus or donut of low-energy in the inner magnetosphere.

To Chappell's surprise, the real find was around the north pole where RIMS measured gases flowing upward from the ionosphere into space as DE-1 arced to about 4.6 Earth radii above the pole (the orbit was 464 x 23,370 km [288 x 14,490 mi]).

"The more we measured it," he said, "the more we realized that this was a big source of material." RIMS measured ions of hydrogen, helium, oxygen, and nitrogen rising at different speeds and on different trajectories.

Apparently, the particles that slammed into the ionosphere to paint the aurora borealis also energized enough atoms to head spaceward. (The same is true of the south pole where DE-1 made similar observations.) The acceleration mechanism is not fully understood, though. Giles said that very low frequency (VLF) radio waves, emitted in the magnetosphere, may contribute energy.

In 1987, Chappell and the RIMS team published a paper describing the polar ion fountain and described the ionosphere as a "fully adequate source of material for the magnetosphere."

That means that scientists did not need to look to the solar wind to fill the magnetosphere. Earth, they contended, does a good job of supplying all the materials.

RIMS, though, was not fully adequate to resolve the challenge that the science team had laid down. Its orbit only went to about 4 Earth radii (about 25,6000 km [16,000 mi]), and it was limited in measuring the environment around it.

RIMS also could not measure the lowest energy ions which would help confirm that the source was the ionosphere and not the solar wind.

It turns out that all spacecraft develop an electrical charge. For high altitude satellites, exposure to sunlight and the passage through plasmas give a satellite a charge of about 5 to 10 eV. A small cloud of ions, a *plasma sheath*, builds around the spacecraft and repels anything with lower energy.

"We needed a device to neutralize that plasma sheath," Chappell said. "Unless you can do that, you won't ever see those particles."

As luck would have it, NASA/Marshall had proposed building a Thermal Ion Dynamics Experiment (TIDE) for the upcoming Polar spacecraft which would be part of the International Solar Terrestrial Physics campaign.

"Because of the new technologies, we're just at the point where we can make these measurements," Giles said. "Controlling spacecraft potential is something that's fairly new."

TIDE works with a Plasma Source Instrument (PSI), a small ion gun that would squirt enough electrified xenon out to neutralize the spacecraft's plasma sheath.

An earlier SSL story explained why Plasmas Can't Hide from Neutralized TIDE.

"That allows the naturally low-energy ions to get to the spacecraft," Chappell said. While TIDE operates continuously, the plasma source has only been used for two months since Polar was launched because the plasma cloud obscures readings for other instruments.

Polar was launched Feb. 24, 1996, into, as the name suggests, a polar orbit, arcing to 8.9 Earth radii (the orbit is 5,558 x 50,420 km [3,446 x 31,149 mi].)

"And lo and behold, there's all this low-energy plasma flowing out of the atmosphere and into the magnetosphere," Chappell said. Some of the material does fall back to Earth, mostly the heavier ions with low energies, but enough apparently is accelerated to reach into the tail of the magnetosphere.

TIDE expanded on RIMS's measurements and showed that the ion flow starts with very low energies, down to a few tenths of an electron-volt, and that its altitude is twice as high as DE-1's orbit.

"The really incredible thing," Chappell said, "is that if you do a very careful model of the magnetic field and electric field, and then put in 10 eV particles, they go into the plasma sheet, and are energized at least a thousandfold."

computer generated paths of hydrogen ion in magnetosphere

computer generated paths of hydrogen ion in magnetosphere

A computer model depicts the path of a hydrogen ion - a free proton - as it arcs away from the northern ionosphere to about as far away as the moon's orbit (384,000 km [240,000 mi]). When it flows into the plasma sheet, it is dramatically energized - from about 5 eV to 5,000 eV. Then it zooms back along the sheet and bounces back and forth (an effect called mirroring) along Earth's magnetic field lines. The simulation only shows about 3-1/2 hours of one of a large number of possible paths. In the top left view (from deep space towards the Earth), the ion would rise out of the screen, dive back, and then zig-zag into the screen as if tracing a series of ribs.

As instrument designs were advancing, so were computers and software which could model what is happening inside the magnetosphere. Dominique Delcourt, then working at NASA/Marshall, developed a program which takes the data from TIDE and shows particles zooming up to 60 Earth radii (almost 400,000 km) into the magnetotail, zipping back along the plasma sheet, and then bouncing back and forth along the Earth's magnetic field lines.

The last effect is called mirroring. As a particle moves along a magnetic field line, its orientation to the field line changes until the particle is repelled towards the other pole. How deep the particle penetrates depends on its angle with respect to the magnetic field. If the angle is aligned with the magnetic field, the particle will hit the upper atmosphere and cause the aurora borealis or australis. If it is at right angles, the particle will mirror and head in the other direction.

"We're not starting with a solar wind particle at 1 keV," Chappell sad, "but with atmospheric particles at a few eV. They make up everything you see up there."

Even if it is not filling the magnetotail, or even contributing a lot of material, the solar wind still plays an important role, Chappell said. It generates the electric field which energizes the tail, and even compresses it so it squirts materials back towards Earth to cause substorms that disrupt satellite communications, interfere with terrestrial power lines, and push the auroras towards lower latitudes.

Part of the challenge in these studies is that most of the magnetotail is made of the same material as the solar wind: protons. Actually, they are called hydrogen ions, but since most hydrogen is made of one proton circled by one electron, a hydrogen ion is just a free proton. And protons don't carry ID cards to tell whether they came from the Sun or the Earth. (Isotope analysis counts the numbers of protons and neutrons in an atom, so it only works on heavier elements.)

Other satellites at present do not carry instruments that can neutralize the plasma sheath as TIDE does. Flights with suborbital rockets can only make limited contributions to solving the question. Giles said: "Rocket flights work on a microscale. TIDE works on a global scale and long-term trends." Rockets also sample the source region for the polar fountain, while TIDE looks at the destinations and what happens to the particles.

Because each orbit measures a single arc over the polar region, Chappell would like to make a series of low-energy measurements, with the plasma source on, to get full coverage of the entire polar region under a range of seasonal conditions and different levels of solar activities.

AGU Fall 1997 Meeting, Moscone Center, San Francisco, Calif., Dec. 8-12. Session SM42D, *Ion Acceleration and Outflow: Recent Progress.* Thursday, Dec. 11, room MC 310.

- Barbara L. Giles, Charles R. Chappell, D. C. Delcourt, T. E. Moore, M. O. Chandler, P. D. Craven. **Magnetospheric Plasmas: A Direct Measurement of the Ionospheric Source**. SM42D-08, 3:55 p.m.
- Charles R. Chappell, Barbara L. Giles, D. C. Delcourt, T. E. Moore, M. O. Chandler, P. D. Craven. **Magnetospheric Plasmas: Flow and Energization of the Ionospheric Source**. SM42D-09, 4:10 p.m.

Join our growing list of subscribers - **sign up for our express news delivery** and you will receive a mail message every time we post a new story!!!

More

Headlines

return to Space Sciences Laboratory Home

Author: Dave Dooling
Curator: Linda Porter
NASA Official: Gregory S. Wilson

Appendix VII

The impossibility of planetary or Solar System's discs accretion to form solid celestial bodies

In his Adams Prize Essay for the year 1856, James Clerk Maxwell took the first step towards a comprehensive theory of the unique system of rings associated with Saturn. The main theme of Maxwell's work was to devise a model for the system of rings which could move in a steady state of motion round Saturn and which could be stable for small disturbances, provided the law of gravitation was applicable to the Saturnian system. Maxwell considered the case a single ring, concentric with Saturn.

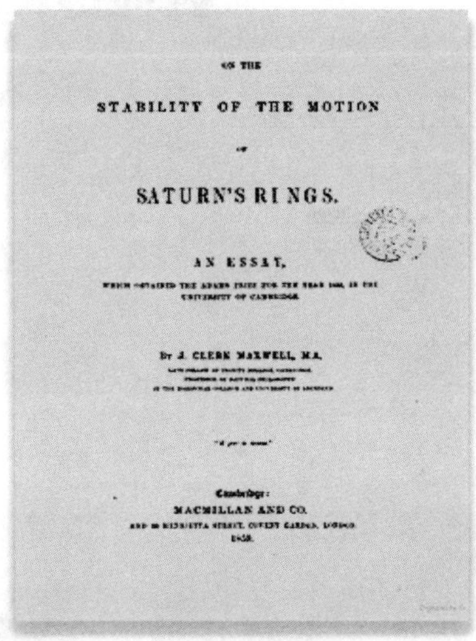

ON THE

STABILITY OF THE MOTION

OF

SATURN'S RINGS.

AN ESSAY,

By J. CLERK MAXWELL, M.A.

MACMILLAN AND CO.

He concluded that (i) the steady motion of a uniform rigid circular ring would be unstable ; and that the steady motion of a non-uniform rigid circular ring could be stable only if the density were very irregular, (ii) the steady motion of a ring composed of discrete particles, of equal mass, moving in a circle round the planet would be stable if a certain condition were satisfied, and (iii) the steady motion of a liquid ring would be unstable.

For Saturn's dense rings important physical processes that crucially influence the steady state properties of the rings: *dissipative collisions* and *shear.* This implies that the rings are in a flat, dynamically cold state where physical collisions of the ring particles, are continuously dissipating energy from the particles' random motion.

These particles are densely packed in the ring plane, continuously forming transient gravitationally bound opaque clumps, that are *disrupted* again by shear on orbital timescales. In this way, neighbouring ring regions are effectively coupled by a drag force: The inner part of the ring tends to *accelerate* (disrupting) the outer part, and vice versa.

That's why an orbiting debris, *never* clumps or coalesce to form a permanent solid unit. What was true of Saturn's rings would also be true for the Solar System. The Sun, planets and moons could *never* have formed from coalescing (accreting) gas and dust because the forces of dissolution (dissipation) are too strong.

Despite that in 2004, the NASA *Cassini* probe to Saturn showed Maxwell's conclusion was *right;* this fact was dismissed by mediocre Western academics. Following the track of Maxwell's work *On the Stability of the Motion of Saturn's Rings*, two Russian professors: Alexei M. Fridman and Nikolai N. Gorkavyi, authored an outstanding book: *Physics of Planetary Rings*.

In it, the authors have made decisive contributions to research into collisional, collective and resonance phenomena in planetary rings. Based on experimental evidence and sound logical reasoning, this Russian team correctly *predicted* the existence of unknown Uranian satellites prior to the Voyager 2 fly-by. This is how *real* science works.

In contrast, we have "research" being carried out in the West, where a bunch of "experts" running silly and flawed computer models, fed with fictitious data, then proceed to claim *pseudoscientific* "facts." These are propagandized trough mainstream media, to attract attention and provoke strong emotional reactions from the naïve public.

Fridman and Gorkavyi have contributed essential ideas to the *full* understanding of planetary rings via the stability analysis of *dynamical continuous systems*. This answers the fundamental question: Why does the matter of rings *remain* uniformly spread along the orbit? (it does NOT clump). This makes *impossible,* the formation of any celestial body, even in protoplanetary discs.

The aforementioned scientific facts, makes *impossible* the current theory of Moon's formation: the "moon-forming impact." An object about the size of Mars, called Theia collided with Earth. The collision was so large that in addition to adding lots of material to the Earth, there was enough energy to vaporize some of the rock and metal from both the proto-Earth and the impacting object.

This vapour formed a *disc* around the Earth that "eventually" cooled and clumped together to become the Moon. As it was already shown, this account of Moon's formation is *false*. It is a wishful thinking. It cannot happen, ever.

.

Appendix VIII

Failure of mainstream science to find out correctly, the MoI of Earth and other planets.

A common and even decisive argument against the Hollow Earth Theory is that the so-called "experts" based on the concept of moment of inertia (MoI), have concluded without a shadow of a doubt, that Earth is a solid body because basically, their *unrealistic* calculations show that Earth's moment of inertia factor is 0.331 MR^2.

Theoretical considerations indicate that a celestial body having a moment of inertia factor below 0.4 means that its density increases with depth, therefore it is a solid body. However, this statement is *false* because it comes from and *idealistic*, a priori conception that all planets must follow. This is not how science is done, by diktat. What do those "scientists" think they are?

Knowledge of the moment of inertia provides an important constraint on possible distributions of the density ρ inside the Earth, which is the point here. Equations are derived connecting the mean I of the principal moments of inertia with observational quantities such as the Earth's mean radius, mass M, and angular velocity n of rotation about the polar axis, the constant G of gravitation, and the precession constant H. The theory also involves the distribution of the ellipticities € of internal surfaces of assumed of constant density.

Estimates of moments of inertia (MoI) require measurement of the J_2 contribution to the gravity field. Usually this is done by observing the rotational responses (precession) of a spacecraft or satellite orbiting a planet, or by measuring the shape (and hence the fattening) of the planet very accurately. It involves indirect estimates of MoI considering the planet as an *isolated body*, neglecting its interaction with other celestial bodies that can determine accurately, the MoI of the planet under scrutiny.

To work out the MoI, scientists consider the moment of inertia of a spherically symmetrical planet, i.e., density is just a function of radius. Spherical symmetry means that the moment of inertia about any axis going through the centre of the planet will be the same. For simplicity, it is chosen the rotation axis to compute the moment of inertia.

For a general non-spherically symmetric density distribution it is customary to calculate the three principal moments of inertia of the planet. These are the moments of inertia about the x, y and z axes. These are called A, B and C respectively. For a spherically symmetric planet it is clear that A = B = C. Here C can be computed.

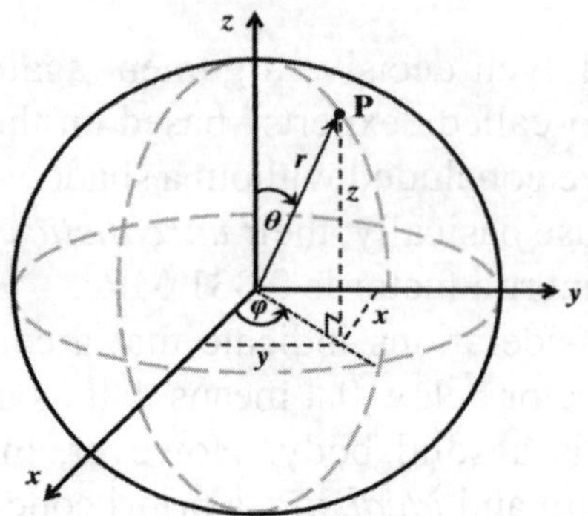

Using a spherical coordinate system:

$x = r \sin\theta \cos\phi$
$y = r \sin\theta \sin\phi$
$z = r \cos\theta$

Where ϕ is longitude and is θ colatitude. An element of volume is given by:

$dV = dxdydz$ in a cartesian coordinate system
$dV = r^2 \sin \theta dr d\theta d\phi$ in a spherical coordinate system

Back to the planet: The mass element centred at point, P is a perpendicular distance *s* from the *z* axis. It has a moment of inertia about the *z* axis given by $s^2\, dm$. The total moment of inertia of the planet about the *z* axis is the *integral* of all such elements over the entire (continuous) mass distribution. So:

$$C = {}_M\int s^2\, dm$$

Where $dm = \rho\,(r) dV$. Substituting *dm* in spherical coordinates:

$$C = {}_V\int r^2 \sin^2 \theta\, V(r)\, dV.$$

Where the integral is over the volume of the planet. Of course, this integral doesn't include any *void* in the planet's volume. This integral can be written as one over r (from 0 to a), over θ (from 0 to π) and over ϕ (from 0 to 2 π). By turning the expression dV into spherical coordinates and after further mathematical analysis, the following equation comes up:

$$C = 8\ \pi/3\ {}_{0}\int^{a} \rho(_{r}) r^{4} dr$$

Where density (ρ) is an *unknown* function of the planet's radius (r) and that's the problem. The moment of inertia tells us something about the density distribution within the planet either it is solid or *hollow*. However, unless we have other information about the planet (such as the mean density, the planets size, etc.) there are many different density distributions that can give us the measured value of C.

This is because C is proportional to the integral of the product (r) and r^4. You cannot get the actual values of A, B and C from the gravity field alone; there is *insufficient information*. You can only get their differences (e.g., C-A).

To fix this issue, the scientific community resort to the traditional theories of studying the figure of the Earth's interior, which usually start from the *assumption* of hydrostatic equilibrium (HSE) and one-dimensional Earth *models*, obtaining the equipotential surface with symmetrical shape.

The first generation of such kind theories is the famous Clairaut equation, in which J_2 term is considered only. A small correction to Clairaut equation, i.e., a J_4 term, is added in the Darwin–de Sitter theory. And the third generation of these theories is proposed and completed by Prof. C. Dennis.

Extensive studies (guesswork) of seismic waves and the application of Snell's Law, which determines ray-path trajectory of waves, provides an internal seismic (estimated) wave-speed and density model of Earth. The model for the internal structure of the Earth is represented by the preliminary reference Earth model (PREM; Dziewonski & Anderson, 1981).

It shows gradual density transitions are due to "increasing" pressure and temperature with depth. However, it *ignores* the research made by the drilling of the Kola Deep Borehole in Russia, which indicates a *decreasing* pressure starting at a depth of 9 kilometres. Nonetheless, perhaps to simplify calculations of the moment of inertia factor, Earth or any planet for that matter, is *assumed* to have uniform density.

Suppose (guess) that a planet has a uniform density ρ* then:

$$C = 8\pi/3\, \rho^* \int_0^a r^4 dr = 8\pi/15\, \rho^*\, a^5$$

The quantity that is used is usually not C but C/Ma^2 which is a dimensionless number (M is the mass of the planet), known as the *moment of inertia factor*. Now the mass of a solid sphere is:

$$M = 4/3\, \pi\, a^3 r$$

So, for a uniform density of an ideal planet:

$$C/Ma^2 = [8\pi/15\, \rho^*\, a^5]x[3/4\, \pi\, a^3\, \rho^*]x[1/a^2] = 2/5 = 0.4$$

If the density is greater near the centre of the planet we find that $C/Ma^2 < 04$. For example, C/Ma^2 for the Earth is 0.3308. Here we can start to see the *fallacy* of this argument. A physical parameter of a planet (MoI) is determined not by reality but by a *fictitious* arrangement of its internal distribution of mass (density), which is endlessly *enforced* by academia. Due to the process of *reification*, scientists are sure their fiction has become a reality, which is further supported by their spurious "scientific research".

How might scientists measure C? A straightforward, but tedious, calculation allows J_2 to be cast in terms of the principal moments of inertia of the body. J_2 is a coefficient in the *spherical harmonic* representation of Earth's gravitational field.

Spherical harmonics are solutions of the spherical Laplace's Equation: They are basically an adaption of Fourier analysis to a *spherical surface.* Just like with Fourier series, the superposition of spherical harmonics can be used to *represent* and analyze physical phenomena distributed on the surface on (or within) the Earth.

In this case the gravity potential (U_G). The $J_{2,3\ldots n}$ coefficients are called *gravitational moments* since they are integrals of the internal density distribution, weighted by progressively higher powers of the radius.

After spherical harmonics calculations have been done, we find that:

$$J_2 = \frac{C - A}{Ma^2}$$

This equation allows us to estimate C - A because J_2 is measured for the Earth, the gravitational attractions of the Sun and the Moon, acting on the equatorial bulge, "cause" a precession of the axis of rotation. From the rate of precession, we can find the dynamical ellipticity, H, which is given by:

$$\frac{C - A}{C} = H$$

The relationship between the precessional quantities and the dynamical ellipticity H can be shown to be (see, for example, De Sitter 1938):

$$(p + p_g) \sec \epsilon_0 + \chi = \frac{3}{2}\left[\frac{n_S^2}{\omega}(1 - e^2)^{-3/2} \cdot \frac{M_S}{M_S + M_E + M_M} + \frac{n_M^2}{\omega} \cdot \frac{M_M}{M_E + M_M} \cdot N\right] \cdot H$$

Where p = general precession in longitude, pg = geodetic precession, ϵ0 = obliquity of Earth's equatorial plane to the ecliptic, χ = planetary precession, ns = mean orbital velocity of the Earth, ω = rotational velocity of the Earth on its axis, e = the eccentricity of the Earth's orbit, MS = mass of the Sun, ME = mass of the Earth, MM = mass of the Moon, nM = mean orbital velocity of the Moon, and N = a function of orbital parameters introduced by Hill (1893) and developed by him with Delaunay's lunar theory. H is estimated to be 1/305.51 and by combining equations we have, for the Earth, the following moment of inertia factor:

$$C = J_2/H\ Ma^2 = 0.\ 3308Ma^2$$

However, the *problem* with this result is that it comes out from a *flawed* argument, because the relationship between the precessional quantities and the dynamical ellipticity H is *wrong*. There is NO such thing as gravitational attractions of the Sun and the Moon, acting on the equatorial bulge, causing a precession of the Earth's axis of rotation.

The cause of apparent motion of precession of equinoxes is attributed *presumptuously* to the movement of Earth's axis. Sir Isaac Newton, assuming a *static* Solar System, explained the movement of Earth's axis as a direct consequence of Earth's wobble, which is due to the gravitational pull on Earth's equatorial bulge by the Moon and Sun respectively (Lunisolar Theory). This North Pole movement in turn is supposed to be the cause of precession of equinox.

The hypothesis that is a differential gravitational force is causing oscillation or wobble of the Earth on its axis, duly shifting the position of North Pole with reference to star Polaris at an equal rate of 50.3 seconds in overhead sky, which is in turn causing precession of equinoxes on equatorial plane. Scientists later on deferred with this version and made *alternative* propositions. However, several scientists claim no definite movement of Earth is identified, which could possibly explain comprehensively the cause of precession of equinoxes; which remained unexplained as on today.

The Moon's gravitational pull on Earth's equator causes tidal waves in the oceans. When a similar presumption that Earth's wobble is due to gravitational pull and pushes on equatorial bulge is due to Moon and Sun, it is conspicuous that there are NO associated *disturbances* on Earth's surface. Hence, the hypothesis of Earth's wobble is not credible.

When Earth spins on its axis in West to East direction (anti clockwise) it is natural that North Pole of the axis moves in the *same* direction. It is *how* North Pole can describe a circle of precession about star Polaris in a clockwise direction, *opposite* to the natural rotation of North Pole of the axis, which conspicuously remains *unexplained.*

The hypothesis of Earth's wobble does not explain the above contradiction. Hence, the hypothetical proposition that the retrograde motion of North Pole is due to Earth's wobble is not credible. Whereas the circle of precision generated by North Pole in 25,772 years in clockwise direction that is *opposite* to the natural direction of Earth's axis as observed, however, indeed is a fact.

This phenomenon is due to *realistic* orbital movement of the Earth not orbiting a static Sun and describing a 2D ellipse, but orbiting a *dynamic* Sun through the Milky Way, were Earth is describing a 3D helicoidal trajectory. Precession is observable in Solar System *motion.*

That it is NOT the Earth that wobbles but rather the motion of the Solar System *curving* through space slowly *changing* the orientation of the Solar System in relation to the *fixed stars* that produces the observable known as the Precession of the Equinox. The annual shift of Earth's orbit on the equatorial plane is due to its 3D orbital spin, which causes precession. Therefore, any calculation based upon the accepted Lunisolar Theory to determine H is *useless* and *misleading*.

Nonetheless of the *pitfalls* already discussed, these traditional methods are still used to determine the MoI and internal mass distribution on planets. Therefore, as by decree, these are unquestionable "solid bodies" and any deviation of this statement is considered ironically, pseudoscience or a conspiracy theory.

In general, to investigate the MoI, bodies fall into two classes: those for which hydrostatic effects dominate (i.e., "rapid rotators") and those for which the rotational bulge is no bigger than the other effects on topography and gravity.

The rapid rotators further subdivide into those for which the Radau-Darwin Approximation for the hydrostatic flattening, is roughly valid (terrestrial bodies and icy satellites) and the gaseous bodies (in which the density variations are too large, and a more complex and detailed theory has been devised). Following this scheme, the models created for Jupiter and Neptune are compared to Earth's model and conclude that Jupiter may have a silica-rich core, whereas Neptune's model has inconclusive density and radius results that should be further investigated.

And now is when things become really interesting. When the Juno Mission to Jupiter was launched in 2011, there were hopes that the MoI of this planet could be "validated" giving credence to the current models.

A paper published the same year (Helled et al, 2011) reads:

"The moment of inertia of a giant planet reveals important information about the planet's internal density structure and this information is not identical to that contained in the gravitational moments. The forthcoming Juno mission to Jupiter might determine Jupiter's normalized moment of inertia NMoI = C/MR2 by measuring Jupiter's pole precession and the Lense–Thirring acceleration of the spacecraft (C is the axial moment of inertia, and M and R are Jupiter's mass and mean radius, respectively). We investigate the possible range of NMoI values for Jupiter based on its measured gravitational field using a simple core/envelope model of the planet assuming that J_2 *and* J_4 *are perfectly known and are equal to their measured values. The model suggests that for fixed values of* J_2 *and* J_4 *a range of NMoI values*

between 0.2629 and 0.2645 can be found. The Radau–Darwin relation gives a NMoI value that is larger than the model values by less than 1%. A low NMoI of ~0.236, inferred from a dynamical model (Ward, W.R., Canup, R.M. [2006]. Astrophys. J. 640, L91–L94) is inconsistent with this range, but the range is model dependent. Although we conclude that the NMoI is tightly constrained by the gravity coefficients, a measurement of Jupiter's NMoI to a few tenths of percent by Juno could provide an important constraint on Jupiter's internal structure."

The above paragraph is much ado about nothing. It lacks substance. Is recycling the same stuff of spherical harmonic coefficients and MoI fancying to find the same computed value, without adding nothing new and meaningful. Nowhere says Juno's mission could prove them wrong. The Juno spacecraft acquired science observations of Jupiter on 27 August 2016 and since then has measured Jupiter's gravitational field to high precision during its first two perijoves. Further data came to put developed *models* to test. Interior models for Jupiter (and Saturn) are based on the phase diagram and a *theoretical* equation of state (P, T) for hydrogen helium mixtures and the assumption of adiabatic temperature variation. They are constrained by the requirement to fit the observed flattening and the gravity coefficients J_2, J_4.

Quantum mechanical calculations suggested that solid or fluid hydrogen becomes *metallic* at high pressure, i.e. has free electrons and becomes a good electrical conductor. A small *rocky core* of a several Earth masses is not required by the geophysical data for Jupiter (but for Saturn), but is likely in any case, because a sizeable solid nucleus may be required to be present first in order to accrete H_2 and He onto it by gravitation.

So, what happened with the Juno Mission survey of Jupiter anyway? By studying the gas giant's gravitational field, researchers can "predict" an abundance of heavy elements in Jupiter's interior, including water, and they *hope* to find evidence of a solid core, as has been predicted by models, and measure its mass and rate of rotation. The gravity data was much *improved*, and the simplest interpretation of the higher order even harmonics implies that the planet have a *diluted* central concentration or core of heavy elements rather than a compact one. This is pretty much, an euphemism for Jupiter being *hollow*, neither more nor less.

Scientists conclude that it is very *challenging* to explain Jupiter's dilute core from standard formation models, based in the obsolete and inaccurate accretion-collision hypothesis of planetary formation. This “dilute core” lacks a well-defined boundary and takes up to a half of the planet’s interior. How such a structure can possibly form, astronomers and astrophysicists don’t yet understand. But, how can they if they don’t have what it takes? Therefore, the customary value of the moment of inertia coefficient for Jupiter:

$$C/Mr^2 = 0.262$$

Is NO longer *valid* because when C < 0.4 the density is concentrated at the centre of the planet and it is considered solid. But, is due to the fact Jupiter’s density is *diluted* at its centre and more concentrated in the *outer layers* of the planet.

Therefore: C > 0.4 indicating a distribution of mass and density in a shape of a *spherical thick shell* for Jupiter, which indeed is the case as this shell extends up to half of the planet’s radius. This shows us that Jupiter like Earth, is practically *hollow*.

Sources

AAS Nova. *Trapping dust to form planets*. https://aasnova.org/2017/10/20/trapping-dust-to-form-planets/

Anchorage Daily News. Confirmed: *Aurora Borealis Makes Sounds*.https://www.adn.com/science/article/confirmed-aurora borealis-makes-sounds/2012/07/10/

Arge, C. N., Mullan, D. J., & Dolginov, A. Z. *Magnetic Moments and Angular Momenta of Stars and Planets*. Astrophysical Journal, Part 1vol.443,no2,p.795-803(1995) http://adsabs.harvard.edu/full/1995ApJ...443..795A

Arizona State University. *Gestalt Principles and Dynamic Symmetry. Nature's Design Connections to Our Built World*. http://www.public.asu.edu/~detrie/msj.uc_daap/article.html

Bennett M., Percy David S. *Dark Moon*. London. Aulis Publishers. 1999

Binary Research Institute. *Precession of the Equinox*. 2015http://binaryresearchinstitute.com/bri/introduction/precession-of-the-equinox/

Bless, A.A. *The Composition of the Interior of the Earth*. Proceedings of the National Academy of Sciences of the United States of America. Vol. 17, No. 4 (Apr. 15, 1931), pp. 225-229

Bligh, Bernard. *The Sun Is Electrically Charged & the Conundrum of the "Missing Mass" Is Explained*. Prespacetime Journal. 2010. http://independent.academia.edu/BlighBernard

Bolt, Bruce. *Inside the Earth. Evidence from Earthquakes*. San Francisco. W.H. Freeman and company, 1982

Bott, Martin H.P. *The Interior of the Earth: its Structure, Constitution and Evolution.* London. Edward Arnold Publishers. 1982

Brian II, William L. Moongate: *Suppressed Findings of the U.S. Space Program*. Future Science Res Pub Co. 1982

Bryangwallace.dreamhosters.com. *The Farce of Physics*. http://bryangwallace.dreamhosters.com/

Bullen, K.E. *An Introduction to the Theory of Seismology*. Cambridge University Press. 1965

Cellularuniverse.org. *The Aether Experiments and the Impact on Cosmology*.http://www.cellularuniverse.org/AA1Aether&Cosmology.htm

Conrad Ranzan. *The Nature of Gravity – How One Factor Unifies Gravity's Convergent, Divergent, Vortex, and Wave Effects*. International Journal of Astrophysics and Space Science. Vol. 6, No. 5, 2018, pp. 73-92

Copi, Irving M. *Introduction to Logic*. Macmillan Library Reference. 1982

Corliss, William. *Handbook of Unusual Natural Phenomena*. Glen Arm. The Sourcebook Project. 1977

Darkecoecology.net. *Drilling Deep/Knowledge from Underground (Kola Deep Borehole)* http://www.darkecology.net/drilling-deep

De Bremaecker, Jean-Claude. *Geophysics: The Earth's Interior*. New York. John Wiley & Sons. 1985

De Meo, James. *A Dynamic and Substantive Cosmological Ether*. Proceedings, Natural Philosophy Alliance. Vol.1, No.1, Spring 2004

Edwards, Matthew R. (editor) *Pushing Gravity. New Perspectives on Le Sage's Theory of Gravitation*. Montreal. Roy Keys Inc. Apeiron. 2002

Eioba.com. *Torsion Fields - Theory of Physical Vacuum*. http://www.eioba.com/a/1tzs/torsion-fields-theory-of-physical-vacuum-shipov-and-heim

Encyclopaedia Britannica. https://www.britannica.com

Eos.org. *The Unsolved Mystery of the Earth Blobs*. https://eos.org/features/the-unsolved-mystery-of-the-earth-blobs

Fleming, Ray. *The Zero Point Universe*. Independently published. 2020

Ferromsky V.I., Denisik S.A., Ferronsky S.V. *Jacobi Dynamics. A Unified Theory with Applications to Geophysics, Celestial Mechanics, Astrophysics and Cosmology*. Second edition. Springer ASSL, 1986

Foulger, Gillian R. et.al. *Caveats on tomographic images*. Terra Nova Journal. Volume 25. Issue 4. Pages 259-281. 2013

Garland, G. D. *The Earth's Shape and Gravity*. Oxford. Pergamon Press. 1977

Gutenberg, Beno. *Physics of the Earth's Interior*. New York. Academic Press Inc. 1959

Harris, Rollin A. *Manual of Tides*. U.S. Government Printing Office. 1904

Hayasaka, Hideo, Takeuchi Sakae. *Anomalous weight reduction on a gyroscope's right rotations around the vertical axis on the Earth*.

Physical Review Letters. 63, 2701 – Published 18 December 1989. https://journals.aps.org/prl/abstract/10.1103/PhysRevLett.63.2701

Haynes, Robert C. *Introduction to Space Science*. New York. John Wiley & Sons, Inc. 1971

Herndon, Marvin J. *Teaching Students to Question Earth-Core Convection.*http://nuclearplanet.com/Question%20EarthCore%20Convection.html

Hess, Wilmot and Mead, Gilbert. *Introduction to Space Science*. New York. Gordon and Breach, Science Publishers, Inc. 1968

Holmes, Arthur. *Principles of Physical Geology. Second Edition Completely Revised.* London. Nelson. 1965

Hoyle, Fred. *Home is Where the Wind Blows*. Calif: Univ Sci Books 1994; p. 413.

Jeffreys, Harold. *Earthquakes and Mountains*. London. The Scientific Book Club. Second edition. 1950

Jeffreys, Harold. *Scientific Inference*. Cambridge University Press. 1931

Jeffreys, Harold. *The Earth. Its Origin, History and Physical Constitution*. Cambridge University Press. 1970

Jordan, Pascual. *The Expanding Earth*. Oxford. Pergamon Press. 1971

Kaplan S.A. and V.N. Tsytovich. *Plasma Astrophysics*. Oxford. Pergamon Press. 1973

Kerridge, J.F., and Vedder, J., *Accretionary Processes in the Early Solar System: An Experimental Approach*. Science 177:161–163, 1972; p. 161. https://www.ncbi.nlm.nih.gov/pubmed/17779914

King, Elbert A. *Space Geology*. An Introduction. New York. John Wiley and Sons, Inc. 1976

Klotz, Frank G. *America on the Ice. Antarctic Policy Issues*. Washington DC. National Defense University Press. 1990.

Kozlovsky, Y. A. *The Superdeep Well of the Kola Peninsula*. Berlin: Springer-Verlag. p. 558. 1987

Kragh, Helge. *Varying Gravity. Dirac's Legacy in Cosmology and Geophysics*. Birkhäuser. Springer International Publishing, Switzerland 2016

Lerner, Eric J. *The Bing Bang Never Happened*. London. Simon & Schuster. 1991

Means, W.D. *Stress and Strain. Basic Concepts of Continuum*

Mechanics for Geologists. Springer-Verlag. New York. 1976

Malakhov, A. *The Mystery of Earth's Mantle*. Moscow. Peace Publishers. 1966

Moulton, Forest R. *An Introduction to Celestial Mechanics*. New York. The Macmillan Company. 1960

NASA Science. *Earth weaves its own invisible cloak - Polar fountains fill magnetosphere with ions*. https://science.nasa.gov/science-news/science-at-nasa/1997/ast09dec97_3

Nature.com. *Earth's Magnetic Field is Acting up and Geologists Don't Know Why* .https://www.nature.com/articles/d41586-019-00007-1

O' Neill, John J. *Prodigal Genius. The Life of Nikola Tesla*. London. Granada Publishing Ltd. 1980

Outerplaces.com. *Mysterious Particles Coming from the Ground in Antarctica Have Physicists Puzzled*. https://www.outerplaces.com/science/item/18894-mysterious-particles-antarctica

Parasnis, D.S. *Principles of Applied Geophysics*. London. Chapman & Hall. 1994

Peng Zhang, R. E. Cohen & K. Haule. *Retraction: Effects of electron correlations on transport properties of iron at Earth's core conditions*.Nature517,605–607(2015) https://www.nature.com/articles/nature17648

Planetaryformation.com. *Planetary Formation Theory in Chaos*. http://planetaryformation.com/nebular-hypothesis/planetary-formation-theory-in-chaos/

Platt, David. *Plate Tectonics: A Paradigm Under Threat*. Journal of Scientific Exploration, Vol. 14, No. 3, pp. 307–352, 2000

Physics.org. *Challenging core belief: Have we misunderstood how Earth's solid centre formed?* https://phys.org/news/2018-02-core-belief-misunderstood-earth-solid.html

Physics.org. *New theory points to 'zombie vortices' as key step in star formation*. https://phys.org/news/2013-08-theory-zombie-vortices-key-star.html

Runcorn, S.K. (editor). *The Application of Modern Physics to the Earth and Planetary Interiors*. London. Wiley-Interscience. 1969

Seligman, Darryl. Laughlin, Gregory. *A Vorticity-preserving Hydrodynamical Scheme for Modelling Accretion Disk Flows*.

The Astrophysical Journal. 2017

Sigurgeirsson, H. Stuart A.M. *A model for preferential concentration*. Physics of Fluids. Volume 14. Issue 12. 2002. https://doi.org/10.1063/1.1517603

Setterfield.org. *Plasma Physics, Astronomy and the ZPE Model.* http://www.setterfield.org/Plasma_Astronomy_ZPE/text.html

Sciencedaily.com. *Planet Formation Could Lie in Stellar Storms RatherThanGravitationalInstability*. https://www.sciencedaily.com/releases/2008/12/081212101543.htm

Scientific American. *Astronomers Spy Planet-Spawning Vortex AroundYoungStar*. https://www.scientificamerican.com/article/astronomers-spy-planet-spawning-vortex-around-young-star/?redirect=1

Scitechdaily.com. *Earthquake Doublets Reveal Changing Speed of the Earth's Inner Core.* https://scitechdaily.com/earthquake-doublets-reveal-changing-speed-of-the-earths-inner-core/

Sheldrake, Rupert. *A New Science of Life.* London. Paladin Grafton Books. 1983

Space.com. *'Totally Wrong' on Jupiter: What Scientists Gleaned from NASA's Juno Mission.* https://www.space.com/39348-juno-jupiter-mission-planet-revelations.html

Sputnik News. *Russian Scientists Reveal Earth's Magnetism is MoreMysteriousThanEver* https://sputniknews.com/environment/201605101039356573-scientists-earth-magnetic-field-mystery/

Stefansson, Vilhjalmur. *Unsolved Mysteries of the Arctic*. New York. Collier Books. 1962

Surdin, Maurice. *Le champ électromagnétique fluctuant de l'univers (Essai d'une électrodynamique stochastique)* Annales de l'I. H. P., section A, tome 15, no 3 (1971), p. 203-241

Surdin, Maurice. *The Magnetic Field of Rotating Bodies*. http://articles.adsabs.harvard.edu//full/1994cmcp.conf..134S/0000134.000.html

Temple, Robert. *A New Science of Heaven*. London. Coronet. 2023

Tombe, F. B. *AetherTheory*. www.cartesio-episteme.net

Vorontsov-Vel'yaminov, B.A. *Astronomical Problems*. Oxford. Pergamon Press. 1969

Witszche Rolf, *The Plasma Core Earth*. Independently published. 2019

Author’s hometown

Get these fascinating books from your nearest bookstore or directly from: Adventures Unlimited Press

www.adventuresunlimitedpress.com

COVERT WARS AND BREAKAWAY CIVILIZATIONS
By Joseph P. Farrell

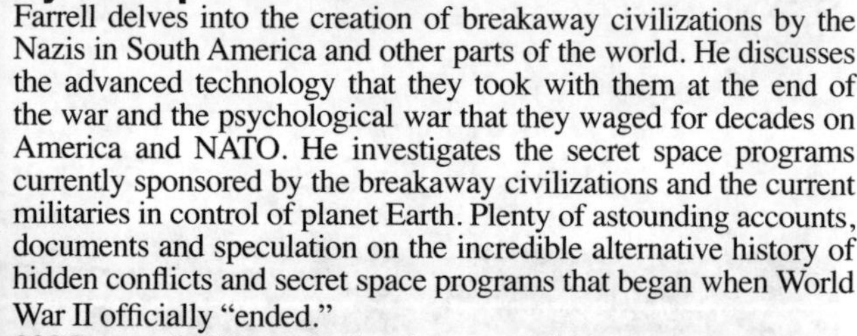

Farrell delves into the creation of breakaway civilizations by the Nazis in South America and other parts of the world. He discusses the advanced technology that they took with them at the end of the war and the psychological war that they waged for decades on America and NATO. He investigates the secret space programs currently sponsored by the breakaway civilizations and the current militaries in control of planet Earth. Plenty of astounding accounts, documents and speculation on the incredible alternative history of hidden conflicts and secret space programs that began when World War II officially "ended."

292 Pages. 6x9 Paperback. Illustrated. $19.95. Code: BCCW

THE ENIGMA OF CRANIAL DEFORMATION
Elongated Skulls of the Ancients
By David Hatcher Childress and Brien Foerster

In a book filled with over a hundred astonishing photos and a color photo section, Childress and Foerster take us to Peru, Bolivia, Egypt, Malta, China, Mexico and other places in search of strange elongated skulls and other cranial deformation. The puzzle of why diverse ancient people—even on remote Pacific Islands—would use head-binding to create elongated heads is mystifying. Where did they even get this idea? Did some people naturally look this way—with long narrow heads? Were they some alien race? Were they an elite race that roamed the entire planet? Why do anthropologists rarely talk about cranial deformation and know so little about it? Color Section.

250 Pages. 6x9 Paperback. Illustrated. $19.95. Code: ECD

ARK OF GOD
The Incredible Power of the Ark of the Covenant
By David Hatcher Childress

Childress takes us on an incredible journey in search of the truth about (and science behind) the fantastic biblical artifact known as the Ark of the Covenant. This object made by Moses at Mount Sinai—part wooden-metal box and part golden statue—had the power to create "lightning" to kill people, and also to fly and lead people through the wilderness. The Ark of the Covenant suddenly disappears from the Bible record and what happened to it is not mentioned. Was it hidden in the underground passages of King Solomon's temple and later discovered by the Knights Templar? Was it taken through Egypt to Ethiopia as many Coptic Christians believe? Childress looks into hidden history, astonishing ancient technology, and a 3,000-year-old mystery that continues to fascinate millions of people today. Color section.

420 Pages. 6x9 Paperback. Illustrated. $22.00 Code: AOG

HESS AND THE PENGUINS
The Holocaust, Antarctica and the Strange Case of Rudolf Hess
By Joseph P. Farrell

Farrell looks at Hess' mission to make peace with Britain and get rid of Hitler—even a plot to fly Hitler to Britain for capture! How much did Göring and Hitler know of Rudolf Hess' subversive plot, and what happened to Hess? Why was a doppleganger put in Spandau Prison and then "suicided"? Did the British use an early form of mind control on Hess' double? John Foster Dulles of the OSS and CIA suspected as much. Farrell also uncovers the strange death of Admiral Richard Byrd's son in 1988, about the same time of the death of Hess.

288 Pages. 6x9 Paperback. Illustrated. $19.95. Code: HAPG

HIDDEN FINANCE, ROGUE NETWORKS & SECRET SORCERY
The Fascist International, 9/11, & Penetrated Operations
By Joseph P. Farrell

Farrell investigates the theory that there were not *two* levels to the 9/11 event, but *three*. He says that the twin towers were downed by the force of an exotic energy weapon, one similar to the Tesla energy weapon suggested by Dr. Judy Wood, and ties together the tangled web of missing money, secret technology and involvement of portions of the Saudi royal family. Farrell unravels the many layers behind the 9-11 attack, layers that include the Deutschebank, the Bush family, the German industrialist Carl Duisberg, Saudi Arabian princes and the energy weapons developed by Tesla before WWII.

296 Pages. 6x9 Paperback. Illustrated. $19.95. Code: HFRN

THRICE GREAT HERMETICA & THE JANUS AGE
By Joseph P. Farrell

What do the Fourth Crusade, the exploration of the New World, secret excavations of the Holy Land, and the pontificate of Innocent the Third all have in common? Answer: Venice and the Templars. What do they have in common with Jesus, Gottfried Leibniz, Sir Isaac Newton, Rene Descartes, and the Earl of Oxford? Answer: Egypt and a body of doctrine known as Hermeticism. The hidden role of Venice and Hermeticism reached far and wide, into the plays of Shakespeare (a.k.a. Edward DeVere, Earl of Oxford), into the quest of the three great mathematicians of the Early Enlightenment for a lost form of analysis, and back into the end of the classical era, to little known Egyptian influences at work during the time of Jesus.

354 Pages. 6x9 Paperback. Illustrated. $19.95. Code: TGHJ

REICH OF THE BLACK SUN
Nazi Secret Weapons & the Cold War Allied Legend
by Joseph P. Farrell

Why were the Allies worried about an atom bomb attack by the Germans in 1944? Why did the Soviets threaten to use poison gas against the Germans? Why did Hitler in 1945 insist that holding Prague could win the war for the Third Reich? Why did US General George Patton's Third Army race for the Skoda works at Pilsen in Czechoslovakia instead of Berlin? Why did the US Army not test the uranium atom bomb it dropped on Hiroshima? Why did the Luftwaffe fly a non-stop round trip mission to within twenty miles of New York City in 1944? Farrel takes the reader on a scientific-historical journey in order to answer these questions. Arguing that Nazi Germany won the race for the atom bomb in late 1944,

352 PAGES. 6x9 PAPERBACK. ILLUSTRATED. $16.95. CODE: ROBS

THE GIZA DEATH STAR REVISITED
An Updated Revision of the Weapon Hypothesis of the Great Pyramid
By Joseph P. Farrell

Join revisionist author Joseph P. Farrell for a summary, revision, and update of his original *Giza Death Star* trilogy in this one-volume compendium of the argument, the physics, and the all-important ancient texts, from the Edfu Temple texts to the Lugal-e and the Enuma Elish that he believes may have made the Great Pyramid a tremendously powerful weapon of mass destruction. Those texts, Farrell argues, provide the clues to the powerful physics of longitudinal waves in the medium that only began to be unlocked centuries later by Sir Isaac Newton and Nikola Tesla's "electro-acoustic" experiments.

360 Pages. 6x9 Paperback. Illustrated. $19.95. Code: GDSR

THE DEMON IN THE EKUR
Angels, Demons, Plasmas, Patristics, and Pyramids
By Joseph P. Farrell

Farrell looks at the Demon in the *Ekur* (the gathering place of the gods in Sumerian tradition) and the Great Pyramid Weapon Hypothesis. He delves deep into the realm of angels, presenting John of Damascus' Angelology and examines the "Immaterial Materiality" of angels; their ability to penetrate ordinary matter and "unlimited" nature; their ability to shapeshift; and the everlasting temporality, or "Sempiternity," of angels. Farrell also explores "Plasma Cosmotheology" and the Plasma Life Hypothesis. He presents intriguing pictures of nuclear detonations and discusses the Hyper-Dimensional Transduction Hypothesis. He includes a discussion of crystals as tuners and transducers, and explores the planetary associations of crystals with angels.

160 Pages. 6x9 Paperback. Illustrated. $16.95. Code: DITE

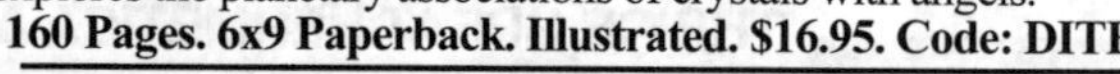

THE ANTI-GRAVITY FILES
A Compilation of Patents and Reports
Edited by David Hatcher Childress

With plenty of technical drawings and explanations, this book reveals suppressed technology that will change the world in ways we can only dream of. Chapters include: A Brief History of Anti-Gravity Patents; The Motionless Electromagnet Generator Patent; Mercury Anti-Gravity Gyros; The Tesla Pyramid Engine; Anti-Gravity Propulsion Dynamics; The Machines in Flight; More Anti-Gravity Patents; Death Rays Anyone?; The Unified Field Theory of Gravity; and tons more. Heavily illustrated. 4-page color section.

216 pages. 8x10 Paperback. Illustrated. $22.00. Code: AGF

ANCIENT TECHNOLOGY IN PERU & BOLIVIA
By David Hatcher Childress

Childress speculates on the existence of a sunken city in Lake Titicaca and reveals new evidence that the Sumerians may have arrived in South America 4,000 years ago. He demonstrates that the use of "keystone cuts" with metal clamps poured into them to secure megalithic construction was an advanced technology used all over the world, from the Andes to Egypt, Greece and Southeast Asia. He maintains that only power tools could have made the intricate articulation and drill holes found in extremely hard granite and basalt blocks in Bolivia and Peru, and that the megalith builders had to have had advanced methods for moving and stacking gigantic blocks of stone, some weighing over 100 tons.

340 Pages. 6x9 Paperback. Illustrated.. $19.95 Code: ATP

ROSWELL AND THE REICH

The Nazi Connection

By Joseph P. Farrell

Farrell has meticulously reviewed the best-known Roswell research from UFO-ET advocates and skeptics alike, as well as some little-known source material, and comes to a radically different scenario of what happened in Roswell, New Mexico in July 1947, and why the US military has continued to cover it up to this day. Farrell presents a fascinating case sure to disturb both ET believers and disbelievers, namely, that what crashed may have been representative of an independent postwar Nazi power—an extraterritorial Reich monitoring its old enemy, America, and the continuing development of the very technologies confiscated from Germany at the end of the War.

540 pages. 6x9 Paperback. Illustrated. $19.95. Code: RWR

SECRETS OF THE UNIFIED FIELD

The Philadelphia Experiment, the Nazi Bell, and the Discarded Theory

by Joseph P. Farrell

Farrell examines the now discarded Unified Field Theory. American and German wartime scientists and engineers determined that, while the theory was incomplete, it could nevertheless be engineered. Chapters include: The Meanings of "Torsion"; Wringing an Aluminum Can; The Mistake in Unified Field Theories and Their Discarding by Contemporary Physics; Three Routes to the Doomsday Weapon: Quantum Potential, Torsion, and Vortices; Tesla's Meeting with FDR; Arnold Sommerfeld and Electromagnetic Radar Stealth; Electromagnetic Phase Conjugations, Phase Conjugate Mirrors, and Templates; The Unified Field Theory, the Torsion Tensor, and Igor Witkowski's Idea of the Plasma Focus; tons more.

340 pages. 6x9 Paperback. Illustrated. $18.95. Code: SOUF

NAZI INTERNATIONAL

The Nazi's Postwar Plan to Control Finance, Conflict, Physics and Space

by Joseph P. Farrell

Beginning with prewar corporate partnerships in the USA, including some with the Bush family, he moves on to the surrender of Nazi Germany, and evacuation plans of the Germans. He then covers the vast, and still-little-known recreation of Nazi Germany in South America with help of Juan Peron, I.G. Farben and Martin Bormann. Farrell then covers Nazi Germany's penetration of the Muslim world including Wilhelm Voss and Otto Skorzeny in Gamel Abdul Nasser's Egypt before moving on to the development and control of new energy technologies including the Bariloche Fusion Project, Dr. Philo Farnsworth's Plasmator, and the work of Dr. Nikolai Kozyrev. Finally, Farrell discusses the Nazi desire to control space, and examines their connection with NASA, the esoteric meaning of NASA Mission Patches.

412 pages. 6x9 Paperback. Illustrated. $19.95. Code: NZIN

ARKTOS

The Polar Myth in Science, Symbolism & Nazi Survival

by Joscelyn Godwin

Explored are the many tales of an ancient race said to have lived in the Arctic regions, such as Thule and Hyperborea. Progressing onward, he looks at modern polar legends: including the survival of Hitler, German bases in Antarctica, UFOs, the hollow earth, and the hidden kingdoms of Agartha and Shambala. Chapters include: Prologue in Hyperborea; The Golden Age; The Northern Lights; The Arctic Homeland; The Aryan Myth; The Thule Society; The Black Order; The Hidden Lands; Agartha and the Polaires; Shambhala; The Hole at the Pole; Antarctica; more.

220 Pages. 6x9 Paperback. Illustrated. Bib. Index. $16.95. Code: ARK

SAUCERS, SWASTIKAS AND PSYOPS
A History of a Breakaway Civilization
By Joseph P. Farrell

Farrell discusses SS Commando Otto Skorzeny; George Adamski; the alleged Hannebu and Vril craft of the Third Reich; The Strange Case of Dr. Hermann Oberth; Nazis in the US and their connections to "UFO contactees"; The Memes—an idea or behavior spread from person to person within a culture—are Implants. Chapters include: The Nov. 20, 1952 Contact: The Memes are Implants; The Interplanetary Federation of Brotherhood; Adamski's Technological Descriptions and Another ET Message: The Danger of Weaponized Gravity; Adamski's Retro-Looking Saucers, and the Nazi Saucer Myth; Dr. Oberth's 1968 Statements on UFOs and Extraterrestrials; more.

272 Pages. 6x9 Paperback. Illustrated. $19.95. Code: SSPY

LBJ AND THE CONSPIRACY TO KILL KENNEDY
By Joseph P. Farrell

Farrell says that a coalescence of interests in the military industrial complex, the CIA, and Lyndon Baines Johnson's powerful and corrupt political machine in Texas led to the events culminating in the assassination of JFK. Chapters include: Oswald, the FBI, and the CIA: Hoover's Concern of a Second Oswald; Oswald and the Anti-Castro Cubans; The Mafia; Hoover, Johnson, and the Mob; The FBI, the Secret Service, Hoover, and Johnson; The CIA and "Murder Incorporated"; Ruby's Bizarre Behavior; The French Connection and Permindex; Big Oil; The Dead Witnesses: Guy Bannister, Jr., Mary Pinchot Meyer, Rose Cheramie, Dorothy Killgallen, Congressman Hale Boggs; LBJ and the Planning of the Texas Trip; LBJ: A Study in Character, Connections, and Cabals; LBJ and the Aftermath: Accessory After the Fact; The Requirements of Coups D'État; more.

342 Pages. 6x9 Paperback. $19.95 Code: LCKK

THE TESLA PAPERS
Nikola Tesla on Free Energy & Wireless Transmission of Power
by Nikola Tesla, edited by David Hatcher Childress

David Hatcher Childress takes us into the incredible world of Nikola Tesla and his amazing inventions. Tesla's fantastic vision of the future, including wireless power, anti-gravity, free energy and highly advanced solar power. Also included are some of the papers, patents and material collected on Tesla at the Colorado Springs Tesla Symposiums, including papers on: •The Secret History of Wireless Transmission •Tesla and the Magnifying Transmitter •Design and Construction of a Half-Wave Tesla Coil •Electrostatics: A Key to Free Energy •Progress in Zero-Point Energy Research •Electromagnetic Energy from Antennas to Atoms

325 PAGES. 8x10 PAPERBACK. ILLUSTRATED. $16.95. CODE: TTP

COVERT WARS & THE CLASH OF CIVILIZATIONS
UFOs, Oligarchs and Space Secrecy
By Joseph P. Farrell

Farrell's customary meticulous research and sharp analysis blow the lid off of a worldwide web of nefarious financial and technological control that very few people even suspect exists. He elaborates on the advanced technology that they took with them at the "end" of World War II and shows how the breakaway civilizations have created a huge system of hidden finance with the involvement of various banks and financial institutions around the world. He investigates the current space secrecy that involves UFOs, suppressed technologies and the hidden oligarchs who control planet earth for their own gain and profit.

358 Pages. 6x9 Paperback. Illustrated. $19.95. Code: CWCC

ANTARCTICA AND THE SECRET SPACE PROGRAM

David Hatcher Childress

David Childress, popular author and star of the History Channel's show *Ancient Aliens*, brings us the incredible tale of Nazi submarines and secret weapons in Antarctica and elsewhere. He then examines Operation High-Jump with Admiral Richard Byrd in 1947 and the battle that he apparently had in Antarctica with flying saucers. Through "Operation Paperclip," the Nazis infiltrated aerospace companies, banking, media, and the US government, including NASA and the CIA after WWII. Does the US Navy have a secret space program that includes huge ships and hundreds of astronauts?

392 Pages. 6x9 Paperback. Illustrated. $22.00 Code: ASSP

HAUNEBU: THE SECRET FILES

The Greatest UFO Secret of All Time

By David Hatcher Childress

Childress brings us the incredible tale of the German flying disk known as the Haunebu. Although rumors of German flying disks have been around since the late years of WWII it was not until 1989 when a German researcher named Ralf Ettl living in London received an anonymous packet of photographs and documents concerning the planning and development of at least three types of unusual craft. Chapters include: A Saucer Full of Secrets; WWII as an Oil War; A Saucer Called Vril; Secret Cities of the Black Sun; The Strange World of Miguel Serrano; Set the Controls for the Heart of the Sun; Dark Side of the Moon: more. Includes a 16-page color section. Over 120 photographs and diagrams.

352 Pages. 6x9 Paperback. Illustrated. $22.00 Code: HBU

ANDROMEDA: THE SECRET FILES

The Flying Submarines of the SS

By David Hatcher Childress

Childress brings us the amazing story of the German Andromeda craft, designed and built during WWII. Along with flying discs, the Germans were making long, cylindrical airships that are commonly called motherships—large craft that house several smaller disc craft. It was not until 1989 that a German researcher named Ralf Ettl, living in London, received an anonymous packet of photographs and documents concerning the planning and development of at least three types of unusual craft—including the Andromeda. Chapters include: Gravity's Rainbow; The Motherships; The MJ-12, UFOs and the Korean War; The Strange Case of Reinhold Schmidt; Secret Cities of the Winged Serpent; The Green Fireballs; Submarines That Can Fly; The Breakaway Civilization; more. Includes a 16-page color section.

382 Pages. 6x9 Paperback. Illustrated. $22.00 Code: ASF

VRIL: SECRETS OF THE BLACK SUN

By David Childress

A remnant of the Nazi military—particularly the SS—continued to operate aircraft and submarines around the world in the decades after the end of the war. This volume closes with how the SS operates today in the Ukraine and how the Wagner second in command, Dimitry Utkin, killed in the fiery crash of Yevgeny Prigozhin's private jet between Moscow and St. Petersburg in August of 2023, had SS tattoos on his shoulders and often signed his name with the SS runes. Chapters include: Secrets of the Black Sun; The Extra-Territorial Reich; The Rise of the SS; The SS Never Surrendered; Secret Submarines, Antarctica & Argentina; The Marconi Connection; Yellow Submarine; Ukraine and the Battalion of the Black Sun; more. Includes an 8-page color section.

382 Pages. 6x9 Paperback. Illustrated. $22.00 Code: VSBS

THE GODS IN THE FIELDS

Michael, Mary and Alice-Guardians of Enchanted Britain

By Nigel Graddon

We learn of Britain's special place in the origins of ancient wisdom and of the "Sun-Men" who taught it to a humanity in its infancy. Aspects of these teachings are found all along the St. Michael ley: at Glastonbury, the location of Merlin and Arthur's Avalon; in the design and layout of the extraordinary Somerset Zodiac of which Glastonbury is a major part; in the amazing stone circles and serpentine avenues at Avebury and nearby Silbury Hill: portals to unimaginable worlds of mystery and enchantment; Chapters include: Michael, Mary and Merlin; England's West Country; The Glastonbury Zodiac; Wiltshire; The Gods in the Fields; Michael, Mary and Alice; East of the Line; Table of Michael and Mary Locations; more.

280 Pages. 6x9 Paperback. Illustrated. $19.95. Code: GIF

AXIS OF THE WORLD

The Search for the Oldest American Civilization

by Igor Witkowski

Polish author Witkowski's research reveals remnants of a high civilization that was able to exert its influence on almost the entire planet, and did so with full consciousness. Sites around South America show that this was not just one of the places influenced by this culture, but a place where they built their crowning achievements. Easter Island, in the southeastern Pacific, constitutes one of them. The Rongo-Rongo language that developed there points westward to the Indus Valley. Taken together, the facts presented by Witkowski provide a fresh, new proof that an antediluvian, great civilization flourished several millennia ago.

220 pages. 6x9 Paperback. Illustrated. $18.95. Code: AXOW

LEY LINE & EARTH ENERGIES

An Extraordinary Journey into the Earth's Natural Energy System

by David Cowan & Chris Arnold

The mysterious standing stones, burial grounds and stone circles that lace Europe, the British Isles and other areas have intrigued scientists, writers, artists and travellers through the centuries. How do ley lines work? How did our ancestors use Earth energy to map their sacred sites and burial grounds? How do ghosts and poltergeists interact with Earth energy? How can Earth spirals and black spots affect our health? This exploration shows how natural forces affect our behavior, how they can be used to enhance our health and well being.

368 pages. 6x9 Paperback. Illustrated. $18.95. Code: LLEE

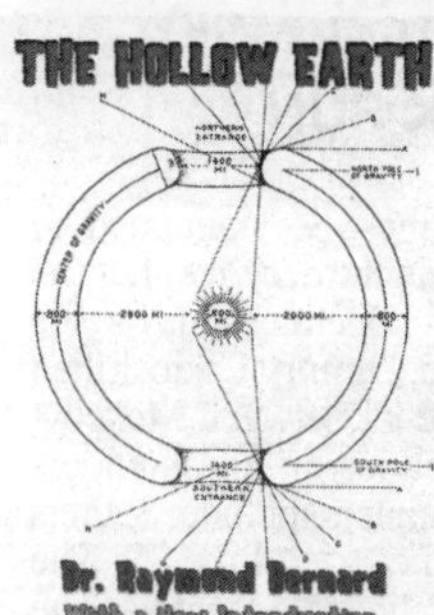

THE HOLLOW EARTH

By Dr. Raymond Bernard, A.B., M.A., Ph.D., Introduction by David Hatcher Childress

This classic of the 60s weird UFO literature touches on UFOs from Antarctica, free energy, tunnels in South America, government cover-ups and the notion that the earth is hollow. Dr. Bernard believes that in the hollow interior of the Earth lives a super-race which wants noting to do with man on the surface. They launched their flying saucers only after man threatened the world with A-bombs. Childress explains much of the unexplainable in. Dr. Bernard's bizarre book, this is a great addition to your library of the strange.

288 Pages. 5x9 Paperback. Illustrated. $16.95. Code: HERT

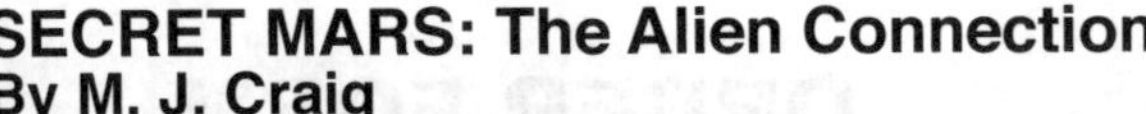

SECRET MARS: The Alien Connection
By M. J. Craig

While scientists spend billions of dollars confirming that microbes live in the Martian soil, people sitting at home on their computers studying the Mars images are making far more astounding discoveries… they have found the possible archaeological remains of an extraterrestrial civilization. Hard to believe? Well, this challenging book invites you to take a look at the astounding pictures yourself and make up your own mind. *Secret Mars* presents over 160 incredible images taken by American and European spacecraft that reveal possible evidence of a civilization that once lived, and may still live, on the planet Mars… powerful evidence that scientists are ignoring! A visual and fascinating book!

352 Pages. 6x9 Paperback. Illustrated. $19.95. Code: SMAR

GODS AND SPACEMEN THROUGHOUT HISTORY
Did Ancient Aliens Visit Earth in the Past?
By W. Raymond Drake

From prehistory, flying saucers have been seen in our skies. As mankind sends probes beyond the fringes of our galaxy, we must ask ourselves: "Has all this happened before? Could extraterrestrials have landed on Earth centuries ago?" Drake spent many years digging through huge archives of material, looking for supposed anomalies that could support his scenarios of space aliens impacting human history. Chapters include: Spacemen; The Golden Age; Sons of the Gods; Lemuria; Atlantis; Ancient America; Aztecs and Incas; India; Tibet; China; Japan; Egypt; The Great Pyramid; Babylon; Israel; Greece; Italy; Ancient Rome; Scandinavia; Britain; Saxon Times; Norman Times; The Middle Ages; The Age of Reason; Today; Tomorrow; more.

280 Pages. 6x9 Paperback. Illustrated. $18.95. Code: GSTH

PYTHAGORAS OF SAMOS
First Philosopher and Magician of Numbers
By Nigel Graddon

This comprehensive account comprises both the historical and metaphysical aspects of Pythagoras' philosophy and teachings. In Part 1, the work draws on all known biographical sources as well as key extracts from the esoteric record to paint a fascinating picture of the Master's amazing life and work. Topics covered include the unique circumstances of Pythagoras' birth, his forty-year period of initiations into all the world's ancient mysteries, his remarkable meeting with a physician from the mysterious Etruscan community, Part 2 comprises, for the first time in a publicly available work, a metaphysical interpretation of Pythagoras' Science of Numbers.

294 Pages. 6x9 Paperback. Illustrated. $18.95. Code: PYOS

VIMANA:
Flying Machines of the Ancients
by David Hatcher Childress

According to early Sanskrit texts the ancients had several types of airships called vimanas. Like aircraft of today, vimanas were used to fly through the air from city to city; to conduct aerial surveys of uncharted lands; and as delivery vehicles for awesome weapons. David Hatcher Childress, popular *Lost Cities* author, takes us on an astounding investigation into tales of ancient flying machines. In his new book, packed with photos and diagrams, he consults ancient texts and modern stories and presents astonishing evidence that aircraft, similar to the ones we use today, were used thousands of years ago in India, Sumeria, China and other countries. Includes a 24-page color section.

408 Pages. 6x9 Paperback. Illustrated. $22.95. Code: VMA

ORDER FORM

10% Discount When You Order 3 or More Items!

One Adventure Place
P.O. Box 74
Kempton, Illinois 60946
United States of America
Tel.: 815-253-6390 • Fax: 815-253-6300
Email: auphq@frontiernet.net
http://www.adventuresunlimitedpress.com

ORDERING INSTRUCTIONS

- ✓ Remit by USD$ Check, Money Order or Credit Card
- ✓ Visa, Master Card, Discover & AmEx Accepted
- ✓ Paypal Payments Can Be Made To:
  info@wexclub.com
- ✓ Prices May Change Without Notice
- ✓ 10% Discount for 3 or More Items

SHIPPING CHARGES

United States

- ✓ POSTAL BOOK RATE
- ✓ Postal Book Rate { $5.00 First Item; 50¢ Each Additional Item
- ✓ Priority Mail { $8.50 First Item; $2.00 Each Additional Item
- ✓ UPS { $9.00 First Item (Minimum 5 Books); $1.50 Each Additional Item

NOTE: UPS Delivery Available to Mainland USA Only

Canada

- ✓ Postal Air Mail { $19.00 First Item; $3.00 Each Additional Item
- ✓ Personal Checks or Bank Drafts MUST BE US$ and Drawn on a US Bank
- ✓ Canadian Postal Money Orders OK
- ✓ Payment MUST BE US$

All Other Countries

- ✓ Sorry, No Surface Delivery!
- ✓ Postal Air Mail { $29.00 First Item; $7.00 Each Additional Item
- ✓ Checks and Money Orders MUST BE US$ and Drawn on a US Bank or branch.
- ✓ Paypal Payments Can Be Made in US$ To:
  info@wexclub.com

SPECIAL NOTES

- ✓ RETAILERS: Standard Discounts Available
- ✓ BACKORDERS: We Backorder all Out-of-Stock Items Unless Otherwise Requested
- ✓ PRO FORMA INVOICES: Available on Request
- ✓ DVD Return Policy: Replace defective DVDs only

ORDER ONLINE AT: www.adventuresunlimitedpress.com

10% Discount When You Order 3 or More Items!

Please check: ✓

☐ This is my first order ☐ I have ordered before

Name
Address
City
State/Province — Postal Code
Country
Phone: Day — Evening
Fax — Email

Item Code	Item Description	Qty	Total

Please check: ✓

Subtotal ▸
Less Discount-10% for 3 or more items ▸

☐ Postal-Surface — Balance ▸
☐ Postal-Air Mail (Priority in USA) — Illinois Residents 6.25% Sales Tax ▸
Previous Credit ▸
☐ UPS (Mainland USA only) — Shipping ▸
Total (check/MO in USD$ only) ▸

☐ Visa/MasterCard/Discover/American Express

Card Number:
Expiration Date: — Security Code:

✓ SEND A CATALOG TO A FRIEND: